Oxidative Eustress in Exercise Physiology

Oxidative Stress and Disease

Enrique Cadenas, MD, PhD
University of Southern California School of Pharmacy
Los Angeles, California

Helmut Sies, PhD
Heinrich Heine University
Dusseldorf, Germany

Oxidative Eustress in Exercise Physiology
edited by James N. Cobley, and Gareth W. Davison

Oxidative Stress in Cancer, AIDS, and Neurodegenerative Diseases
edited by Luc Montagnier, René Olivier, and Catherine Pasquier

Understanding the Process of Aging
The Roles of Mitochondria, Free Radicals, and Antioxidants,
edited by Enrique Cadenas and Lester Packer

Redox Regulation of Cell Signaling and Its Clinical Application,
edited by Lester Packer and Junji Yodoi

Antioxidants in Diabetes Management
edited by Lester Packer, Peter Rösen, Hans J. Tritschler, George L. King, and Angelo Azzi

Free Radicals in Brain Pathophysiology
edited by Giuseppe Poli, Enrique Cadenas, and Lester Packer

Nutraceuticals in Health and Disease Prevention
edited by Klaus Krämer, Peter-Paul Hoppe, and Lester Packer

Environmental Stressors in Health and Disease
edited by Jürgen Fuchs and Lester Packer

Handbook of Antioxidants
Second Edition, Revised and Expanded
edited by Enrique Cadenas and Lester Packer

Flavonoids in Health and Disease
Second Edition, Revised and Expanded
edited by Catherine A. Rice-Evans and Lester Packer

Redox–Genome Interactions in Health and Disease
edited by Jürgen Fuchs, Maurizio Podda, and Lester Packer

Thiamine
Catalytic Mechanisms in Normal and Disease States
edited by Frank Jordan and Mulchand S. Patel

Phytochemicals in Health and Disease
edited by Yongping Bao and Roger Fenwick

For more information about this series, please visit:
https://www.crcpress.com/Oxidative-Stress-and-Disease/book-series/CRCOXISTRDIS

Oxidative Eustress in Exercise Physiology

Edited by
James N. Cobley and Gareth W. Davison

CRC Press is an imprint of the
Taylor & Francis Group, an **informa** business

First edition published 2022
by CRC Press
6000 Broken Sound Parkway NW, Suite 300, Boca Raton, FL 33487-2742

and by CRC Press
2 Park Square, Milton Park, Abingdon, Oxon, OX14 4RN

CRC Press is an imprint of Taylor & Francis Group, LLC

Library of Congress Cataloging-in-Publication Data
Names: Cobley, James N., editor. | Davison, Gareth W., editor.
Title: Oxidative stress in exercise physiology / edited by James N. Cobley and Gareth W. Davison.
Description: First edition. | Boca Raton : CRC Press, 2022. |
Series: Oxidative stress and disease; vol 47 | Includes bibliographical references and index. |
Summary: "This book aims to unravel key physiological responses and adaptations to exercise paradigms regulated at the cell, tissue, and whole-body level in model systems and in humans both with respect to health and disease. By unravelling and contextualising how fundamental chemistry impacts biology at the physiological level, this book offers enduring appeal for scientists working in disparate disciplines. The present work recruited world-leading contributors, known as experts in their chosen field, to provide a comprehensive resource for new and established investigators working in the field, as well as, serving as a primer for those interested in the field"– Provided by publisher.
Identifiers: LCCN 2021048060 (print) | LCCN 2021048061 (ebook) |
ISBN 9780367508760 (hardback) | ISBN 9780367508777 (paperback) | ISBN 9781003051619 (ebook other)
Subjects: LCSH: Oxidative stress. | Exercise–Physiological aspects.
Classification: LCC RB170 .O9546 2022 (print) | LCC RB170 (ebook) |
DDC 615.8/2–dc23/eng/20211115
LC record available at https://lccn.loc.gov/2021048060
LC ebook record available at https://lccn.loc.gov/2021048061

ISBN: 9780367508760 (hbk)
ISBN: 9780367508777 (pbk)
ISBN: 9781003051619 (ebk)

DOI: 10.1201/9781003051619

Typeset in Joanna
by codeMantra

CONTENTS

Series Preface / vii
Editors / ix
Contributors / xi

1 • **INTRODUCTION TO OXIDATIVE (EU)STRESS IN EXERCISE PHYSIOLOGY / 1**
Gareth W. Davison and James N. Cobley

2 • **MEASURING OXIDATIVE DAMAGE AND REDOX SIGNALLING: PRINCIPLES, CHALLENGES, AND OPPORTUNITIES / 11**
James N. Cobley and Gareth W. Davison

3 • **EXERCISE REDOX SIGNALLING: FROM ROS SOURCES TO WIDESPREAD HEALTH ADAPTATION / 23**
Ruy A. Louzada, Jessica Bouviere, Rodrigo S. Fortunato and Denise P. Carvalho

4 • **OXYGEN TRANSPORT: A REDOX O_2DYSSEY / 41**
P.N. Chatzinikolaou, N.V. Margaritelis, A.N. Chatzinikolaou, V. Paschalis, A.A. Theodorou, I.S. Vrabas, A. Kyparos and M.G. Nikolaidis

5 • **MITOCHONDRIAL REDOX REGULATION IN ADAPTATION TO EXERCISE / 59**
Christopher P. Hedges and Troy L. Merry

6 • **BASAL REDOX STATUS INFLUENCES THE ADAPTIVE REDOX RESPONSE TO REGULAR EXERCISE / 71**
Ethan L. Ostrom and Tinna Traustadóttir

7 • **TIME TO 'COUPLE' REDOX BIOLOGY WITH EXERCISE IMMUNOLOGY / 85**
Alex J. Wadley and Steven J. Coles

8 • **EXERCISE AND RNA OXIDATION / 95**
Emil List Larsen, Kristian Karstoft and Henrik Enghusen Poulsen

9 • **EXERCISE AND DNA DAMAGE: CONSIDERATIONS FOR THE NUCLEAR AND MITOCHONDRIAL GENOME / 103**
Josh Williamson and Gareth W. Davison

10 • **NUTRITIONAL ANTIOXIDANTS FOR SPORTS PERFORMANCE / 115**
Jamie N. Pugh and Graeme L. Close

11 • **ANTIOXIDANT SUPPLEMENTS AND EXERCISE ADAPTATIONS / 123**
Shaun A. Mason, Lewan Parker, Adam J. Trewin and Glenn D. Wadley

12 • **NITRIC OXIDE BIOCHEMISTRY AND EXERCISE PERFORMANCE IN HUMANS: INFLUENCE OF NITRATE SUPPLEMENTATION / 137**
Stephen J. Bailey and Andrew M. Jones

13 • **(POLY)PHENOLS IN EXERCISE PERFORMANCE AND RECOVERY: MORE THAN AN ANTIOXIDANT? / 153**
Tom Clifford and Glyn Howatson

14 • **EXERCISE: A STRATEGY TO TARGET OXIDATIVE STRESS IN CANCER / 167**
Amélie Rébillard, Cindy Richard and Suzanne Dufresne

15 • **OXIDATIVE STRESS AND EXERCISE TOLERANCE IN CYSTIC FIBROSIS / 183**
Cassandra C. Derella, Adeola A. Sanni and Ryan A. Harris

16 • **AGEING, NEURODEGENERATION AND ALZHEIMER'S DISEASE: THE UNDERLYING ROLE OF OXIDATIVE DISTRESS / 193**
Richard J. Elsworthy and Sarah Aldred

17 • **EXERCISE, METABOLISM AND OXIDATIVE STRESS IN THE EPIGENETIC LANDSCAPE / 209**
Gareth W. Davison and Colum P. Walsh

Index / 223

SERIES PREFACE

Oxidative eustress, a form of physiological oxidative stress, is a concept that addresses the roles of hydrogen peroxide – the major redox-sensing molecule – in redox signaling under physiological conditions. Oxidative eustress differs from oxidative stress (or oxidative distress) inasmuch as the latter is associated with damage to biomolecules. From this perspective, the present book *Oxidative Eustress in Exercise Physiology* is timely and innovative and is a new perspective to address the redox-sensitive exercise paradigms: nutritional antioxidants that might blunt the adaptive responses to exercise and redox-sensitive glucose uptake by skeletal muscle.

This book acquires further significance because it dissects the fundamental redox biochemistry of exercising skeletal muscle, defines the underlying mechanism, and – importantly – advances the translational impact in relation to health and disease. To this end, the editors convened contributors with a wide expertise spectrum: examples are the role of antioxidants in personalized nutrition and exercise, carbohydrate energy metabolism in exercise, the vaso-protective functions of physical exercise, neuromuscular integrity and function, genetic factors and frailty, nutrition and physical exercise, mitochondrial function, clinical trials on high-intensity interval training, exercise-induced redox signaling, and many more. We congratulate Drs Cobley and Davison on this.

HELMUT SIES
ENRIQUE CADENAS
Oxidative Stress & Disease Series Editors

EDITORS

James N. Cobley, PhD, is a Senior Lecturer in Free Radicals at the University of the Highlands and Islands (Inverness, UK). His doctoral work, completed in 2013, focused on the redox regulation of molecular exercise adaptations in young and old human skeletal muscle. Since then, Dr Cobley has focused on developing methods to measure protein thiol redox state, which has resulted in the development of two new methods: ALISA and RedoxiFluor. Dr Cobley intends, in collaboration with others, to use both technologies to determine if and how protein thiol-defined redox signalling regulates exercise adaptations and responses.

Gareth W. Davison, PhD, is Professor of Exercise Biochemistry and Physiology and Director of Research at the Sport and Exercise Sciences Research Institute at Ulster University in the UK. He holds a BA, MSc, and an MSt in Genomic Medicine from the University of Cambridge and was awarded his PhD in Biochemistry and Physiology in 2002. Professor Davison is a Fellow of the American College of Sports Medicine and currently serves on several editorial boards, holding Editor roles with the *Journal of Sports Sciences, Frontiers in Physiology* (Redox Physiology Section) and *Antioxidants.* His research interests are aligned to exercise, DNA damage, and antioxidant function. Recently, his laboratory has focused on bridging the gap between intracellular redox metabolism and DNA methylation in health and disease.

CONTRIBUTORS

SARAH ALDRED
The Centre for Human Brain Health (CHBH)
University of Birmingham
Birmingham, United Kingdom

STEPHEN J. BAILEY
School of Sport, Exercise and Health Sciences
Loughborough University
Loughborough, United Kingdom

JESSICA BOUVIERE
Miller School of Medicine
University of Miami
Coral Gables, Florida
and
Institute of Biophysics Carlos Chagas Filho
Federal University of Rio de Janeiro
Rio de Janeiro, Brazil

DENISE P. CARVALHO
Institute of Biophysics Carlos Chagas Filho
Federal University of Rio de Janeiro
Rio de Janeiro, Brazil

A.N. CHATZINIKOLAOU
Department of Mathematics
National and Kapodistrian University of Athens
Athens, Greece

P.N. CHATZINIKOLAOU
Department of Physical Education and
Sport Science at Serres
Aristotle University of Thessaloniki
Thessaloniki, Greece

TOM CLIFFORD
School of Sport, Exercise and Health Sciences
Loughborough University
Loughborough, United Kingdom

GRAEME L. CLOSE
School of Sport and Exercise Sciences
Liverpool John Moores University
Liverpool, United Kingdom

STEVEN J. COLES
Institute of Science & the Environment
University of Worcester
Worcester, United Kingdom

CASSANDRA C. DERELLA
Department of Medicine
Augusta University
Augusta, Georgia

SUZANNE DUFRESNE
M2S Laboratory
University of Rennes
Rennes, France

RICHARD J. ELSWORTHY
School of Sport, Exercise and Rehabilitation Sciences
College of Life and Environmental Sciences
University of Birmingham
Birmingham, United Kingdom

RODRIGO S. FORTUNATO
Institute of Biophysics Carlos Chagas Filho
Federal University of Rio de Janeiro
Rio de Janeiro, Brazil

RYAN A. HARRIS
Department of Medicine
Augusta University
Augusta, Georgia

CHRISTOPHER P. HEDGES
Discipline of Nutrition, School of Medical Sciences
The University of Auckland
Auckland, New Zealand

GLYN HOWATSON
Department of Sport Exercise and Rehabilitation
Northumbria University
Newcastle–upon–Tyne, United Kingdom

ANDREW M. JONES
Sport and Health Sciences
College of Life and Environmental Sciences
University of Exeter
Exeter, United Kingdom

KRISTIAN KARSTOFT
Department of Clinical Pharmacology
Copenhagen University Hospital –
Bispebjerg and Frederiksberg
Copenhagen, Denmark
and
Centre for Physical Activity Research
Copenhagen University Hospital – Rigshospitalet
Copenhagen, Denmark

A. KYPAROS
Department of Physical Education and
Sport Science at Serres
Aristotle University of Thessaloniki
Thessaloniki, Greece

EMIL LIST LARSEN
Department of Clinical Pharmacology
Copenhagen University Hospital –
Bispebjerg and Frederiksberg
Copenhagen, Denmark

RUY A. LOUZADA
Miller School of Medicine
University of Miami
Coral Gables, Florida
and
Institute of Biophysics Carlos Chagas Filho
Federal University of Rio de Janeiro
Rio de Janeiro, Brazil

N.V. MARGARITELIS
Department of Physical Education and
Sport Science at Serres
Aristotle University of Thessaloniki
Thessaloniki, Greece
and
Dialysis Unit
424 General Military Hospital of Thessaloniki
Thessaloniki, Greece

SHAUN A. MASON
Institute for Physical Activity and Nutrition
School of Exercise and Nutrition Sciences
Deakin University
Geelong, Australia

TROY L. MERRY
Discipline of Nutrition,
School of Medical Sciences
The University of Auckland
Auckland, New Zealand

M.G. NIKOLAIDIS
Department of Physical Education and
Sport Science at Serres
Aristotle University of Thessaloniki
Thessaloniki, Greece

ETHAN L. OSTROM
Department of Radiology, School of Medicine
University of Washington
Seattle, Washington

LEWAN PARKER
Institute for Physical Activity and Nutrition
School of Exercise and Nutrition Sciences
Deakin University
Geelong, Australia

V. PASCHALIS
School of Physical Education and Sport Science
National and Kapodistrian University of Athens
Athens, Greece

HENRIK ENGHUSEN POULSEN
Department of Clinical Pharmacology
Copenhagen University Hospital –
Bispebjerg and Frederiksberg
Copenhagen, Denmark
and
Department of Clinical Medicine
University of Copenhagen
Copenhagen, Denmark
and
Department of Cardiology
Copenhagen University Hospital – North Zealand
Hillerød, Denmark

JAMIE N. PUGH
School of Sport and Exercise Sciences
Liverpool John Moores University
Liverpool, United Kingdom

AMÉLIE RÉBILLARD
M2S Laboratory
University of Rennes
Rennes, France

CINDY RICHARD
M2S Laboratory
University of Rennes
Rennes, France

ADEOLA SANNI
Department of Medicine
Augusta University
Augusta, Georgia

A.A. THEODOROU
Department of Life Sciences, School of Sciences
European University Cyprus
Nicosia, Cyprus

TINNA TRAUSTADÓTTIR
Department of Biological Sciences
Northern Arizona University
Flagstaff, Arizona

ADAM J. TREWIN
Institute for Physical Activity and Nutrition
School of Exercise and Nutrition Sciences
Deakin University
Geelong, Australia

I.S. VRABAS
Department of Physical Education and
Sport Science at Serres
Aristotle University of Thessaloniki
Thessaloniki, Greece

ALEX J. WADLEY
School of Sport, Exercise & Rehabilitation Sciences
University of Birmingham
Edgbaston, United Kingdom

GLENN D. WADLEY
Institute for Physical Activity and Nutrition
School of Exercise and Nutrition Sciences
Deakin University
Geelong, Australia

COLUM P. WALSH
Faculty of Life and Health Sciences
Ulster University
Northern Ireland, United Kingdom

JOSH WILLIAMSON
Faculty of Life and Health Sciences
Ulster University
Northern Ireland, United Kingdom

CHAPTER ONE

Introduction to Oxidative (Eu)stress in Exercise Physiology

Gareth W. Davison and James N. Cobley

CONTENTS

Introduction / 1
Oxygen and Its Derivatives / 2
Superoxide Anion / 2
Hydrogen Peroxide / 3
Hydroxyl Radical / 3
Reactive Nitrogen Species / 4
Brief Overview of Indirect and Direct Biomarkers of Exercise-Induced Oxidative Stress / 4
Brief Overview of ROS as Eustress Regulators of Skeletal Muscle / 6
Conclusion / 7
Acknowledgement / 7
References / 7

INTRODUCTION

Exercise-induced oxidative (eu)stress has received considerable attention over the last four decades, with *circa* 900 original investigations published since 1978. Indeed, redox research *per se* is a contemporary feature of biomedical journals and conferences, with a plethora of exciting new information relating to health and disease being presented.

The discovery that free radicals (i.e., molecules capable of existing independently, and containing one or more unpaired electrons) exist in living biological organisms occurred in 1954 when Commoner and colleagues (Commoner et al., 1954) first used electron paramagnetic resonance (EPR) spectroscopy to detect free radicals in growing seeds. Around this time, Gerschman et al. (1954) proposed that free radicals damage cells, and second, when mice are exposed to hyperoxia, the biological injury that ensues is likely related to increased free radical production. Following the publication of this work in the mid-1950s, a series of seminal studies confirmed the biological importance (Powers et al., 2016). For example, the work of Chance and Williams (1956) demonstrated that respiring mitochondria can generate hydrogen peroxide, which can diffuse in cells. Then, McCord and Fridovich (1969) discovered the metalloenzyme superoxide dismutase (SOD), showing in the process that superoxide radicals are spontaneously degraded to hydrogen peroxide. The milestone discovery of SOD not only provided much of the basis of our current understanding of antioxidant defence systems, but it is also credited with providing the first convincing evidence that biological reactive oxygen species (ROS) exist and that they are likely salient regulators of cell biology (Powers et al., 2016).

Since these landmark discoveries, the exercise physiology and redox field have grown

DOI: 10.1201/9781003051619-1

exponentially with prominent work enhancing our understanding of the role that free radicals (e.g., superoxide) and non-radicals (e.g., hydrogen peroxide), collectively termed ROS, play in systemic and intracellular muscle biology. It is now known, for example, that both different exercise modalities generate ROS and their production is associated with molecular biomarkers of oxidative stress. To the contrary, exercise-induced low-levels of ROS activate key signalling pathways leading to exercise-induced skeletal muscle adaptations. This chapter is designed to introduce the basic concept of oxidative (eu)stress, and in doing so, a brief overview of the key studies pertaining to exercise-induced oxidative stress and those identified as eustress regulators of skeletal muscle will be highlighted. A brief introduction to ground state molecular dioxygen and its derivatives (i.e., ROS) will also provide context to the general area of interest.

'Oxidative stress' is a popular term among exercise physiologists and is widely used in the literature, but rarely defined. Sies and Cadenas (1985) defined the term oxidative stress as 'a disturbance in the prooxidant–antioxidant balance in favour of the former'. Although this definition was widely accepted for over two decades, this description of oxidative stress has undergone scrutiny (Powers et al., 2016), and subsequently, Jones and Sies revised the definition of oxidative stress to 'an imbalance between oxidants and antioxidants in favour of the oxidants, leading to a disruption of redox signalling and control and/or molecular damage' (Sies & Jones, 2007). On the latter, the basic premise is that, in an open metabolic system, a steady-state redox balance is maintained at a particular setpoint, providing a basal redox tone, and any deviation away from the steady-state redox balance is regarded as a stress response. This oxidative distress concept, which explains supra-physiological disruption to redox signalling and/or oxidative stress damage to biomolecules, however, only explains one end of the continuum. At the other end, the term oxidative eustress may be used to connect low levels of oxidative stress to intracellular redox signalling and regulation (Sies 2020a).

OXYGEN AND ITS DERIVATIVES

Arguably the most important free radicals in biological systems are radical derivatives from oxygen, and although oxygen is necessary for survival, it has the potential to become toxic when supplied at concentrations higher than normally encountered (i.e., the toxicity of molecular oxygen is primarily due to the production of ROS). Ground state molecular oxygen is, in fact, a di-radical, with two unpaired electrons located in a different antibonding orbital with the same directional spin. Consequently, oxygen can only react with non-radicals by accepting a pair of electrons that spin in an anti-parallel manner (McCord, 1979).

TABLE 1.1

Reactive Oxygen Species Present in Biological Systems

Free Radical	Chemical Formula	Half-Life (seconds)
Hydroxyl	$OH^{\cdot}$	1×10^{-9}
Superoxide	$O_2^{\cdot-}$	1×10^{-6}
Singlet oxygen	$^1O_2$	1×10^{-6}
Alkoxyl	$RO^{\cdot}$	1×10^{-6}
Molecular oxygen	O_2	$> 10^2$
Hydrogen peroxide	H_2O_2	10
Peroxyl	$ROO^{\cdot}$	17
Hydroperoxyl	$HO_2^{\cdot}$	?
Carbonate	$CO_3^{\cdot-}$	?
Semiquinone	Q^{-}	Days

Free radicals in biological organisms include, but not limited to superoxide ($O_2^{\cdot-}$), hydroperoxyl ($HO_2^{\cdot}$), hydroxyl ($OH^{\cdot}$), carbonate ($CO_3^{\cdot-}$), peroxyl ($RO_2^{\cdot}$), alkoxyl ($RO^{\cdot}$) and nitric oxide (NO) radicals (Table 1.1). Hydrogen peroxide (H_2O_2) and hypochlorous acid (HOCL) have no unpaired electrons, and by definition they are non-radicals, but nevertheless these ROS can be powerful oxidants that are often involved in free radical reactions.

SUPEROXIDE ANION

Superoxide $(O_2^{\cdot-})$ is a commonly known oxygen-centred free radical. The reduction of a single electron to an O_2 molecule produces the superoxide anion. $O_2^{\cdot-}$ is relatively unreactive with non-radical species in comparison to other radical types (e.g., hydroxyl radical); however, if $O_2^{\cdot-}$ is generated near the site of any biochemical molecule it can be damaging. The reactivity of $O_2^{\cdot-}$ in aqueous solutions is more likely to occur in vivo (Halliwell and Gutteridge, 2015). In this environment, $O_2^{\cdot-}$ can act as a base, accepting a proton to form the hydroperoxyl radical ($HO_2^{\cdot}$). The pKa for this reaction is 4.8, indicating that approximately only 1% of $O_2^{\cdot-}$ is in the $HO_2^{\cdot}$ form at physiological pH (Pryor, 1986). However, there may be more $HO_2^{\cdot}$ at comparatively acidic membrane nanodomains. $O_2^{\cdot-}$ under a normal metabolic response is dismutated by SOD, which can increase the rate of intracellular dismutation by a factor of 10^{-9} $M^{-1}s^{-1}$ to form hydrogen peroxide (H_2O_2) (Chance et al., 1979), as shown in equation 1.1:

$$O_2^{\cdot-} + O_2^{\cdot-} + 2H^+ \rightarrow H_2O_2 + O_2 \qquad (1.1)$$

This would suggest that $O_2^{\cdot-}$ is deliberately produced in vivo, as the presence of SOD has no other cell purpose than to dismutate $O_2^{\cdot-}$ to H_2O_2.

HYDROGEN PEROXIDE

Any biological system generating $O_2^{\cdot-}$ produces H_2O_2 as a direct result of the dismutation reaction as above. H_2O_2 is a relatively long-lived species, and the longer half-life allows H_2O_2 to pass freely through biological membranes (via aquaporins), which $O_2^{\cdot-}$ generally cannot do without the aid of a negatively charged chloride (Cl^-) channel (Matsuo and Kaneko, 2000). H_2O_2 can therefore act as a passage to transmit free radical-induced damage across cellular compartments (Young and Woodside, 2001).

H_2O_2 is produced by a variety of intracellular reactions, although SOD is the main method. Several enzymatic reactions, including those catalysed by D-amino-acid oxidase in the peroxisomes and monoamine oxidase in the mitochondrial outer membrane, may directly produce H_2O_2 (Sen et al., 2000). H_2O_2 in the presence of transition metal ions (e.g., copper and iron; Fenton chemistry) can produce OH, which is the most reactive free radical known to chemistry (Halliwell, 1989).

HYDROXYL RADICAL

The highly oxidising hydroxyl radical is the final intermediary product to be formed prior to tissue damage (Lloyd et al., 1997). The aforementioned reactive species ($O_2^{\cdot-}$, H_2O_2) can exert pathological effects by giving rise to hydroxyl radical formation (Young and Woodside, 2001). The reason being that the hydroxyl radical has the ability, once formed, to react immediately and abstract a hydrogen atom from many biological molecules, including sugars, DNA, lipids and thiols at an extremely high diffusion-controlled rate (Halliwell, 1991; Young and Woodside, 2001). The degradation of these compounds may produce damaged products and a range of secondary organic radicals of variable reactivity (Halliwell and Gutteridge, 2015) that may possibly be measured directly by ESR spectroscopy (*e.g.*, peroxyl, alkoxyl or alkyl radicals). However, due to the extremely high reactivity and short half-life of the hydroxyl radical, it may be unable to react with any molecule beyond five molecular diameters from the site of formation (Pryor, 1986; Matsuo and Kaneko, 2000). Thus, the concept of site specificity is important, since if this radical is formed near DNA, it may damage any of the DNA bases, which may progress to the development of pathology.

Hydroxyl radicals can be formed in various ways in vivo with homolytic and heterolytic bond fission of water being well-known sources (Halliwell and Gutteridge, 2015). However, the most widely known mechanism of hydroxyl radical formation in an aqueous biological system is the transition metal-catalysed decomposition of $O_2^{\cdot-}$ and H_2O_2.

Fritz Haber and Joseph Weiss first proposed, in 1932, that hydroxyl radicals are produced when superoxide binds with hydrogen peroxide (Haber and Weiss, 1932), as outlined in equation 1.2:

$$O_2^{\cdot-} + H_2 \rightarrow O_2 + OH^{\cdot} + OH^- \qquad (1.2)$$

However, this reaction, which created a great deal of attention at the time, was found to be far, far too slow to be biologically important. This underlines the importance of considering kinetics, as well as the products or course of a reaction (Davison, 2010). Similarly, Henry Fenton, in 1894, first discovered that by adding H_2O_2 to the reducing agent, ferrous iron (Fe^{2+}), an organic compound like tartaric acid could be oxidised (Fenton, 1894). Partly as a result of studies conducted in the 1940s, the reaction is now known to involve the generation of the hydroxyl radical and was subsequently named the Fenton reaction (equation 1.3).

$$Fe^{2+} + H_2O_2 \rightarrow Fe^{3+} + OH^{\cdot} + OH^{-} \quad (1.3)$$

Copper can also catalyse the above Fenton reaction, and as such the intracellular levels of these transition metal ions are critical in defining the extent of hydroxyl radical production from superoxide and H_2O_2. However, it is debatable whether Fenton chemistry occurs in vivo due to the presence of the storage proteins, ferritin and caeruloplasmin, which tightly bind iron and copper, respectively, inhibiting free radical generation (Davison, 2010). It is, however, pertinent that storing iron in ferritin can oxidise thiols and produce H_2O_2. This 'sequestration' of metal ions has been reported to be an important antioxidant defence mechanism (Halliwell, 1996).

REACTIVE NITROGEN SPECIES

Reactive nitrogen species (RNS) is a term used to explain the process of an unpaired electron residing on a nitrogen molecule (Halliwell and Gutteridge, 2015). Common RNS include nitrogen dioxide $\left(NO_2^{\cdot}\right)$, nitrous acid ($HNO_2$), dinitrogen trioxide ($N_2O_3$), nitroxyl anion ($NO^-$), nitric oxide $\left(NO^{\cdot}\right)$ and peroxynitrous acid (ONOOH). Of these, perhaps the most widely known is $NO^{\cdot}$.

Nitric oxide $\left(NO^{\cdot}\right)$ [*nitrogen monoxide*] is by definition a free radical due to the presence of a single unpaired electron (Sen et al., 2000). $NO^{\cdot}$ is produced in several mammalian cells and can control blood flow, thrombosis and neural activity (Beckman and Koppenol, 1996). It has the ability to rapidly diffuse between cells and bind to $O_2^{\cdot-}$ to produce the peroxynitrite anion ($ONOO^-$), which although not a free radical has the potential to attack and damage cellular membranes (Sen et al., 2000). One-electron reduction of NO would yield the nitroxyl anion, which is a relatively unreactive and short-lived species (Halliwell and Gutteridge, 2015).

BRIEF OVERVIEW OF INDIRECT AND DIRECT BIOMARKERS OF EXERCISE-INDUCED OXIDATIVE STRESS

Much of our current understanding of exercise and oxidative stress from the perspective of cell and substrate damage has been made possible by improvements in measurement technology. Popular techniques used to assess macromolecular damage include the by-products of protein, lipid and/or DNA oxidation.

Pentane gas is derived from *n*-6 polyunsaturated fatty acids and can be detected in expired breath following polyunsaturated fatty acid oxidation. In a landmark study, pentane was used by Dillard and colleagues (1978) to provide the first evidence that oxidative stress is associated with 60 minutes of exercise at 50% VO_{2max}. The source of pentane was attributed to the peroxidation of liver lipid membranes. Since this early study, many investigators have demonstrated that several forms of exercise inclusive of prolonged endurance or short-duration, high-intensity exercise can result in increased lipid peroxidation (Ashton et al., 1998; Alessio, 2000; Davison et al., 2006).

Common assays used to indicate exercise-induced lipid peroxidation include lipid hydroperoxides, malondialdehyde or thiobarbituric acid reactive substances (TBARS) and isoprostanes. On the latter, F_2-isoprostances, a prostaglandin-like compound generated by non-enzymatic peroxidation of arachidonic acid, is probably regarded as the most reliable (due to increased specificity) method to quantify free radical-mediated lipid peroxidation in exercise (Fisher-Wellman and Bloomer, 2009). The TBARS assay is no longer recommended for use in an exercise context, due to its lack of specificity reacting with a variety of substrates at 532 nm to form malondialdehyde. HPLC combined with fluorometric detection can improve MDA quantification (Cobley et al., 2017), however, this biochemical tool and expertise may inaccessible to many. It is generally recommended that two or more assays are required to confirm exercise-induced lipid peroxidation.

Damage to DNA by exercise produced free radicals, leads to the formation of a variety of base and sugar modification products. The appearance of these molecular by-products generally indicates an oxidative stress response, as they are not present during normal nucleotide metabolism (Fisher-Wellman and Bloomer, 2009). In a seminal study, Hartmann et al. (1994) demonstrated that exhaustive exercise can induce blood DNA strand breaks as measured by the single-cell gel electrophoresis assay (comet) 24 h post-exercise. Consistent with this, others have shown that exhaustive running, cycling or rowing exercise can lead to single-strand breaks to blood DNA (Niess et al., 1996; Davison et al., 2005; Fogarty et al., 2013a). Indeed, over the years, many investigators have used the comet assay with or without modification (enzyme digestion) to confirm that high-intensity exercise damages DNA (Tsai et al., 2001; Mastaloudis et al., 2004; Brown et al. 2019). Relatively few studies have examined DNA oxidation directly in muscle tissue. Fogarty et al. (2013b) showed that following high-intensity isolated muscle contractions, 8-hydroxy-2-deoxyguanosine (marker of DNA base oxidation) increased in mitochondria, and more recently, Williamson et al. (2020a) observed an increase in global mitochondrial DNA damage (as measured by Long Amplicon qPCR) following repetitive bouts of exercise at 90%–95% heart rate max. While Williamson et al. (2020b) demonstrate that DNA single (comet) and double-strand breaks (γ-H2AX co-localised with 53BP1) are increased as a function of high intensity, all DNA damage seems to be repaired within a 48-hour time frame.

A major target substrate for oxidative attack is proteins due to their steer quantity in biological systems (Davies, 2012). Oxidative damage to proteins can occur directly by interaction of the protein with a free radical species or indirectly by interaction of the protein with a secondary product. Oxidative stress-related protein modification occurs via peptide backbone cleavage, cross-linking and/or modification of the side chain of an amino acid. Most protein damage is recognised as irreparable, and oxidative protein structure modification can lead to loss of enzymatic, contractile or structural function in the affected proteins, thus making them increasingly susceptible to proteolytic degradation (Fisher-Wellman and Bloomer, 2009). Although free radical-mediated protein damage leads to multiple oxidation by-products, the presence of carbonyl groups and/or derivatives have been commonly used as biomarkers (measured via immunochemical and spectrophotometric based assays, or through a mass spectrometry approach) of oxidative damage in an exercise setting. Reznick et al. (1992) were the first group to report an increase in reactive carbonyl derivatives in skeletal muscle of rats following a single bout of maximal exercise, while Witt et al. (1992) conducted the first reported study on the effects of regular exercise on the accumulation of reactive carbonyl derivatives and observed an increase in protein carbonyls following 8 weeks of moderate-intensity training. Over the years, an increase in protein oxidation evident by accumulation of protein carbonyls has been reported by several investigators (Radak et al., 2003; Lamprecht et al., 2008; Vezzoli et al. 2016). However, others have shown no change in protein oxidation with exercise (Gaeini et al., 2006; Kabasakalis et al., 2011), which may be reflective of insufficient sampling times, the difference in training status of participants and/or duration of exercise methodology employed (Fisher-Wellman and Bloomer, 2009).

Traditionally, exercise-induced oxidative stress is quantified by measuring the by-products of macromolecular damage as outlined. That said, others examining the relationship between exercise and free radical generation have used electron paramagnetic resonance (EPR) spectroscopy to directly detect primary and/or secondary free radical species. EPR spectroscopy is a physical method of observing molecules with an unpaired electron in a magnetic field (Ikeya, 1993) and is arguably the most sensitive, specific and direct method of measuring free radical species (Ashton et al., 1999). EPR spectroscopy was first used in an exercise context by Davies et al. (1982) showing a two to threefold increase in free radical production in rat muscle and liver homogenates following exercise to exhaustion. This study is arguably the most quoted study within the exercise and oxidative stress literature. Following this work, Jackson et al. (1985) used EPR to examine skeletal muscle tissue from mice and rats after exercise, and observed an increase in free production in all intact and homogenate muscle samples measured at 77 K. A *gauss* (*g*) value of 2.004 was ascribed to the EPR signals detected, comparing favourably to that previously reported by Davies et al. (1982) and attributed to a free radical species generated in the ubiquinone region of the mitochondrial electron transport chain. The work of Jackson et al. (1985) also highlighted the inability to detect an EPR signal in aqueous samples measured at room temperature, and this may be due to water having a high dielectric constant (Voet and Voet, 1995), which would favour the ability of water molecules to strongly absorb the microwaves used to detect a magnetic field. To overcome this, Ashton et al. (1998) used an organic solvent (toluene) combined with the spin trap α-phenyl-*tert*-butylnitrone (PBN) to detect room temperature lipid-derived (cell membrane) free radical species in blood following exhaustive exercise. This EPR spin trapping approach has been replicated by others showing an exercise-induced free radical production (alkoxyl, alkyl and/or peroxyl species) in healthy and diseased models (Ashton et al., 1999; Davison et al., 2002, 2006, 2008a,b). While Davies et al. (1982) and Jackson et al. (1985) were amongst the first to discover that contracting animal muscle generates oxidants, Bailey and colleagues (2007) provided direct evidence for intramuscular free radical (ubisemiquinone) accumulation following exercise in humans. Although EPR spectrometers can be expensive, and the detection and analysis of the acquired spectra require practical experience by the user, EPR spectroscopy remains an indispensable technique to study exercise-induced oxidative stress.

BRIEF OVERVIEW OF ROS AS EUSTRESS REGULATORS OF SKELETAL MUSCLE

Although ROS were generally considered cell damaging agents, a substantial body of evidence has now emerged indicating that intracellular ROS are also essential signalling molecules (Figure 1.1). Exercise-induced oxidative eustress provokes a series of beneficial hormetic responses as controllers of intracellular muscle adaptation (Cobley, 2020). While the boundary between oxidative eustress and distress is difficult to discern (and dependent on exercise intensity and duration), inferences can only be made when the radical species, intracellular location and a specific tissue are considered for any given biological context (Cobley, 2020).

The concept that exercise regulates tissue-specific adaptation through ROS derived mainly through work conducted by the Vina, Jackson and Ristow research groups. In an early study, Gomez-Cabrera et al. (2005) ascertained that inhibiting xanthine oxidase activity during exercise prevents the exercise-induced activation of signalling pathways involved in the cell adaptative process. Numerous studies have relied on an antioxidant interventional approach to examine the connection between ROS and signalling-induced adaptation. Jackson et al. (2004) have shown that expression of heat shock protein 72 is suppressed by vitamin E supplementation, while Gomez-Cabrera and colleagues (2008) have demonstrated in rats that vitamin C supplementation inhibits the exercise-induced increase in PGC-1α and mitochondrial biogenesis in muscle. Using a human model, Ristow et al. (2009) confirmed the findings of Gomez-Cabrera et al. (2008) observing that ingestion of 1 g of vitamin C combined with 400 IU of vitamin E per day prevents mitochondrial biogenesis in muscle.

The specific mechanisms involved in the oxidative eustress–muscle adaptation relationship are

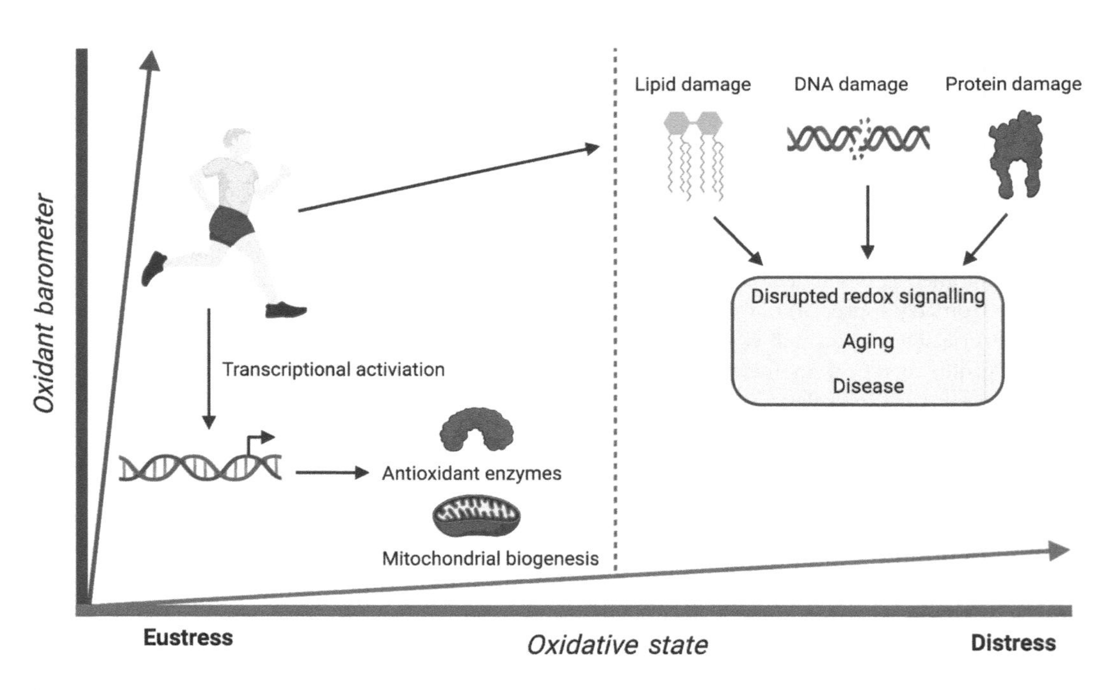

Figure 1.1 Exercise-induced oxidative eustress and distress.

fully reviewed elsewhere; however, at this juncture, the transcription factor, nuclear erythroid 2-related factor (Nrf-2), is highlighted as a main protagonist aligned to the upregulation of enzymatic antioxidant (e.g., SOD, GPx) gene transcription, while other redox-sensitive transcription factors such as NF-KB and the forkhead family of transcription factors (FoxO) may regulate PGC-1α and mitochondrial biogenesis. These transcription factors, which are central to several redox-sensitive signalling pathways, are in part controlled by $O_2 .^-$ and/or H_2O_2 (Ji, 2015). Central to this idea is the current working paradigm that ROS, notably H_2O_2, exert their effects, in part, by directly or indirectly oxidising protein thiols (i.e., cysteine residues) to distinct reversibly oxidised chemotypes (e.g., from a sulphhydryl [reduced] to a sulphenic acid [oxidised]), which is a central protein regulatory mechanism. Therefore, any manipulation of these pathways via antioxidant ingestion, for example, may interfere with the muscle redox control of exercise-specific adaptation.

As outlined by Powers et al. (2016), interest in redox controls systems has rapidly grown to incorporate a new paradigm that redox switches regulate protein function by cysteine modifications. To this end, advances in redox proteomics and the understanding of oxidisable thiols will provide new insights into the role of ROS in mediating exercise-induced adaptations (McDonagh et al., 2014).

Oxidant barometer represents ROS (e.g., $O_2^{\cdot -}$, H_2O_2) production. Oxidative state denotes the eustress to distress continuum. Dashed line represents an oxidative threshold.

CONCLUSION

The concept that exercise-induced ROS can damage important macromolecules, such as DNA, lipids and proteins, has now been circulating for over 30 years. More recently, however, the emphasis has been towards defining the role of ROS in exercise training adaptations. The hormesis theory generally refers to an environmental and beneficial effect on a cell or organism at low doses that can otherwise be harmful at a high dose, creating a bell-shaped curve (Mattson, 2008). In the context of exercise and low physiological doses of ROS, redox-sensitive transcription factors are activated leading to cell adaption through upregulation of several molecular pathways; termed exercise-induced oxidative eustress. Oxidative eustress is a ripe area for further investigation in the context of exercise with emerging advances in redox proteomics coupled with new genomic techniques designed to interrogate the redox-sensitive transcriptome. As such, determining the interplay between exercise and oxidative eustress will pave the way for the next 30 years of robust, scientific inquiry, in this exciting field of exercise biochemistry.

ACKNOWLEDGEMENT

Figure created using BioRender.

REFERENCES

Alessio, H.M. (2000) Lipid peroxidation in healthy and diseased models: Influence of different types of exercise. In: Sen, C.K., Packer, L., Hanninen, O. (eds.) *Handbook of Oxidants and Antioxidants in Exercise.* Elsevier, Amsterdam.

Ashton, T., Rowlands, C.C., Jones, E., Young, I.S., Jackson, S.K., Davies, B., Peters, J.R. (1998) Electron spin resonance spectroscopic detection of oxygen-centred radicals in human serum following exhaustive exercise. *Eur J Appl Physiol.* 77, 498–502.

Ashton, T., Young, I.S., Peters, J.R., Jones, E., Jackson, S.K., Davies, B., Rowlands, C.C. (1999) Electron spin resonance spectroscopy, exercise, and oxidative stress: An ascorbic acid intervention study. *J Appl Physiol.* 87, 2032–2036.

Bailey, D.M., Lawrenson, L., McEneny, J., Young, I.S., James, P.E., Jackson, S.K., Henry, R.R., Mathieu-Costello, O., McCord, J.M., Richardson, R.S. (2007) Electron paramagnetic spectroscopic evidence of exercise-induced free radical accumulation in human skeletal muscle. *Free Radic Res.* 41, 182–190.

Beckman, J.S., Koppenol, W.H. (1996) Nitric oxide, superoxide, and peroxynitrite: The good, the bad, and the ugly. *Am J Physiol.* 271, C1424–C1437.

Brown, M., McClean, C.M., Davison G., Brown, J., Murphy, M.H. (2019) Preceding exercise and postprandial hypertriglyceridemia: Effects on lymphocyte cell DNA damage and vascular inflammation. *Lipids in Health and Disease.* 18, 125.

Chance, B., Sies, H., Boveris, A. (1979) Hydroperoxide metabolism in mammalian organs. *Physiol Rev.* 59, 527–605.

Chance, B., Williams, G.R. (1956) The respiratory chain and oxidative phosphorylation. *Adv Enzymol Relat Subj Biochem*. 17, 65–134.

Cobley, J.N. (2020) How exercise indices oxidative eustress. In: Sies, H. (ed.) *Oxidative Stress: Eustress and Distress*. Elsevier, Amsterdam.

Cobley, J.N., Close, G.L., Bailey, D.M., Davison, G.W. (2017) Exercise redox biochemistry: conceptual, methodological and technical recommendations. *Redox Biol*. 12, 540–548,

Commoner, B., Townsend, J. Pake, G.E. (1954) Free radicals in biological materials. *Nature*. 174, 689–691.

Davies, M.J. (2012) Oxidative damage to proteins. In: Chatgilialoglu, C., Studer, A. (eds.) *Encyclopedia of Radicals in Chemistry, Biology and Materials*. John Wiley & Sons, Hoboken, NJ.

Davies, K.J.A., Quintanilha, A.T., Brooks, G.A., Packer, L. (1982) Free radicals and tissue damage produced by exercise. *Biochem Biophy Res Comm*. 107, 1198–1205.

Davison, G.W. (2010) *Exercise and Oxidative Stress in Health and Disease*. Lambert Academic Publishing, USA.

Davison, G.W., Ashton, T., Davies, B., Bailey, D.M. (2008a) In vitro electron paramagnetic resonance characterisation of free radicals: relevance to exercise-induced lipid peroxidation and implications of ascorbate prophylaxis. *Free Radical Res*. 42, 379–386.

Davison, G.W., Ashton, T., George, L., Young, I.S., McEneny, J., Davies, B., Jackson, S.K., Peters, J.R., Bailey, D.M. (2008b) Molecular detection of exercise-induced free radicals following ascorbate prophylaxis in type 1 diabetes mellitus: A randomised controlled trial. *Diabetologia*. 51, 2049–2059.

Davison, G.W., George, L., Jackson, S.K., Young, I.S., Davies, B., Bailey, D.M., Peters, J.R., Ashton, T. (2002) Exercise, free radicals, and lipid peroxidation in type 1 diabetes mellitus. *Free Rad Biol Med*. 33, 1543–1551.

Davison, G.W., Hughes, C.M., Bell, R.A. (2005) Exercise and mononuclear cell DNA damage: The effects of antioxidant supplementation. *Int J Sport Nutr Exerc Metab*. 15, 480–92.

Davison, G.W., Morgan, R.M., Hiscock, N., Garcia, J.M., Grace, F., Boisseau, N., Davies, B., Castell, L., McEneny, J., Young, I.S., Hullin, D., Ashton, T., Bailey, D.M. (2006) Manipulation of systemic oxygen flux by acute exercise and normobaric hypoxia: Implications for reactive oxygen species generation. *Clin Sci*. 110, 133–141.

Dillard, C.J., Litov, R.E., Savin, W.M., Dumelin, E.E., Tappel, A.L. (1978) Effects of exercise, vitamin E, and ozone on pulmonary function and lipid peroxidation. *J Appl Physiol Respir Environ Exerc Physiol*. 45(6), 927–32. https://doi.org/10.1152/jappl.1978.45.6.927.

Fenton, H.J.H. (1894) Oxidation of tartaric acid in presence of iron. *J Chem Soc Trans*. 65, 899–911.

Fisher-Wellman, K., Bloomer, R.J. (2009) Acute exercise and oxidative stress: A 30 year history. *Dynamic Medicine*. 8, 1.

Fogarty, M.C., Hughes, C.M., Burke, G., Brown, J.C., Davison, G.W. (2013a) Acute and chronic watercress supplementation attenuates exercise- induced peripheral mononuclear cell DNA damage and lipid peroxidation. *Br J Nutr*. 109, 293–301.

Fogarty, M.C., De Vito, G., Hughes, C.M., Burke, G., Brown, J., McEneny, J., Brown, D., McClean, C., Davison, G.W. (2013b) Effects of α-lipoic acid on mtDNA damage after isolated muscle contractions. *Med Sci Sports Exerc*. 45, 1469–1477.

Gaeini, A.A., Rahnama, N., Hamedinia, M.R. (2006) Effects of vitamin E supplementation on oxidative stress at rest and after exercise to exhaustion in athletic students. *J Sports Med Phys Fitness*. 46, 458–461.

Gerschman, R., Gilbert, D.L., Nye, S.W., Dwyer, P., Fenn, W.O. (1954) Oxygen poisoning and x irradiation: A mechanism in common. *Science*. 119, 623–626.

Gomez-Cabrera, M.C., Borras, C., Pallardo, F.V., Sastre, J., Ji, L.L., Vina, J. (2005) Decreasing xanthine oxidase-mediated oxidative stress prevents useful cellular adaptations to exercise in rats. *J Physiol*. 567, 113–120.

Gomez-Cabrera, M.C., Domenech, E., Romagnoli, M., Arduini, A., Borras, C., Pallardo, F.V., Sastre, J., Vina, J. (2008) Oral administration of vitamin C decreases muscle mitochondrial biogenesis and hampers training-induced adaptations in endurance performance. *Am J Clin Nutr*. 87, 142–149.

Haber, F., Weiss, J. (1932) Über die Katalyse des Hydroperoxydes. *Naturweissenschaften*. 20, 948–950.

Halliwell, B. (1989) Superoxide, iron, vascular endothelium and reperfusion injury. *Free Rad Res Comms*. 5, 315–318.

Halliwell, B. (1991) Reactive oxygen species in living systems: Source, biochemistry, and role in human disease. *Am J Med*. 91(3C), 14S.

Halliwell, B. (1996) Antioxidants in human health and disease. *Annu Rev Nutr*. 16, 33–50.

Halliwell, B., Gutteridge, J.M.C. (2015) *Free Radicals in Biology and Medicine*. Oxford University Press, Oxford.

Hartmann, A., Plappert, U., Raddatz, K., Griinert-Fuchs, M., Speit, G. (1994) Does physical activity induce DNA damage? *Mutagenesis*. 9, 269–72.

Ikeya, M. (1993) *New Applications of Electron Spin Resonance: Dating Dosimetry and Microscopy*. World Scientific Publishing, London.

Jackson, M.J., Edwards, R.H.T., Symons, M.C.R. (1985) Electron spin resonance studies of intact mammalian skeletal muscle. *Biochim Biophy Acta*. 847, 185–190.

Jackson, M.J., Khassaf, M., Vasilaki, A., McArdle, F., McArdle, A. (2004) Vitamin E and the oxidative stress of exercise. *Ann N Y Acad Sci*. 1031, 158–168.

Ji, L.L. (2015) Redox signalling in skeletal muscle: Role of aging and exercise. *Adv Physiol Educ*. 39, 352–359.

Kabasakalis, A., Kyparos, A., Tsalis, G., Loupos, D., Pavlidou, A., Kouretas, D. (2011) Blood oxidative stress markers after ultramarathon swimming. *J Strength Cond Res*. 25, 805–811.

Lamprecht, M., Greilberger, J.F., Schwaberger, G., Hofmann, P., Oettl, K. (2008) Single bouts of exercise affect albumin redox state and carbonyl groups on plasma protein of trained men in a workload-dependent manner. *J Appl Physiol*. 104, 1611–1617.

Lloyd, R.V., Hanna, P.M., Mason, R.P. (1997) The origin of the hydroxyl radical oxygen in the fention reaction. *Free Rad Biol Med*. 22, 885–8.

Mastaloudis, A., Yu, T.W., O'Donnell, R.P., Frei, B., Dashwood, R.H., Traber, M.G. (2004) Endurance exercise results in DNA damage as detected by the comet assay. *Free Rad Biol Med*. 36, 966–75.

Mattson, M.P. (2008) Hormesis defined. *Ageing Res Rev*. 7, 1–7.

Matsuo, M., Kaneko, T. (2000) The chemistry of reactive oxygen species and related free radicals. In: Radak, Z. (ed.) *Free Radicals in Exercise and Aging*. Human Kinetics, Champaign, IL.

McCord, J.M. (1979) Superoxide, superoxide dismutase and oxygen toxicity. *Rev Biochem Toxicol*. 1, 109–124.

McCord, J.M., Fridovich, I. (1969) Superoxide dismutase. An enzymic function for erythrocuprein (hemocuprein). *J Biol Chem*. 244, 6049–6055.

McDonagh, B., Sakellariou, G.K., Jackson, M.J. (2014) Application of redox proteomics to skeletal muscle aging and exercise. *Biochem Soc Trans*. 42, 965–970.

Niess, A.M., Hartmann, A., Grünert-Fuchs, M., Poch, B., Speit, G. (1996) DNA damage after exhaustive treadmill running in trained and untrained men. *Int J Sports Med*. 17, 397–403.

Powers, S.K., Radak, Z., Ji, L.L. (2016) Exercise-Induced oxidative stress: Past, present and future. *J Physiol*. 594, 5081–5092.

Pryor, W.A. (1986) Oxy-radicals and related species: Their formation, lifetimes, and reactions. *Ann Rev Physiol*. 48, 657–667.

Radak, Z., Ogonovszky, H., Dubecz, J., Pavlik, G., Sasvari, M., Pucsok, J., Berkes, I., Csont, T. Ferdinandy, P. (2003) Super-marathon race increases serum and urinary nitrotyrosine and carbonyl levels. *Eur J Clin Investig*. 33, 726–30.

Reznick, A.Z., Witt, E., Matsumoto, M., Packer, L. (1992) Vitamin E inhibits protein oxidation in skeletal muscle of resting and exercised rats. *Biochem Biophys Res Commun*. 189, 801–806.

Ristow, M., Zarse, K., Oberbach, A., Kloting, N., Birringer, M., Kiehntopf, M., Stumvoll, M., Kahn, C.R., Bluher, M. (2009) Antioxidants prevent health-promoting effects of physical exercise in humans. *Proc Natl Acad Sci USA*. 106, 8665–8670.

Sen, C.K., Roy, S., Packer, L. (2000) Exercise-induced oxidative stress and antioxidant nutrients. In: Maughan, R.J. (eds.) *Nutrition in Sport*. Blackwell Science, Oxford.

Sies, H. (2020a) On the history of oxidative stress: concept and some aspects of current development. *Curr Opin Toxicol*. 7, 122–126.

Sies, H. (2020b) Oxidative stress: Concept and some practical aspects. *Antioxidants*. 9, 852.

Sies, H., Cadenas, E. (1985) Oxidative stress: Damage to intact cells and organs. *Philos Trans R Soc Lond B Biol Sci*. 311, 617–631.

Sies, H., Jones, D.P. (2007) *Oxidative Stress*. Elsevier, Amsterdam.

Tsai, K., Hsu, T.G., Hsu, K.M., Cheng, H., Liu, T.Y., Hsu, C.F., et al. (2001) Oxidative DNA damage in human peripheral leukocytes induced by massive aerobic exercise. *Free Rad Biol Med*. 31, 1465–1472.

Vezzoli, A., Dellanoce, C., Mrakic-Sposta, S., Montorsi, M., Moretti, S., Tonini, A., Pratali, L., Accinni, R. (2016) Oxidative stress assessment in response to ultraendurance exercise: thiols redox status and ROS production according to duration of a competitive race. *Oxidative Medicine and Cellular Longevity*. 2016, 1–13.

Voet, D., Voet, D. (1995) *Biochemistry*. John Wiley and Sons, Hoboken, NJ.

Witt, E.H., Reznick, A.Z., Viguie, C.A., Starke-Reed, P., Packer, L. (1992) Exercise, oxidative damage and effects of antioxidant manipulation. *J Nutr.* 122, 766–773.

Williamson, J., Hughes, C.M., Burke, G., Davison, G.W. (2020b) A combined γ-H2AX and 53BP1 approach to determine the DNA damage-repair response to exercise in hypoxia. *Free Rad Biol Med.* 154, 9–17.

Williamson, J., Hughes, C.M., Cobley, J.N., Davison, G.W. (2020a) The mitochondria-targeted antioxidant MitoQ, attenuates exercise-induced mitochondrial DNA damage. *Redox Biology.* 36, 101673.

Young, I.S., Woodside, J.V. (2001) Antioxidants in health and disease. *J Clin Path.* 54, 176–186.

CHAPTER TWO

Measuring Oxidative Damage and Redox Signalling

PRINCIPLES, CHALLENGES, AND OPPORTUNITIES

James N. Cobley and Gareth W. Davison

CONTENTS

The Importance of Measuring Oxidative Eustress / 11
Blink and You'll Miss It: The Challenge of Directly Measuring Reactive Species / 12
Novel Approaches to Measure Oxidative Damage / 13
Redox Signalling: The Promise of Novel Immunological Assays / 16
Systemic Redox Analysis: Moving beyond Measuring Antioxidant Enzyme Activity Towards PRDX Isoform Dimers / 19
Concluding Recommendations / 19
Acknowledgements / 20
References / 20

THE IMPORTANCE OF MEASURING OXIDATIVE EUSTRESS

Understanding oxidative eustress relies on measuring reactive species directly and/or indirectly using oxidative damage and redox signalling reporters (Murphy et al., 2011). One example suffices to demonstrate the importance of measuring oxidative eustress: a lack of appropriate assays (e.g., redox signalling reporters) means whether vitamin C and E blunt molecular exercise adaptations by scavenging free radicals is ambiguous (Cobley et al., 2015a). Disambiguating the underlying mechanisms is essential for unravelling the molecular basis of oxidative eustress (Cobley, 2020). For example, if either vitamin impaired redox signalling (e.g., decreased KEAP1 thiol oxidation), it would imply an underappreciated role for free radicals, lipid peroxidation, and/or reversal of S-nitrosation or sulphenic acid formation (Niki, 2012; Winterbourn, 2015; Cobley, 2020). In particular, the ability of vitamin C to reduce protein sulphenic acids ($k \sim 10^3\ M^{-1}s^{-1}$) could attenuate hydrogen peroxide (H_2O_2) mediated redox signalling (Anschau et al., 2020). Appropriate assays can, therefore, transform our understanding of oxidative eustress.

Measuring reactive species, oxidative damage, and redox signalling presents several formidable challenges. For example, measuring superoxide using dihydroethidium (DHE) poses kinetic challenges (see below). Further, interpreting DHE fluorescence is complex (i.e., several non-selective reactions contribute) and appropriate controls are difficult (i.e., superoxide dismutase (SOD) mimetics are chemically promiscuous) (Sawyer and Valentine, 1981; Zielonka and Kalyanaraman,

DOI: 10.1201/9781003051619-2

2010; Batinic-Haberle et al., 2014). In exercise physiology, most studies attempt to chemically footprint oxidative damage and/or redox signalling. Too often overtly flawed assays like the thiobarbituric acid-reactive substance (TBARS) assay for lipid peroxidation are used (see Table 2.1). Taking the previous example, using TBARS to determine whether vitamin C and/or E decrease lipid peroxidation to blunt exercise adaptations is inappropriate (Cobley et al., 2017). Insufficient specificity allied to assay-induced lipid peroxidation invalidates TBARS (Forman et al., 2015). Even when a valid assay is used (e.g., an OxyBlot to measure protein carbonylation), it is usually confined to the global scale (i.e., the identity of the carbonylated proteins remains unknown). More broadly, the literature has focused on a narrow subset of oxidative damage biomarkers and has largely neglected chemical surrogates of redox signalling. After briefly reviewing direct strategies, we critique how to measure oxidative damage and redox signalling in humans. The present chapter is delimited to thiol-based redox signalling and oxidative damage to proteins and lipids.

BLINK AND YOU'LL MISS IT: THE CHALLENGE OF DIRECTLY MEASURING REACTIVE SPECIES

Their short half-life (e.g., 10^{-9} seconds for hydroxyl radical) due to kinetically efficient endogenous biological reactions (e.g., diffusion-controlled reaction of superoxide with SOD isoforms) makes measuring reactive species directly using fluorophores or electron paramagnetic resonance (EPR) spin trapping challenging (Zielonka and Kalyanaraman, 2018). As Zielonka and Kalyanaraman remark, superoxide even at a supraphysiological 1 μM (physiological = 10–200 pM) level will decrease to 500 nM within 2 seconds at pH 7.4 due to spontaneous dismutation alone. It is, therefore,

TABLE 2.1
Flawed redox assays and/or approaches.

Assay/Approach	Rationale	Flaws/Issues	Alternative (s)
TBARS	A purported measure of lipid peroxidation	The assay procedure itself artificially generates MDA. Non-lipids (e.g., sugars and proteins) react to produce MDA	Measure MDA by HPLC. Measure F_2-isoprostanes or lipid hydroperoxides
2′,7′-dichlorofluorescin (DCF)	To measure reactive species directly	Fails to react appreciably with several key species. For example, DCF is chemically unable to react with H_2O_2 as is often claimed. DCF redox cycles with O_2 to produce superoxide. Sensitive to peroxidase mediated oxidation	Measure free radical levels using DHE
*Total antioxidant capacity (TAC)**	Purports to measure the antioxidant capacity of a biological system	No total marker of antioxidant capacity exists—the redox environment is simply too complex! Confounded by exposure to 21% O_2 in the assay. Orthogonal to many two-electron species like H_2O_2 and key enzymes (e.g., peroxiredoxins) that set intracellular buffering capacity	Measure vitamin C radical by EPR or total thiols. In cells, measure antioxidant enzyme activity
Glutathione redox state or antioxidant enzyme activity in plasma/serum	To report systemic redox state and antioxidant enzyme activity	Insufficient activity plasma and serum to make the assay meaningful. For example, in contrast to cells (mM levels), glutathione is present at μM levels in plasma. Amplified by failing to alkylate glutathione to prevent ex vivo oxidation. Many assays are confounded by haemolysis	Measure thioredoxin/peroxiredoxin content in plasma/serum or their redox state in erythrocytes

* Applies TAC variants like oxygen radical absorbance capacity (ORAC)

difficult for any fluorophore or spin trap to intercept superoxide because they must compete with kinetically efficient sinks (i.e., SOD reacts with superoxide at least 10,000 times faster than DHE). Notwithstanding, EPR spin trapping can measure free radicals. Free radicals and non-radicals can also be measured using fluorophores (e.g., MitoNeoD, Shchepinova et al., 2017). Typically, a spin trap or fluorophore is added to a sample before rapid ex vivo analysis (e.g., EPR, Davison et al., 2008). Seminal studies have used direct measurement to show that contracting skeletal muscle fibres unequivocally produce free radicals (EPR) and that much contraction-induced free radical production emanates from the cytosol (fluorophores) (Davies et al., 1982; Sakellariou et al., 2013).

While direct measurement approaches are invaluable consider that: (1) ex vivo oxidation induced by exposure to 21% ground state molecular dioxygen (O_2) coupled to transition metal ion (e.g., Fe^{2+}) release could be considerable (Halliwell and Gutteridge, 2015); (2) discerning reactive species identity is difficult (Zielonka and Kalyanaraman, 2010); (3) probes and spin traps can have off-target effects, especially when in molar excess; and (4) measures must be performed rapidly to prevent one from simply measuring ex vivo oxidation (i.e., recall the micro and nanosecond lifetime of most reactive species). Nevertheless, provided appropriate controls are used (Spasojević, 2011), direct measures can address several pressing questions (e.g., reactive species providence). Researchers may, therefore, wish to consider: (1) immuno-spin-trapping (i.e., using an antibody to detect spin trap protein adducts that are diagnostic of protein radicals) because it can be coupled to proteomic analysis; and (2) high-performance liquid chromatography-based detection of diagnostic probe products to discern reactive species identity (e.g., 2-OH-E for superoxide mediated DHE oxidation (Zielonka and Kalyanaraman, 2018)). Finally, use Amplex Red to infer mitochondrial H_2O_2 release in isolated skeletal muscle fibres with caution since many of the aforementioned limitations and several more apply (Murphy, 2009; Munro and Pamenter, 2019). For example, the ability of the Amplex red to permeate the inner mitochondrial membrane can distort the assay indirectly by altering the proton motive force and directly by spurious carboxylesterase mediated resorufin oxidation (Miwa et al., 2016).

NOVEL APPROACHES TO MEASURE OXIDATIVE DAMAGE

Oxidative damage is usually assessed to determine the ability of a given acute or chronic training stimulus (e.g., high-intensity interval training) and/or nutritional antioxidant (e.g., vitamin C) to modify a key redox parameter like lipid peroxidation (Cobley et al., 2014). Measuring oxidative damage is no trivial task (Halliwell and Whiteman, 2004; Murphy et al., 2011; Halliwell and Gutteridge, 2015). The inherent complexity of the redox environment coupled to the diverse spectrum of chemically heterogenous oxidative damage adducts coexisting in biological systems means no ideal biomarker exists. Nuanced hierarchical spatiotemporal complexity means no all-encompassing redox biomarker exists: it cannot simply be distilled to a binary logic gate (Cobley et al., 2017). Moreover, the choice of redox biomarker varies according to the research question. For example, to appraise the ability of mitochondria-targeted quinone (MitoQ) to decrease exercise-induced cardiolipin peroxidation in the inner mitochondrial membrane by scavenging alkoxyl and peroxyl radials it is unwise to measure cytosolic glutathione (GSH) because it is spatially and chemically orthogonal to cardiolipin. Chemically defined hypotheses are, therefore, essential for selecting appropriate assays (e.g., targeted lipidomic-based cardiolipin oxidation product profiling). To select appropriate assays (see Table 2.2), we present novel analytical approaches before delineating general interpretational considerations.

Oxidative macromolecule damage yields a panoply of chemically heterogenous products (Halliwell and Gutteridge, 2015). For example, in DNA alone, OH-mediated guanine oxidation can yield over 20 chemically unique products. Analogously, the same adduct can form in a myriad of disparate targets. For example, protein carbonyls can form on a proteome-wide scale with diverse functional consequences. The general approach is, however, to assess a single oxidation adduct at the global level (e.g., global protein carbonylation by ELISA). Omics workflows could be used to unravel the scale, diversity, and targets of exercise-induced oxidative damage for several adducts from carbonylation to nitration (Margaritelis et al., 2016a). For example, to identify the targets of exercise-induced protein damage protein carbonyls could be derivatised

TABLE 2.2
A step-wise guide to selecting an appropriate assay.

Step	Process	Comments	Example
1	Identify the biological redox end-point (e.g., oxidative damage)	Decide whether oxidative damage, redox signalling, or measuring reactive species is important. If redox signalling is important, we can exclude oxidative damage assays (e.g., F_2-isoprostanes) at the outset	Measure redox signalling to assess whether a selected antioxidant (e.g., MitoQ) impacts exercise adaptations
2	Select one or more assays	Select an assay able to chemically report the process of interest. Use multiple assays of one process of interest instead of orthogonal assays (e.g., MDA and F_2-isoprostanes if lipid peroxidation is of interest). Carefully consider if the assay is appropriate for the biological material available (e.g., it is difficult to assess redox signalling systemically). Avoid the flawed assays detailed in Table 2.1	Use cardiolipin oxidation to assess whether mitochondria-targeted vitamin E impacts lipid peroxidation
3	Select the level of analysis required	Decide whether global or targeted analysis is required. Often the level required will depend on how chemically precise the experimental hypothesis is. It may often be sufficient to assess a given process at the global level	Shotgun redox proteomics to investigate the scale of exercise-induced redox signalling
4	Select the analytical workflow	Determine the analytical workflow to be used with reference to the redox biology literature. If selecting between an ELISA or Western blot for a given marker (e.g., protein carbonylation, with reference to step 3, consider whether quantitative (ELISA) or spatial information (MW of carbonylated proteins) is relevant	Targeted lipidomic approaches to assess cardiolipin oxidation products
5	Select appropriate controls	Judicious analysis of the available controls is often required to prevent confounding effects	Include N-ethylmaleimide in the lysis buffer to prevent artificial glutathione oxidation
6	Review what the assay can and can't do	Before proceeding check steps 1–4 and review what the assay can and cannot do. This can often identify an inappropriate approach	Incorporate a mass spec step (e.g., MRM) may need to identify the carbonylation sites if a targeted immunological approach was used to measure a target protein but identifying the residue was key

with hydrazine biotin before being purified with immobile streptavidin supports for high-performance liquid chromatography mass spectrometry (HPLC-MS-MS). Systematic, hypothesis-free, and high-throughput (i.e., multiplexed) isobaric tandem mass tag coupled LC-MS-MS analysis could identify and quantify the occupancy (i.e., percent carbonylation) of each site on the proteome-wide scale at the tissue or systemic level. Functional significance could be determined using protein activity surrogates (e.g., enzyme assays). Appropriate controls (e.g., preventing lysis induced artefacts) are, however, required. Additionally, omics analysis is complicated by several factors including data-dependent acquisition, tryptic digestion issues, and difficulties assessing hydrophobic membrane proteins (Cobley, Sakellariou, et al., 2019). Overall, omics workflows could be used to achieve unprecedented insight into the scale, diversity, and functional significance of exercise-induced oxidative macromolecule damage (see Figure 2.1).

Using omics workflows to measure exercise-induced oxidative damage is rate-limited by the time, expertise, and resources required. To expand current non-omic approaches to measure oxidative damage, we propose targeted immunological analysis. For example, manganese superoxide dismutase (MnSOD) activity is inhibited by tyrosine 34 (Y34) nitration to 3-nitrotyrosine (3-NT). Y34 nitration impairs catalysis by disrupting the hydrogen bond network required to protonate manganese bound superoxide (Abreu and Cabelli, 2010). MnSOD immunocapture followed by 3-NT blotting can assess a functionally annotated oxidative damage biomarker. However, appropriate steps to safeguard faithful immunocapture are required (e.g., covalently immobilising the antibody to a solid support). Additionally, reciprocal immunocapture is useful (e.g., confirming a 3-NT antibody elutes MnSOD). Further, immunoblotting is semi-quantitative. In certain cases, fluorophores could be used to achieve quantitative in-gel analysis. For example, a heterobifunctional hydrazine carbonyl reactive warhead with a short polyethylene glycol (PEG) spacer between a fluorescent moiety for quantitative in-gel fluorescent target protein carbonylation analysis.

Targeted functional analysis has been used to assess how exercise impacts ryanodine receptor carbonylation in skeletal muscle (Place et al., 2015) and can readily be extended to several adducts (e.g., 4-hydroxynoneneal). If the function of a particular oxidative damage signature is known (e.g., tyrosine nitration for MnSOD), targeted analysis simplifies interpretational complexity. For example, if acute exercise increased MnSOD nitration, one can infer decreased enzyme activity and readily confirm it with an enzyme activity assay. The rich omics literature on oxidative damage enables one to select novel targets for analysis at the systemic (i.e., secreted proteins) and tissue level. In contrast, global assays seldom

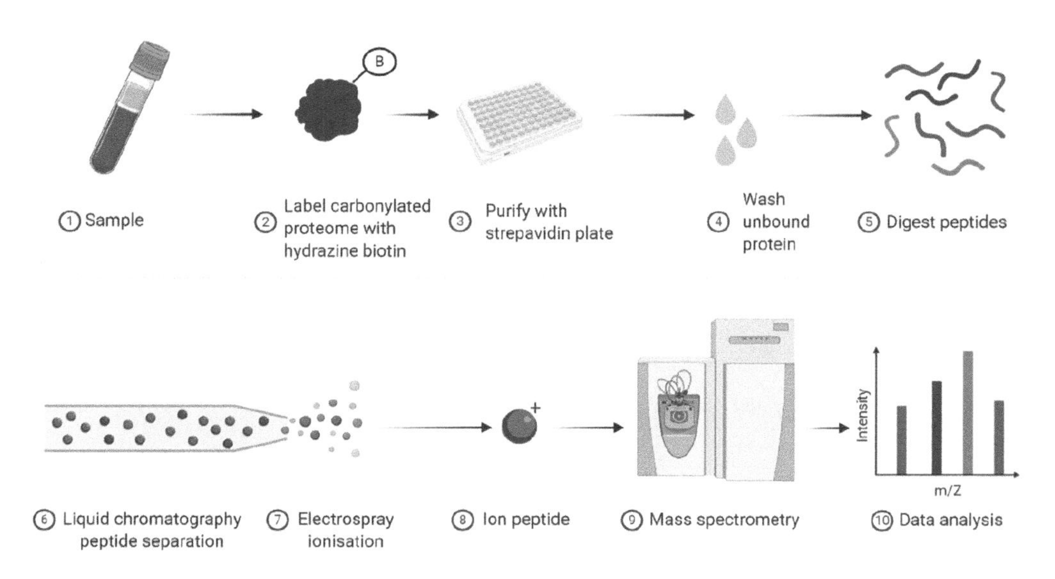

Figure 2.1 Proteomic-based identification of carbonylated proteins. (1) A biological sample is obtained. (2) Carbonylated proteins are labelled with hydrazine conjugated biotin. (3) The carbonylated proteome is separated from the rest of the sample via streptavidin solid support (e.g., 96 well plate) or bead. (4) Unbound protein is removed by extensive washing. (5) Trypsin is used to digest the bound proteins. (6) HPLC is used to separate the peptides before electrospray ionisation (7–8) and subsequent mass spectrometry analysis (9–10). Multiple variations are possible a carbonylated protein antibody could be used, and proteins can be separated in gel. Purification can be omitted if a direct label is used (i.e., direct carbonylation conjugation with a mass spec appropriate hydrazine label).

yield defined chemical information, which makes functional interpretation difficult. For example, inferring whether increased global tyrosine nitration is damaging is difficult because it fails to disclose the modified proteins. Targeted analysis, therefore, establishes a technically sound platform to expand knowledge of exercise-induced oxidative damage beyond global analysis of an oxidative damage adduct by Western blot or ELISA (see Figure 2.2).

Key interpretational points are threefold. First, the absolute steady state of a biomarker is in dynamic equilibrium with formation and repair mechanisms (Murphy et al., 2011). A change could simply reflect differential repair (Cobley et al., 2017), as opposed to increased damage, especially if repair consumes ATP. For example, exercise-induced ATP demand could deprive sulphiredoxin of the ATP needed to reverse sulphinic (SO_2) acid oxidation, which could lead to some proteins becoming irrevocably inactivated (Akter et al., 2018). Likewise, any change at the tissue or systemic level could simply reflect differential efflux. Second, far from being inert end-points, many oxidative macromolecule adducts are bioactive (Niki, 2012). For example, 4-HNE, a lipid peroxidation metabolite, can oxidise protein thiols by Michael addition or Schiff base formation, which could enact redox signalling (Zhang and Forman, 2017). Oxidative damage and redox signalling can, therefore, interact (Cobley et al., 2015a, b; Margaritelis et al., 2016a, b). Third, without functional annotation or appropriate reporters (e.g., enzyme assays ideally with and without selective removal of the modification), it is difficult to infer the biological meaning of any oxidative damage observed. Without corroborating evidence, it is unwise to conclude, especially from a single surrogate, that exercise-induced oxidative damage is functionally significant.

REDOX SIGNALLING: THE PROMISE OF NOVEL IMMUNOLOGICAL ASSAYS

Troublingly, much exercise-induced redox signalling is inferred without the process being

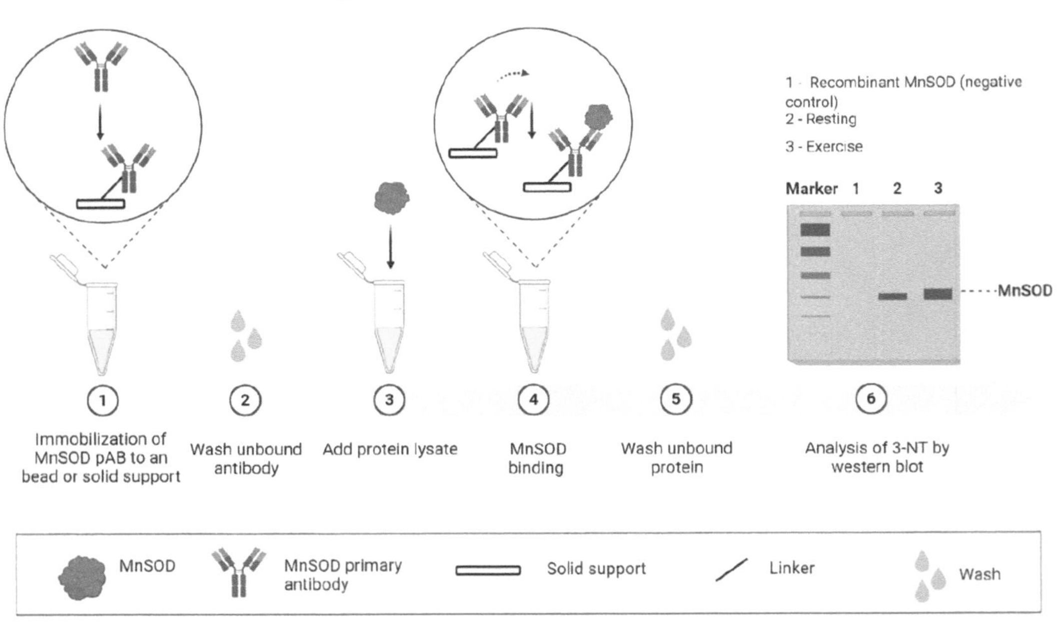

Figure 2.2 Analysis of MnSOD nitration using a targeted immunological approach. (1) An MnSOD primary antibody is immobilised to an agarose bead to solid support (e.g., via epoxy chemistry). (2) Unbound antibody is removed by washing. (3) A protein lysate containing MnSOD is added. (4) MnSOD binds the primary antibody. (5) Non-selectively bound material is removed by extensive washing. (6) MnSOD nitration is assessed by Western blot using a 3-NT antibody. A recombinant protein can be loaded as a negative control. The basic approach described above can be adapted for other proteins or inverted (e.g., using the immobile 3-NT antibody as bait to determine whether it binds MnSOD).

measured! Using an increase in antioxidant enzyme (i.e., activity or abundance), molecular chaperone (e.g., heat shock protein), or nuclear content of Nrf-2 (or related gene transcription) to infer redox signalling is problematic because they respond to other inputs. For example, cytokines can activate nuclear kappa beta signalling to increase antioxidant enzyme abundance and impaired ubiquitin proteasome activity could release Nrf-2 (Cobley et al., 2014). Likewise, inferring redox signalling has occurred, without corroborating evidence, because 2-cysteine peroxiredoxin (PRDX) isoforms dimerise is unwise because it can be coupled to antioxidant defence as opposed to redox signalling (Winterbourn, 2018). In any case, interpreting dimer assays can be complicated by comigration of SO_2/SO_3 species in the "reduced" monomeric band (Cobley and Husi, 2020). Given accessibility concerns rate-limit redox proteomics approaches (see Box 1), immunological assays are required to assess exercise-induced thiol mediated redox signalling. Protein thiol redox state is difficult to assess immunologically because of the inability to distinguish reduced and reversibly oxidised thiols by Western blot (Cobley and Husi, 2020). To address an unmet need, we advocate using catalyst-free Click-PEG to assess protein thiol redox state.

Click-PEG uses catalyst-free, selective, and bioorthogonal inverse electron demand Diels–Alder (IEDDA) click chemistry to distinguish reduced and reversibly oxidised thiols by selectively ligating a low molecular weight PEG moiety to the redox state of interest (Cobley, Noble, et al., 2019; Cobley et al., 2020). The resultant ability to disambiguate reduced and reversibly oxidised species by Western blotting enables Click-PEG to assess protein thiol redox state (van Leeuwen et al., 2017). The propensity of the PEG to impede antibody binding can be bypassed using novel in-gel antibody-first workflows (Cobley and Husi, 2020). Together canonical and antibody-first Click-PEG establishes a route to measure exercise-induced redox signalling (see Figure 2.2). Using the rich redox proteomic literature together with resources like the Oximouse database—details tissue-specific reversible thiol oxidation occupancy by site—one can select appropriate targets to assess redox signalling (Held, 2020; Xiao et al., 2020). As Sies and Jones remark (Sies and Jones, 2020), oxidising a catalytic thiol in GAPDH (Cys152) initiates self-correcting adaptive metabolic fluxes through the pentose phosphate pathway (PPP) to provide NADPH for the GSH and thioredoxin systems (Peralta et al., 2015). Oximouse reveals GAPDH is 5% (i.e., weighted mean of 5 thiols) oxidised in

BOX 1. Redox Proteomic Approaches: A Brief Overview

Redox proteomics is the gold-standard analytical approach to measure reversible thiol oxidation-based redox signalling in a systematic, unbiased (i.e., hypothesis-free), and high-throughput manner (Xiao et al., 2020). While several preparatory workflows exist, the essence of redox proteomics is to selectively enrich a redox state of interest (e.g., S-glutathionylated proteins using antibody enrichment) or use a "switch" approach wherein reduced and reversibly oxidised thiols are labelled with different tags (e.g., light vs. heavy N-ethylmaleimide) (Cobley, Sakellariou, et al., 2019). Regardless of the workflow, the terminal steps are usually the same: enzyme-mediated (i.e., trypsin) digestion of proteins to peptides and subsequent high-performance liquid chromatography followed by mass spectrometry (commonly using electron spray ionisation)-based proteomics (HPLC-MS-MS). Unique tryptic digest fragments are used to assign protein identity before systems biology analysis is used to identify enriched biological pathway (e.g., protein folding) (Held, 2020). Optionally, targeted multiple reaction monitoring (MRM)-based approaches can be used to interrogate specific hypotheses (Held et al., 2010). The developments in tandem mass tags for better quantification, mean redox proteomics approaches offer unprecedent quantitative redox proteome coverage. If possible, therefore, redox proteomics can, and indeed should, be used to investigate exercise-induced redox signalling. If it is possible to use redox proteomics, some of the associated caveats should be considered: (1) it may fail to detect some cysteines if the peptides fail to ionise well or are too short or long; (2) often low copy number proteins and hydrophobic proteins are undetected; and (3) additional experiments are required to determine if the modification is functional.

quiescent skeletal muscle (https://oximouse.hms.harvard.edu/sites.html). Low quiescent oxidation makes GAPDH an attractive target to determine exercise-induced metabolic redox signalling. Reversible thiol oxidation also decreases catalytic reactivity towards D-glyceraldehyde-3-phosphate, which can readily be assessed with and without thiol reductants (e.g., 1,4-dithiothreitol) to determine the functional significance. Unless a protein contains a single thiol, Click PEGylation cannot disclose the identity of the site modified without orthogonal MRM (Cobley et al., 2016; Cobley and Husi, 2020). Click-PEG could be used to decipher exercise-induced redox signalling by enabling investigators to measure the redox state of key proteins (Margaritelis et al., 2020) (Figure 2.3).

Key interpretational points are fivefold. First, a change in thiol oxidation can reflect altered reduction, especially if exercise reroutes NADPH from the GSH and TRX systems to NADPH oxidase isoforms allied to a parallel flux through glycolysis at the expense of the PPP (Henríquez-Olguin et al., 2019). Second, if a thiol protein is oxidised, consider direct peroxomonocarbonate ($HOOCO_2^-$) mediated oxidation (Dagnell et al., 2019). Further, free radicals, peroxynitrite, redox relays, and local PRDX inactivation can all contribute (Woo et al., 2010; Sobotta et al., 2015; Winterbourn, 2015). That is, multiple often overlapping mechanisms can result in target protein thiol oxidation. Third, unless a catalytic thiol is oxidised (e.g., Cys^{373} in Cdc25c), it is difficult to infer function without parallel enzyme activity and/or reporter (e.g., nuclear translocation) analysis. On a related note, the thiol oxidation event could play no role or serve antioxidant defence (Murphy, 2012). Fourth, even if an intervention (e.g., vitamin E) does decrease thiol oxidation (e.g., mitogen-activated protein kinase oxidation), redundancy means compensatory mechanism (e.g., calcium signalling) can transduce the signal (Cobley et al., 2015a,b). Finally, redox signalling extends far beyond reversible thiol oxidation. Other mechanisms including methionine, tyrosine, and Fe-S cluster oxidation all contribute (Sies and Jones, 2020).

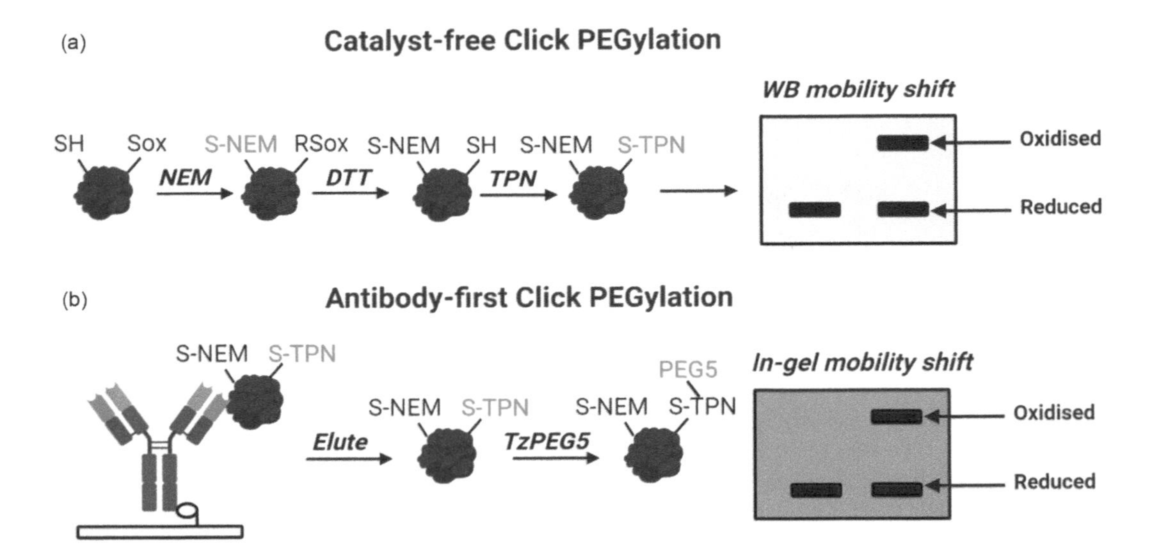

Figure 2.3 Click-PEG workflows to assess exercise-induced redox signalling. (a) Catalyst-free Click-PEG from left to right. Reduced thiols are alkylated with N-ethylmaleimide (NEM) to prevent artificial oxidation. After passing samples through a spin column to remove excess NEM (omitted for clarity), reversibly oxidised thiols are reduced with DTT. After removing excess DTT, newly reduced thiols are labelled with trans-cyclooctene PEG4 maleimide (TPN). The catalyst-free inverse electron demand Diels Alder (IEDDA) reaction is initiated by adding 6-methyltetrazine conjugated PEG5000 (TzPEG5). Reversibly oxidised thiols within the target protein are then detected as mass shifted bands by Western blot (WB). (b) Antibody-first Click-PEG (for when the primary antibody is unable to recognise the PEG-conjugated target protein) from left to right. A primary antibody is covalently immobilised to a solid support (e.g., a glass slide, omitted for clarity) to capture the reversibly oxidised target protein (prepared as above). After removing non-specifically bound material by extensive washing, the target is eluted (e.g., in 1% SDS with heating) before being treated with TzPEG5. Reversibly oxidised thiols within the target protein are then detected in gel as mass shifted bands following Coomassie or silver staining.

SYSTEMIC REDOX ANALYSIS: MOVING BEYOND MEASURING ANTIOXIDANT ENZYME ACTIVITY TOWARDS PRDX ISOFORM DIMERS

Many studies measure oxidative eustress at the systemic level by assessing antioxidant enzyme content or activity in plasma/serum (see Table 2.1). If you do, please consider that: (1) there is little to no value in assessing antioxidant enzyme activity or GSH in plasma; (2) antioxidant enzyme activity or content are poor redox signalling proxies; and (3) many assays fail to capture the activity of the protein as it was in vivo. As an aside, total antioxidant capacity (TAC) and related iteratives are often technically flawed, conceptually fraught, and methodologically ill-defined and should, therefore, be discontinued in an exercise setting (Cobley et al., 2017) (see Table 2.1). New systemic assays are, therefore, required.

Measuring cytosolic 2-Cys PRDX isoform activity using the dimer assay particularly in erythrocytes can meet the need for a new systemic approach to assess oxidative eustress (Low et al., 2007). Provided ex vivo oxidation is controlled (e.g., using N-ethylmaleimide), the PRDX1/PRDX2 dimer assay is a useful proxy of erythrocyte H_2O_2 metabolism and NADPH availability. That is, a change in the dimer likely reflects increased H_2O_2 production/influx and/or altered thioredoxin reductase activity likely secondary to insufficient NADPH supply. PRDX isoforms also react with peroxynitrite and protein-bound peroxides, as well as certain free radicals (De Armas et al., 2019). That is, their oxidation can seldom be exclusively ascribed to H_2O_2; a caveat that incidentally applies to protein thiol-based genetically encoded redox indicators (Müller et al., 2017). Furthermore, PRDX and TRDX isoform content is also a useful assay given their abundance in plasma/serum correlates with many diseases (Cobley et al., 2017). Moreover, selective glutaredoxin-mediated reduction coupled to catalyst-free Click PEGylation could assess S-glutathionylated PRDX2 in plasma. PRDX2 S-glutathionylation is a danger signal known to alter the redox state of cell surfaces proteins (e.g., toll-like receptors) on inflammatory cells (Salzano et al., 2014). Circulating S-glutathionylated PRDX2 may prove an exception to the general rule that redox signalling cannot be assessed systemically (Margaritelis et al., 2016a,b); especially if plasma depleted of S-glutationylated PRDX2 by immunocapture or DTT failed to elicit an inflammatory response (e.g., TNF-α release) in cultured macrophages. The promise of Click PEGylation is underscored by a recent report of increased exercise-induced albumin oxidation in fingertip blood samples (Lim et al., 2020). Additionally, in many cases, it may be informative to supplement the PRDX dimer assay by quantitively measuring the loss of maleimide conjugated fluorophore signal in a plate reader and in-gel to monitor the global thiol redox state. Underappreciated opportunities exist to better understand oxidative eustress in humans at the systemic level (see Figure 2.4).

CONCLUDING RECOMMENDATIONS

Building on previous work (Cobley et al., 2017; Cobley, 2020), ten concluding recommendations are provided for investigators wishing to measure reactive species, oxidative damage, and redox signalling:

1. Rationally select the assay depending on the biological questions (e.g., select a signalling reporter for measuring redox signalling).
2. Abandon flawed assays (e.g., TBARS).
3. Include the relevant controls (e.g., NEM for preventing ex vivo thiol oxidation) and prepare buffers carefully (e.g., pH range to prevent off-target NEM labelling).
4. Use multiple, ideally redox selective, markers to assess oxidative damage (e.g., 4-HNE and 8-iso-prostaglandin $F_{2\alpha}$ isomers to assess lipid peroxidation).
5. Use quantitative omics approaches to assess oxidative damage and redox signalling (e.g., lipidomics to assess lipid peroxidation).
6. Use novel immunological approaches to assess redox signalling (e.g., Click PEGylation to assess target protein thiol redox state).
7. Consider measuring the redox state of secreted proteins.
8. Refrain from assessing GSH, antioxidant enzyme activity, or total antioxidant capacity in plasma to assess oxidative damage or redox signalling.

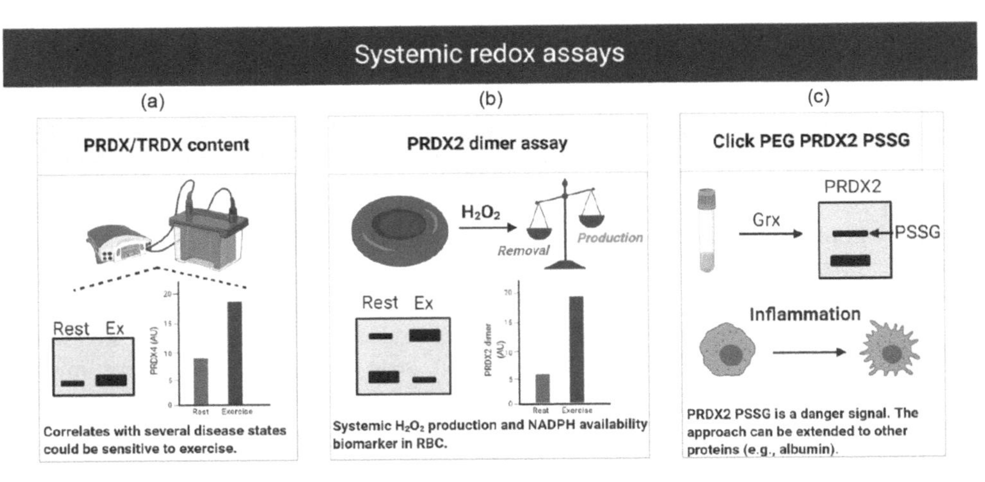

Figure 2.4 Novel systemic redox assays. (a) Western blotting can be used to determine whether PRDX/TRDX isoform content changes in plasma/serum following exercise. Provided RBC haemolysis is carefully controlled for, the approach may provide a useful biomarker of a particular state. For example, acute exercise may increase their content, but chronic training may mitigate their systemic increase in disease. (b) The dimer assay (i.e., non-reducing Western blotting) can be used to assess the monomer (reduced) to dimer (disulphide, RSSR) form of PRDX2 in response to exercise. At the systemic level, the assay reports H_2O_2 metabolism and is sensitive to changes in production and/or removal. Equally, it can be used in tissues (e.g., to infer local changes in H_2O_2 metabolism). (c) Catalyst-free Click PEGylation (see Figure 2.2) with a selective glutaredoxin (GRX) reduction step (as opposed to a generic reductant like DTT) can be used to selectively mass shift S-glutathionylated PRDX2 by Western blot. S-glutathionylated is a danger signal that can activate an inflammatory response; depicted as macrophage activation.

9. Consider assessing diagnostic probe products using HPLC to assess reactive species directly (e.g., HPLC analysis of 2-OH-E for superoxide).
10. Consider metabolism and repair when interpreting any marker (e.g., a change in NADPH availability could increase reversible thiol oxidation by decreasing TRDX reduction).

ACKNOWLEDGEMENTS

Figures 2.1–2.4 were created with Biorender.com.

REFERENCES

Abreu, I. A. and Cabelli, D. E. (2010) 'Superoxide dismutases-a review of the metal-associated mechanistic variations', *Biochimica et Biophysica Acta - Proteins and Proteomics*. Elsevier B.V., 1804(2), pp. 263–274.

Akter, S. et al. (2018) 'Chemical proteomics reveals new targets of cysteine sulfinic acid reductase', *Nature Chemical Biology*. Nature Publishing Group, 14(11), pp. 995–1004.

Anschau, V. et al. (2020) 'Reduction of sulfenic acids by ascorbate in proteins, connecting thiol-dependent to alternative redox pathways', *Free Radical Biology and Medicine*. Pergamon, 156, pp. 207–216.

Batinic-Haberle, I. et al. (2014) 'SOD therapeutics: Latest insights into their structure-activity relationships and impact on the cellular redox-based signaling pathways', *Antioxidants & Redox Signaling*, 20(15), pp. 2372–2415.

Cobley, J. N. et al. (2020) 'Reversible thiol oxidation inhibits the mitochondrial ATP synthase in xenopus laevis oocytes', *Antioxidants*. MDPI AG, 9(3), p. 215.

Cobley, J. N. et al. (2014) 'Lifelong training preserves some redox-regulated adaptive responses after an acute exercise stimulus in aged human skeletal muscle', *Free Radical Biology and Medicine*. Elsevier, 70, pp. 23–32.

Cobley, J. N. et al. (2015a) 'Influence of vitamin C and vitamin E on redox signalling: Implications for exercise adaptations', *Free Radical Biology and Medicine*. Elsevier, 84, pp. 65–76.

Cobley, J. N. et al. (2015b) 'The basic chemistry of exercise-induced DNA oxidation: oxidative damage, redox signaling, and their interplay', *Frontiers in Physiology*, 6, pp. 1–8.

Cobley, J. N. et al. (2016) 'Age- and activity-related differences in the abundance of myosin essential and regulatory light chains in human muscle', *Proteomes*, 4(15), pp. 1–15.

Cobley, J. N. et al. (2017) 'Exercise redox biochemistry: Conceptual, methodological and technical recommendations', *Redox Biology*. Elsevier B.V., 12, pp. 540–548.

Cobley, J. N., Noble, A., et al. (2019) 'Catalyst-free Click PEGylation reveals substantial mitochondrial ATP synthase sub-unit alpha oxidation before and after fertilisation', *Redox Biology*. Elsevier B.V., 26, p. 101258.

Cobley, J. N., Sakellariou, G. K., et al. (2019) 'Proteomic strategies to unravel age-related redox signalling defects in skeletal muscle', *Free Radical Biology and Medicine*. Elsevier Inc., pp. 24–32.

Cobley, J. N. (2020) 'How exercise induces oxidative eustress', in Sies, H. (ed.) *Oxidative Stress*. Elsevier, 132, pp. 447–462.

Cobley, J. N. and Husi, H. (2020) 'Immunological techniques to assess protein thiol redox state: Opportunities, challenges and solutions', *Antioxidants*, 9, p. 315.

Dagnell, M. et al. (2019) 'Bicarbonate is essential for protein-tyrosine phosphatase 1B (PTP1B) oxidation and cellular signaling through EGF-triggered phosphorylation cascades', *Journal of Biological Chemistry*, 294(33), pp. 12330–12338.

Davies, K. J. A. et al. (1982) 'Free radicals and tissue damage produced by exercise', *Biochimica et Biophysica Acta (BBA)*, 107, pp. 1198–1205.

Davison, G. W. et al. (2008) 'In vitro electron paramagnetic resonance characterization of free radicals: Relevance to exercise-induced lipid peroxidation and implications of ascorbate prophylaxis', *Free Radical Research*, 42(4), pp. 379–86.

De Armas, M. I. et al. (2019) 'Rapid peroxynitrite reduction by human peroxiredoxin 3: Implications for the fate of oxidants in mitochondria', *Free Radical Biology and Medicine*. Elsevier Inc., 130, pp. 369–378.

Forman, H. J. et al. (2015) 'Even free radicals should follow some rules: A Guide to free radical research terminology and methodology', *Free Radical Biology and Medicine*. Elsevier, 78, pp. 233–235.

Halliwell, B. and Gutteridge, J. M. C. (2015) *Free Radicals in Biology & Medicine*. Fifth Edit. Oxford University Press, Oxford.

Halliwell, B. and Whiteman, M. (2004) 'Measuring reactive species and oxidative damage *in vivo* and in cell culture: How should you do it and what do the results mean?' *British Journal of Pharmacology*, 142(2), pp. 231–255.

Held, J. M. et al. (2010) 'Targeted quantitation of site-specific cysteine oxidation in endogenous proteins using a differential alkylation and multiple reaction monitoring mass spectrometry approach', *Molecular and Cellular Proteomics*. American Society for Biochemistry and Molecular Biology Inc., 9(7), pp. 1400–1410.

Held, J. M. (2020) 'Redox systems biology: Harnessing the sentinels of the cysteine redoxome', *Antioxidants & Redox Signaling*. NLM (Medline), 32(10), pp. 659–676.

Henríquez-Olguin, C. et al. (2019) 'Cytosolic ROS production by NADPH oxidase 2 regulates muscle glucose uptake during exercise', *Nature Communications*. Nature Publishing Group, 10(1), pp. 1–11.

Lim, Z. X. et al. (2020) 'Oxidation of cysteine 34 of plasma albumin as a biomarker of oxidative stress', *Free Radical Research*. Taylor and Francis Ltd, 54(1), pp. 91–103.

Low, F. M. et al. (2007) 'Peroxiredoxin 2 functions as a noncatalytic scavenger of low-level hydrogen peroxide in the erythrocyte', *Blood*, 109(6), pp. 2611–2617.

Margaritelis, N. V. et al. (2020) 'Redox basis of exercise physiology', *Redox Biology*. Elsevier BV, 35, p. 101499.

Margaritelis, N. V. et al. (2016a) 'Going retro: Oxidative stress biomarkers in modern redox biology', *Free Radical Biology and Medicine*. Elsevier, 98(2), pp. 2–12.

Margaritelis, N. V. et al. (2016b) 'Principles for integrating reactive species into in vivo biological processes : Examples from exercise physiology', *Cellular Signalling*. Elsevier Inc., 28(4), pp. 256–271.

Miwa, S. et al. (2016) 'Carboxylesterase converts Amplex red to resorufin: Implications for mitochondrial H_2O_2 release assays', *Free Radical Biology and Medicine*. Elsevier, 90, pp. 173–183.

Müller, A. et al. (2017) 'Fluorescence spectroscopy of roGFP2-based redox probes responding to various physiologically relevant oxidant species in vitro', *Free Radical Biology & Medicine*. Elsevier B.V., 106, pp. 329–338.

Munro, D. and Pamenter, M. E. (2019) 'Comparative studies of mitochondrial reactive oxygen species in animal longevity: Technical pitfalls and possibilities', *Aging Cell*, 18(5), pp. 1–16.

Murphy, M. P. (2009) 'How mitochondria produce reactive oxygen species', *Biochemical Journal*, 417(1), pp. 1–13.

Murphy, M. P. et al. (2011) 'Unraveling the biological roles of reactive oxygen species', *Cell metabolism*, 13(4), pp. 361–366.

Murphy, M. P. (2012) 'Mitochondrial thiols in antioxidant protection and redox signaling: Distinct roles for glutathionylation and other thiol modifications', *Antioxidants & Redox Signaling*, 16(6), pp. 476–495.

Niki, E. (2012) 'Do antioxidants impair signaling by reactive oxygen species and lipid oxidation products?', *FEBS Letters*. Federation of European Biochemical Societies, 586(21), pp. 3767–3770.

Peralta, D. et al. (2015) 'A proton relay enhances H_2O_2 sensitivity of GAPDH to facilitate metabolic adaptation', *Nature Chemical Biology*, 11(2), pp. 156–163.

Place, N. et al. (2015) 'Ryanodine receptor fragmentation and sarcoplasmic reticulum Ca^{2+} leak after one session of high-intensity interval exercise', *Proceedings of the National Academy of Sciences*, 112(50), p. 201507176.

Sakellariou, G. K. et al. (2013) 'Studies of mitochondrial and nonmitochondrial sources implicate nicotinamide adenine dinucleotide phosphate oxidase(s) in the increased skeletal muscle superoxide generation that occurs during contractile activity', *Antioxidants & Redox Signaling*, 18(6), pp. 603–621.

Salzano, S. et al. (2014) 'Linkage of inflammation and oxidative stress via release of glutathionylated peroxiredoxin-2, which acts as a danger signal', *Proceedings of the National Academy of Sciences of the United States of America*, 111(33), pp. 12157–12162.

Sawyer, D. and Valentine, J. (1981) 'How super is superoxide?', *Accounts of Chemical Research*, 14(12), pp. 393–400.

Shchepinova, M. M. et al. (2017) 'MitoNeoD: A mitochondria-targeted superoxide probe', *Cell Chemical Biology*. Elsevier Ltd., 3, pp. 8–21.

Sies, H. and Jones, D. P. (2020) 'Reactive oxygen species (ROS) as pleiotropic physiological signalling agents', *Nature Reviews Molecular Cell Biology*. Springer US, 21(7), pp. 1–21.

Sobotta, M. C. et al. (2015) 'Peroxiredoxin-2 and STAT3 form a redox relay for H_2O_2 signaling', *Nature Chemical Biology*, 11, pp. 64–70.

Spasojević, I. (2011) 'Free radicals and antioxidants at a glance using EPR spectroscopy', *Critical Reviews in Clinical Laboratory Sciences*, 48(3), pp. 114–142.

van Leeuwen, L. A. G. et al. (2017) 'Click-PEGylation – A mobility shift approach to assess the redox state of cysteines in candidate proteins', *Free Radical Biology and Medicine*. Elsevier B.V., 108(January), pp. 374–382.

Winterbourn, C. C. (2015) 'Are free radicals involved in thiol-based redox signaling?', *Free Radical Biology and Medicine*. Elsevier, 80, pp. 164–170.

Winterbourn, C. C. (2018) 'Biological production, detection, and fate of hydrogen peroxide', *Antioxidants and Redox Signaling*, 20, pp. 541–551.

Woo, H. A. et al. (2010) 'Inactivation of peroxiredoxin i by phosphorylation allows localized H_2O_2 accumulation for cell signaling', *Cell*. Elsevier Ltd, 140(4), pp. 517–528.

Xiao, H. et al. (2020) 'A quantitative tissue-specific landscape of protein redox regulation during aging', *Cell*. Cell Press, 180(5), pp. 968–983.

Zhang, H. and Forman, H. J. (2017) 'Signaling by 4-hydroxy-2-nonenal: Exposure protocols, target selectivity and degradation', *Archives of Biochemistry and Biophysics*. Elsevier Inc, 617, pp. 145–154.

Zielonka, J. and Kalyanaraman, B. (2010) 'Hydroethidine- and MitoSOX-derived red fluorescence is not a reliable indicator of intracellular superoxide formation: Another inconvenient truth', *Free Radical Biology and Medicine*. Elsevier Inc., 48(8), pp. 983–1001.

Zielonka, J. and Kalyanaraman, B. (2018) 'Small-molecule luminescent probes for the detection of cellular oxidizing and nitrating species', *Free Radical Biology and Medicine*. Elsevier Inc., 128, pp. 3–22.

CHAPTER THREE

Exercise Redox Signalling

FROM ROS SOURCES TO WIDESPREAD HEALTH ADAPTATION

Ruy A. Louzada, Jessica Bouviere,
Rodrigo S. Fortunato and Denise P. Carvalho

CONTENTS

Introduction / 23
Exercise Creates an ROS-Rich Environment / 23
Main Sources of ROS in Contracting Muscle / 24
ROS-Generating System and Antioxidant Capacity in the Conventional Muscle-Type Classification / 25
Redox-Mediated Signalling Mainly Occurs via Targeted Modifications of Specific Residues in Proteins / 26
ROS and Contractile Function / 27
Exercise-Generated ROS Is Crucial to Muscle Glucose Uptake / 28
Muscle Adaptations to Exercise Training Rely on ROS-Mediated Signalling Pathways / 28
Mitochondrial Biogenesis and Antioxidant Defence / 29
Hypertrophy / 30
Exercise Creates ROS-Rich Environments by Inducing Both Local and Systemic ROS Waves / 31
Conclusion / 33
Acknowledgements / 33
Conflicts of Interest / 33
References / 33

INTRODUCTION

Exercise Creates an ROS-Rich Environment

The beneficial effects of exercise are related to the activation of multiple integrated and occasionally redundant cellular signalling pathways, which occur in muscle and other tissues simultaneously (Atherton et al., 2005; Bouviere et al., 2021; Hawley et al., 2014; Louzada et al., 2020; Pedersen and Saltin, 2015). Physical activity perturbs redox homeostasis by creating a prooxidative muscular environment (Jackson, 2011; Louzada et al., 2020; Powers and Jackson, 2008). Reactive oxygen species (ROS), such as superoxide $\left(O_2^{\cdot-}\right)$, hydroxyl $\left(OH^{\cdot}\right)$, and the non-radical species H_2O_2, are small radical or non-radical molecules that can

DOI: 10.1201/9781003051619-3

react with a wide spectrum of molecules, changing their structures in a reversible or irreversible way and modifying their function. To prevent, delay or remove ROS-mediated effects, ROS levels are modulated by antioxidant activity, which can be enzymatic (e.g., superoxide dismutase, glutathione peroxidase, catalase, and peroxiredoxins) or non-enzymatic (e.g., glutathione, vitamins C and E). Therefore, the availability of ROS and the extent of their effects depend on a fine balance between their generation and elimination.

Initially, exercise-induced ROS generation was believed to have deleterious effects on health. However, a new concept has been evolving over the last three decades that identifies ROS as critical mediators of signal transduction pathways (Jackson et al., 2016; Powers et al., 2016). Robust evidence that antioxidant supplementation hampers many beneficial effects of exercise (Gomez-Cabrera et al., 2012, 2008; Ristow et al., 2009) has expanded the interest in understanding how exercise induces ROS generation and in uncovering the mechanisms by which ROS-mediated signalling promotes health adaptation to exercise.

There are several sources of ROS production and scavenging in resting and working muscle. Mitochondria, xanthine oxidase, and nicotinamide adenine dinucleotide phosphate (NADPH) oxidase seem to be the major sources of ROS/reactive nitrogen species in skeletal muscle (Henríquez-Olguin et al., 2019; Sakellariou et al., 2013). Physical exercise can activate redox-sensitive intracellular signalling pathways through ROS-related mechanisms (Bjørnsen et al., 2016; Braakhuis et al., 2014; Gomez-Cabrera et al., 2008; Ristow et al., 2009), leading to muscle-function modification through both genomic and non-genomic mechanisms. In addition, mitochondrial function, calcium handling, and force production are causally related to muscle redox signalling.

Beyond intramuscular adaptation to exercise, recent discoveries about the mechanisms by which contracting muscle "communicates" with other organs are emerging as an exciting new field of exercise biology. It has been proposed that many of these widespread beneficial effects of exercise might not only require a complex ROS-dependent intramuscular signalling cascade, but also an integrated network with many remote tissues (reinforcing the hypothetical model of "synchronized ROS waves induced by exercise" (Louzada et al., 2020). The present chapter provides an overview of the main sources of ROS following muscle contraction, highlighting its importance for contractile function, and underlining the roles of ROS in the widespread health benefits of exercise.

Main Sources of ROS in Contracting Muscle

In the onset of muscle contraction, a high demand for ATP drives the cellular metabolism to provide energy for ATPases, calcium handling events, and promotion of the crossbridge interaction between actin and myosin filaments. Within milliseconds, ATP demand rapidly induces an orchestrated metabolic flow to match the demand for ATP. Initially, mitochondria were believed to be the main source of ROS following exercise based on an obvious correlation between increased oxygen consumption and increased ATP production. It has been shown that at least 11 mitochondrial sites can generate ROS in mammals that depends on the bioenergetic state (Wong et al., 2017). Remarkably, during exercise mitochondria operate in state 3 (also known as the maximal ADP stimulated respiration) and thereby reduce ROS generation, contrary to what is observed in the basal conditions, where mitochondria operate in respiration state 4 (Goncalves et al., 2015). Elegantly, when mitochondria are exposed to a condition that mimics exercise, H_2O_2 is considerably reduced (Goncalves et al., 2015; Jackson et al., 2016) and also confirmed using an in vivo mice model of exercise (Henríquez-Olguin et al., 2019).

The NADPH oxidase enzymes (NOXs) are unique proteins specialized in ROS production and are composed of seven members: NOX1 to NOX5 and DUOX1 and 2. Most of the NOXs have been reported to be present in skeletal muscle in vivo (Hori et al., 2011; Loureiro et al., 2016) (Figure 3.1). NOX2 is constitutively associated with the protein p22phox in the biological membrane. Activation of the NOX2-p22phox complex requires the translocation of cytosolic factors such as p47phox, p67phox, and p40phox to the membrane (Figure 3.1) (Henriquez-Olguin et al., 2020; Henríquez-Olguín et al., 2019a). NOX4 activity seems to be mainly dependent on its expression level, as well as the partial oxygen pressure (Ambasta et al., 2004; Ameziane-El-Hassani et al., 2016; Sun et al., 2011). Moreover, ATP can directly bind and negatively regulate NOX4 activity in the inner mitochondrial membrane, suggesting that subcellular redistribution of ATP levels from the mitochondria might act as an allosteric switch to activate NOX4 (Shanmugasundaram et al., 2017). While NOX2 and 4 produce O_2^{-}, DUOX1/2 require maturation factors (DUOXA1/A2) to be targeted at the plasma membrane and generate H_2O_2 (Ameziane-El-Hassani et al., 2016; Carré et al., 2015; Louzada et al., 2018).

Based on mouse fibre isolates, NADPH oxidase appears to be the major contributor of cytosolic ROS in skeletal muscle rather than mitochondria (Michaelson et al., 2010; Sakellariou et al., 2013). NOX2, NOX4, and p22phox were found in sarcolemma and transverse tubules of isolated fibres, whereas the other partners such as p40Phox and p67phox proteins that were found in cytoplasm were translocated to the sarcolemma during muscular contraction (Sakellariou et al., 2013). The emerging role of NOX2 has been reinforced by some elegant in vivo studies (Henríquez-Olguin et al., 2019; Henríquez-Olguín et al., 2016). Henriquez-Olguin et al. showed that NOX2 levels are increased in skeletal muscle immediately after one physical exercise session, returning to basal levels after 1 hour (Henríquez-Olguín et al., 2016). Interestingly, whole-body and endothelial-specific NOX4 deletion disrupted some metabolic responses and adaptations to exercise in mice (Specht et al., 2021), suggesting that NOX4 also contributes to ROS production after exercise.

Another important source of O_2^{-} following contractile activity is the enzyme XO (Gomez-Cabrera et al., 2010, 2005). Xanthine oxidoreductase (XOR) is related to both XO and xanthine dehydrogenase (XDH), which are enzymes that catalyse the reduction of hypoxanthine and xanthine to uric acid. While XDH transfers electrons from the substrate to NAD+, XO transfers it to O_2 and generates O_2^{-} . The conversion of XDH to XO occurs when tissue is injured (Gomez-Cabrera et al., 2010; Ryan et al., 2011) (Figure 3.1b). XO inhibition prevents extracellular superoxide production following an isometric protocol of contraction in the gastrocnemius muscle (Gomez-Cabrera et al., 2005), pointing to XO as an important source of ROS following muscle contraction (Figure 3.1).

The muscular contribution of XO could be debated due to its high expression in endothelial cells. However, studies using allopurinol, a non-specific inhibitor of XO, show that some classical markers of muscle adaptation seem to be dependent on XO activity, which are discussed further in this chapter. NOX2 and NOX4 are likely to compose the predominant oxidase system that contributes to cytosolic superoxide/H_2O_2 generation in muscle fibres during contraction. However, the muscular compartmentalization of ROS production is still a debated topic (Henriquez-Olguin et al., 2020; Henríquez-Olguín et al., 2019a).

ROS-Generating System and Antioxidant Capacity in the Conventional Muscle-Type Classification

Different skeletal muscle fibre types have been identified along with their contractile characteristics and metabolic properties. Briefly, type I myofibres are typically referred to as "slow-twitch oxidative" fibres because they have a slow contraction time to peak tension that is predominantly supported by mitochondrial ATP. Type II myofibres are termed "fast-twitch glycolytic" fibres and have quicker contraction time but a rapid fatigue profile. They are supported by faster ATP recycling, which is provided by glycolysis and the phosphate creatine system (Schiaffino and Reggiani, 2011). Type IIa fibres are intermediate and have a higher contraction time compared to type I fibres, and they have a higher oxidative capacity than type II fibres. These are called fast-twitch oxidative fibres.

Metabolically, types I and IIa fibres exhibit a high oxidative potential and capillary supply, while the fastest fibre in rats (type IIx), and its equivalent in humans (type IIb) are primarily glycolytic. The velocity of shortening and the fibre's twitch duration depend on the myosin composition and the speed of Ca^{2+} release and uptake by the sarcoplasmic reticulum. This is dependent on sequestering systems such as the sarcoplasmic reticulum Ca^{2+} ATPases (SERCAs), which have differentially expressed isoforms in different fibre types.

The traditional classification based on size, metabolic characteristics, and contractile function can be enriched with more recent information on the differences in redox homeostasis of each myofibre type. For instance, skeletal muscle fibres express NOX2, NOX4, and the DUOX1 transcript. NOX2, its regulatory subunits, and NOX4 are present in the sarcolemma, sarcoplasmic reticulum, and T tubules of skeletal muscle fibres (Espinosa et al., 2006; Xia et al., 2003). NOX4 is also detectable in skeletal-muscle mitochondria (Sakellariou et al., 2013). Interestingly, NOX enzymes are differently expressed in slow-twitch

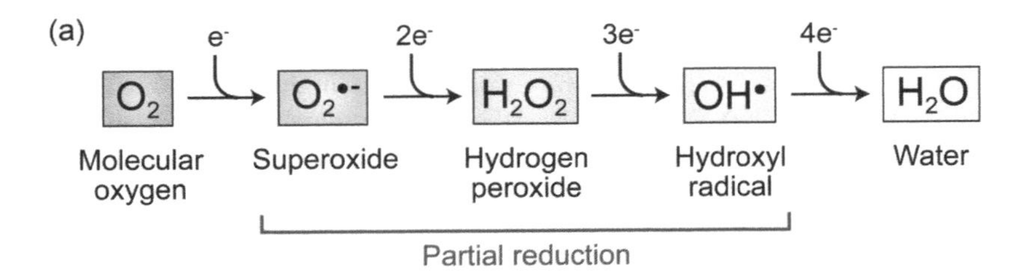

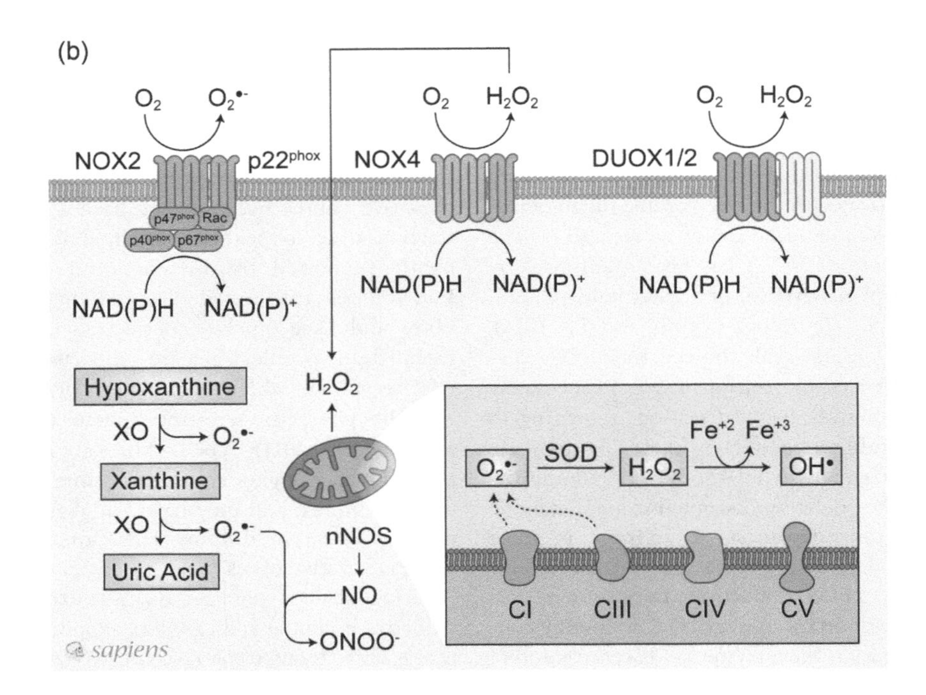

Figure 3.1 Pathways for the formation of ROS from oxygen to H_2O. (a) Formation of ROS through the partial reduction of molecular oxygen. (b) Main sources of ROS in skeletal muscle cells. NAD(P)H oxidases transfer electrons across biological membranes from NAD(P)H to molecular oxygen producing superoxide or H_2O_2 when one or two electrons are transferred, respectively. NOX2 produces superoxide that depends on the assembly of cytoplasmic regulators proteins (e.g., p47phox, RAC, p40phox, p67phox) to NOX2-p22phox complex. NOX4 produces H_2O_2 and its activity is mainly dependent on its expression levels. DUOX1/2 produces H_2O_2, and their translocation from endoplasmic reticulum to the plasma membrane is related to their association with DUOXA1/A2. Xanthine oxidase is a superoxide-producing enzyme that catalyses the oxidation of hypoxanthine to xanthine and can further catalyse the oxidation of xanthine to uric acid. XO expression in skeletal muscle is still debated and is more likely present in the adjacent endothelial cells. Mitochondrial respiratory chain and oxidative phosphorylation system are known to leak electrons to oxygen to produce superoxide and/or H_2O_2.

oxidative muscle compared to fast-twitch glycolytic muscle (Loureiro et al., 2016; Osório Alves et al., 2020).

NOX2 and NOX4 mRNA and their activity are higher in slow-twitch compared with fast-twitch fibres (Loureiro et al., 2016). In addition, the antioxidant activity is directly correlated with NOX activity of the different muscle fibre types. Thus, this mechanism seems to be crucial for maintaining muscle redox homeostasis and function, thereby preventing cellular oxidative stress (Ji, 2015, 2008; Powers and Jackson, 2008).

REDOX-MEDIATED SIGNALLING MAINLY OCCURS VIA TARGETED MODIFICATIONS OF SPECIFIC RESIDUES IN PROTEINS

Redox-mediated signalling is a post-translational regulation of protein function that occurs principally through the modification of protein cysteine residues. Sulphhydryl groups on cysteines are the preferred targets for oxidation or for the formation of disulphide bonds. Therefore, it has been debated that redox-mediated signals are not likely a "switch on/off" mechanism. Instead, redox modification of the quaternary structure can change protein function by activating or inhibiting its enzymatic activity. Allosteric disulphides can also control protein function when they undergo redox change. Moreover, protein redox modifications can affect its lifetime, cellular location. To illustrate what occurs during muscle contraction, redox modification of calcium handling proteins and modulation of the sensitivity to Ca^{2+} are some examples of how exercise-induced ROS affect contractile function.

ROS and Contractile Function

Briefly, skeletal muscle contraction is a complex event that involves rapid communication between electrical stimulation in the plasma membrane and Ca^{2+} release from the sarcoplasmic reticulum (SR). This leads to a transient increase of Ca^{2+} concentration in the cytoplasm and the formation of actin and myosin cross-bridges, leading to sarcomere shortening. Then, Ca^{2+} disappearance from the cytoplasm is mainly mediated by its reuptake by the SR through SR Ca^{2+} adenosine triphosphatase (SERCA) (Schiaffino and Reggiani, 2011), as illustrated in Figure 3.2.

Under physiological conditions, NOX4 seems to be involved in the skeletal muscle excitation-contraction coupling that controls calcium handling during muscular contraction. This occurs through redox modifications of ryanodine receptor-Ca^{2+} release channel 1 (RyR1), facilitating the rapid release of Ca^{2+} into the cytosol (Sun et al., 2011). Multiple cysteine residues within RyR1 are targets of S-oxidation or S-nitrosylation, where at least 21 cysteine residues out of 100 are redox modified in a NOX-dependent manner (Sun et al., 2013). As illustrated in Figure 3.2, RyR1 can also be modified by S-glutathionylation by NOX2-generated ROS at T-tubule membranes (Henríquez-Olguín et al., 2019a; Hidalgo et al., 2006) and S-nitrosylation of Cys3635 of the cysteine-containing CaM-binding domain, which underlies the mechanism of CaM-dependent regulation of RyR1 by NO (Sun et al., 2001). Further studies using animal models submitted to exercise are necessary to confirm the evidence mentioned above. Additionally, transient receptor potential (TRP) channels can be activated by NOX2-derived ROS, thereby increasing the

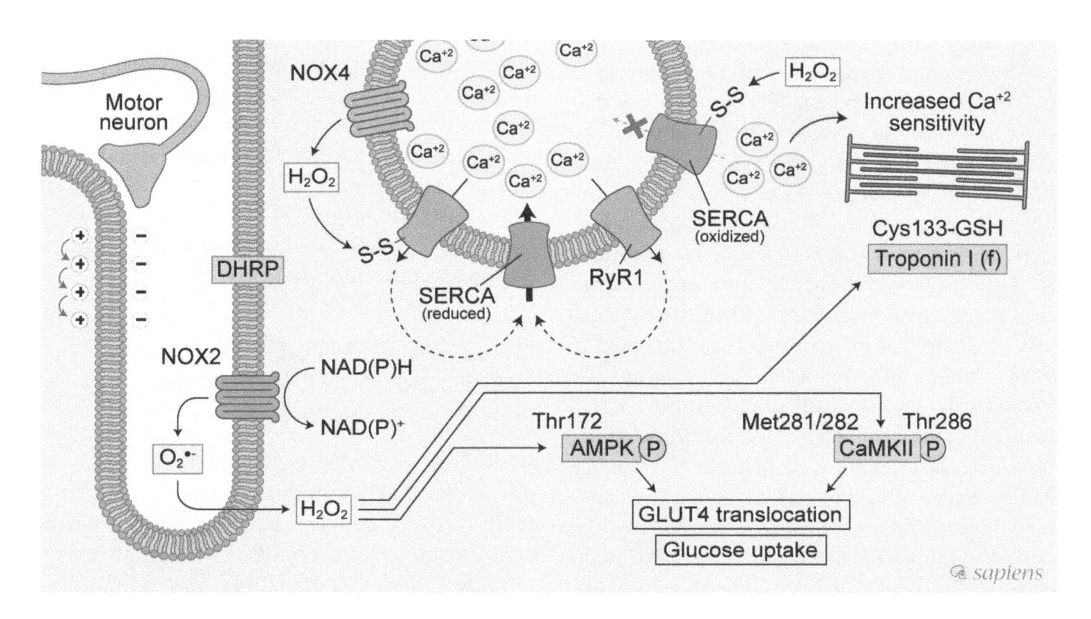

Figure 3.2 Involvement of exercise-generated ROS in crucial steps of calcium handling, force production, and glucose uptake. At the sarcoplasmic reticulum, NOX4-derived H_2O_2 oxidizes RyR cysteines, stimulating calcium release in response to sarcolemma action potential. Exercise-induced ROS increases force production by increasing Troponin I calcium sensitivity. Exercise-induced ROS activate CaMK by increasing its Thr286 phosphorylation, which culminates in multiple effects, such as glucose uptake and mitochondrial biogenesis.

influx of Ca^{2+} through the skeletal muscle sarcolemma (Brinkmeier, 2011).

For redox modification of the contractile apparatus, in vitro experiments mimicking physiological conditions showed that ROS can increase the force production of skeletal muscle (Smith and Reid, 2006), but the doses used are now considered non-physiological bringing some of this into question. S-glutathionylation increases its sensitivity to Ca^{2+} in fast-twitch fibres, most likely at Cys133 of the fast troponin I isoform (TnI(f)) (Mollica et al., 2012), but the relevance of this mechanism on in vivo models remain to be elucidated. The role of NOX-derived ROS in excitation-contraction coupling is summarized in Figure 3.2.

EXERCISE-GENERATED ROS IS CRUCIAL TO MUSCLE GLUCOSE UPTAKE

Skeletal muscle is one of the crucial tissues involved in glucose homeostasis. Regardless of exercise stimulation, upon insulin action, NOX-derived ROS is required for the increase of intracellular Ca^{2+}, which is mediated by the IP3 receptor. This process is involved in the transport of vesicles containing GLUT4 to the plasma membrane (da Justa Pinheiro et al., 2010; Silveira et al., 2006). Adding H_2O_2 directly to a muscle cell increases glucose uptake through the PI3K signalling pathway. Moreover, pretreatment with ROS scavengers (e.g., catalase and superoxide dismutase (SOD)) blunted the muscle glucose uptake induced by muscle contraction (Higaki et al., 2008; Katz, 2007), showing that the ROS produced during contractile activity is involved in glucose uptake.

Exercise is one of the best strategies to control glucose blood levels, mostly because of the capacity of contracting muscle to stimulate glucose uptake. Exercise can increase glucose transport to about 50-fold in humans (Katz, 2007). Despite the direct stimulatory effect of exercise-induced ROS on muscle glucose uptake via PI3K (Higaki et al., 2008), other studies showed that Akt2 deficiency did not affect the exercise-stimulated glucose uptake (Sakamoto et al., 2006), supporting the idea that insulin-stimulated glucose uptake pathways might be dissociated from the exercise-stimulated glucose uptake. Several signalling pathways are simultaneously activated following contraction and evoke glucose uptake. Among them, Ca^{2+}/calmodulin-dependent kinase II (CaMKII) and AMPK are the most studied candidates for playing a role in exercise-stimulated glucose uptake (Figure 3.2).

The calcium released with muscular contraction increases CaMKII activity, which induces GLUT4 translocation and consequently glucose uptake. CaMKII inhibitors can reduce glucose uptake in vitro following contractile activity (Erickson et al., 2011, 2008). Interestingly, CaMKII activity is increased by some oxidative modifications in the methionine pair 281/282, which are located in its regulatory domain (Erickson et al., 2011, 2008). This suggests that CAMKII could be an effector that links ROS and glucose uptake through post-translational modifications, but to the best of our knowledge no direct evidence about this mechanism are available (Figure 3.2).

AMPK is a metabolic sensor implicated in glucose uptake following exercise. After activation, this kinase phosphorylates many other proteins, thereby activating molecular pathways that are involved in ATP synthesis, GLUT4 translocation to the plasmatic membrane, and mitochondrial activity. AMPK is also a redox-sensitive protein and oxidative modification of AMPK cysteine residues can increase its kinase activity regardless of ATP depletion (Zmijewski et al., 2010). A pretreatment with the antioxidant enzyme catalase prevents both glucose uptake and AMPK activation during contractile activity (Higaki et al., 2008), thus linking these three elements in a plausible mechanism that could control muscle glucose uptake (Figure 3.2). However, interdependency of those events was not confirmed using an in vivo model of specific deletion of AMPK in muscle, showing that AMPK and its downstream target TBC1D1 are involved in the post exercise-stimulated glucose uptake but not during exercise (Kjøbsted et al., 2019).

MUSCLE ADAPTATIONS TO EXERCISE TRAINING RELY ON ROS-MEDIATED SIGNALLING PATHWAYS

Several responses that are generally reversible take place at local and systemic levels during and after acute physical exercise, which require a permanent stimulus that is provided by regular exercise training (Hawley et al., 2014). Skeletal muscle is a specialized tissue with remarkable plasticity. Successive muscle contractions at each exercise session led to a variety of physiological adaptations, which are closely related to the type of exercise. Briefly, physical exercise can be divided into two main types: endurance exercise (aerobic resistance training) and muscular strength exercise (anaerobic resistance training) (Schiaffino and Reggiani, 2011). The molecular machinery involved in physical exercise adaptations are closely related to the exercise type (Atherton et al., 2005; Hawley et al., 2014) and are summarised in Figure 3.3.

Most of the signalling pathways activated during and after physical exercise are potentially affected by ROS within contracting muscle. As above, phosphatase activities that control many crucial steps alongside the signalling cascade could be inhibited by ROS, while several kinases involved in exercise responses have redox-sensitive sites that could be activated by ROS. (Kjøbsted et al., 2019; Shao et al., 2014; Zmijewski et al., 2010). In this way, exercise-induced ROS production leads to an increase in the net phosphorylation levels, which fits perfectly well with the proposed model of how the modification of redox-sensitive protein directs many exercise responses and adaptations (Barbieri and Sestili, 2012). It is believed that by simultaneously controlling the levels of phosphatase and kinase, exercise-induced ROS production controls many of the signalling pathways, as illustrated in Figure 3.3.

Mitochondrial Biogenesis and Antioxidant Defence

The central player in endurance exercise adaptations is the peroxisome proliferator-activated receptor-γ coactivator-1α (PGC-1α) (Handschin and Spiegelman, 2008, 2011). Exercise is a powerful stimulus that induces PGC-1α mRNA and PGC-1α protein expression in rodents and humans (Egan et al., 2010; Little et al., 2010; Miura et al., 2007; Russell et al., 2003). PGC-1α has emerged as a mechanistic link to a wide range of beneficial effects of exercise in muscle, such as mitochondrial biogenesis, angiogenesis, fatty acid utilization, type switching of fibres (Atherton et al., 2005; Narkar et al., 2008), and antioxidant defence (St-Pierre et al., 2006) (Figure 3.3). Importantly, studies in vitro and in vivo have used antioxidant supplementation and specific inhibitors of ROS production, and the results indicated that PGC-1α activity is influenced by ROS, but it seems to be indirect through the activation of p38 MAPK and NF-κB, as well as Ca^{2+} signalling by these molecules (Nalbandian et al., 2019; Yan, 2009). As antioxidants can prevent or delay any ROS production during contraction in an

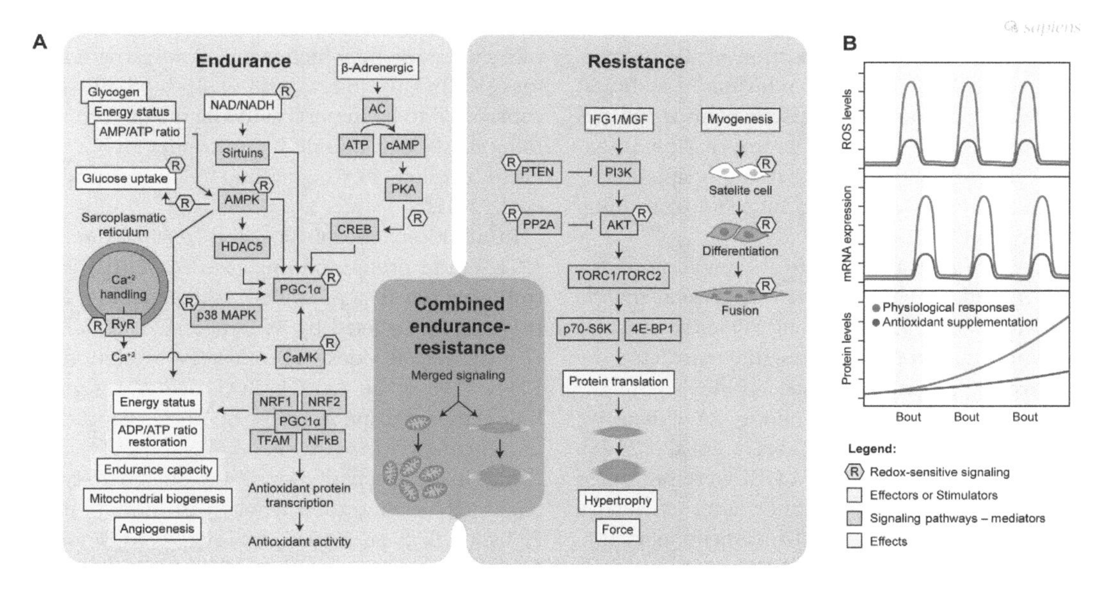

Figure 3.3 Redox-sensitive signalling pathways involved in physical exercise and physiological adaptations. (a) ROS-mediated signalling of endurance and resistance exercise related to the stimulation of mitochondrial levels, metabolic responses, and hypertrophy. The insertion of muscle progenitor cells at the site of skeletal muscle lesions is a crucial event to accomplish hypertrophy. NOX expression is differentially regulated during myogenesis. (b) The cumulative effect of exercise bouts leads to mRNAs and proteins accumulation linked to the physiological responses related to physical exercise. Repeated transient ROS bursts precede mRNA increases at each exercise bout, coupling ROS-mediated signalling to exercise adaptations.

unspecified manner, more mechanistic insights about the sources of ROS during exercise derived from genetically modified animal models are currently available and have filled the knowledge gap about how health adaptation to exercise are reliant on ROS-mediated signalling.

The importance of NOX2 in many physiological adaptations induced by physical exercise has been reported (Henríquez-Olguín et al., 2019a). Increased SOD2 and glutathione peroxidase mRNA in response to acute exercise were blocked by pharmacological inhibition of NOX2, as was the expression of mitochondrial biogenesis genes (e.g., TFAM and citrate synthases mRNA) linking NOX2-derived ROS and the activation of the NF-κB and NRF-2 pathway (Henríquez-Olguín et al., 2016). Adaptations to high-intensity interval training, such as maximal running capacity and mitochondrial adaptation to 6 weeks of training (e.g., Complex I, III, IV and PDH levels), also occur via a NOX2-dependent mechanism. Moreover, increased SOD2 and catalase expression were also shown to be dependent on NOX2 activation in contracting muscle (Henríquez-Olguín et al., 2019b).

Interestingly, exercise training for 3 weeks increased NOX2 expression and activity in slow-twitch fibres, while NOX4 expression was increased in only the red portion of the gastrocnemius (fast-twitch oxidative fibres). In addition, the white portion of the gastrocnemius did not respond to this exercise protocol (low-intensity exercise), suggesting that the adaptation is dependent on the fibre and intensity of training (Loureiro et al., 2016).

XO expression and activity are much more evident in endothelial cells than in skeletal muscle (Hellsten-Westing, 1993), but XO seems to play an important role in signalling mediated by reactive oxygen nitrogen species (RONS) in the gastrocnemius muscle following an exhaustive exercise session (Gomez-Cabrera et al., 2005). Activation of p38 MAPK and ERK were prevented by allopurinol, an XO inhibitor, after exhaustive exercise. Interestingly, dual-specificity phosphatases (DUSPs) counteract MAPK activity, and its Cys 258 might be involved in deactivating DUSP phosphatase activities and triggering their proteasomal degradation (Figure 3.3) (Chen et al., 2019; Kamata et al., 2005). There is a lack of a direct evidence showing that XO-generated ROS hampers MAPK signalling by controlling this phosphatase. Regarding the antioxidant response, it seems that the absence of NF-κB activation led to blunted SOD expression (Gomez-Cabrera et al., 2005). Remarkably, XO inhibition attenuated some ROS-mediated signalling (e.g., P38 MAPK and ERK phosphorylation) in acute exercise, but it did not prevent the increase of PGC-1α, NRF2, GLUT4, and SOD mRNA levels and mitochondrial adaptation to exercise training (Wadley et al., 2013). Noteworthy, as abovementioned, the expression of XO in muscle is still debated and further studies must be done.

Hypertrophy

It is well established that activation of the PI3K/Akt/mTOR pathway increases protein synthesis, but how this pathway is activated remains poorly understood (Spangenburg et al., 2008). Recently, a redox control of this pathway during exercise has been discussed (Gomez-Cabrera et al., 2020). One alternative to activate this signalling pathway during exercise is through extracellular ligands secreted in the muscle microenvironment. IGF-1 is believed to be the most important hypertrophic stimulus. It is produced locally by muscle cells and acts in an autocrine and paracrine manner. IGF-1 post-transcriptional regulation creates distinct molecules, among which is mechano-growth factor (MGF) (Musaró et al., 2004). Both proteins are considered to be important mediators of exercise-induced skeletal muscle hypertrophy, which can occur through ROS signalling (Handayaningsih et al., 2011).

In addition, phosphatase and tensin homolog (PTEN) and protein phosphatase 2A (PP2A) control two crucial steps in PI3K/Akt signalling, and they can be targeted by ROS-mediated signalling (Figure 3.3). Besides the transitory inhibition of PTEN (Lee et al., 2002), ROS-mediated signalling causes phosphorylation of PTEN, triggering its degradation (Abraham and O'Neill, 2014). For PP2A, it has been demonstrated that peroxynitrite-mediated nitration resulted in inhibition of PP2A, which promotes sustained activation of PI3/Akt signalling (Low et al., 2014). Additionally, Akt alone is also a redox-sensitive protein that has a disulphide bond between Cys297 and Cys311 (Murata et al., 2003) and other two cysteine residues in the pleckstrin homology domain (Cys60 and Cys77). Thus, the net balance toward PI3K/Akt/mTORC pathways might be caused by a pro-oxidant environment after exercise and thereby sustaining the activation of an anabolic cascade to mediate hypertrophy (Figure 3.3).

Additionally, a well-recognized way of stimulating mass gain following resistance training is through the insertion of muscle progenitor cells at the site of skeletal muscle lesions. Microlesions induced by exercise generally involve eccentric or lengthening contraction that causes excessive sarcomere strain, leading to cellular membrane perturbation and altering the myofibrillar structure. Type II fibres are more susceptible to this damage than type I fibres. Conversely, type I fibres contain more satellite cells. This higher susceptibility to damage is attributed to the structure and metabolic differences between the two fibre types (Qaisar et al., 2016). Upon exercise-induced microlesions, impaired regeneration after local depletion of satellite cells is observed (Bellamy et al., 2014), showing that these cells are crucial for tissue repair (Sambasivan et al., 2011). Upon skeletal muscle damage, satellite cells are activated and proliferate to become myogenic precursor cells and ultimately differentiate to myoblasts. Multiple signals from injured muscle and supportive cells (local and recruited cells) stimulate migration of myogenic progenitors to the damaged area, as well as its differentiation and fusion to form multinucleated muscle fibres (Le Moal et al., 2017). Many in vitro studies have shown that this process can be unfavourably affected by either excessive or insufficient ROS signalling (Kozakowska et al., 2015; Le Moal et al., 2017) and reactive nitrogen species (RNS) (McCormick et al., 2016).

NOX2 and its subunits p22(phox), p47(phox), and p67(phox), as well as NOX4, are expressed in human and murine skeletal muscle precursor cells. Their activity seems to play an important role in promoting proliferation in these cells (Mofarrahi et al., 2008). Regarding myoblast differentiation in vitro, ERK1/2 activation and myogenic genes in C2C12 cells were affected by both absence and excess of NOX4, suggesting an important role of NOX4-produced ROS signalling during differentiation (Acharya et al., 2013). During muscular differentiation, ROS production increases gradually, and the blockade of the PI3K/p38MAPK cascade inhibited the activation of the NF-κB pathway mediated by NOX2, which is a crucial event in muscular differentiation (Piao et al., 2005). However, the role of NOX4 in this process is not well understood, but it has been shown that NOX4 can decrease myoblast differentiation (Mofarrahi et al., 2008).

Additionally, the physiological role of DUOX isoforms in adult muscle function has not been assessed, but the deletion of DUOXA1 (a partner of DUOX1) in myoblasts enhanced its differentiation, while overexpression of DUOXA1 increased apoptosis and reduced differentiation markers in these cells. The expressions of both DUOX1 and DUOXA1 seem to be reduced in myotubes compared to myoblasts (Sandiford et al., 2014). Patients with Duchenne muscular dystrophy have elevated oxidative stress in skeletal muscle (Rodriguez and Tarnopolsky, 2003), in which NOXs are probably involved. DUOX1 is upregulated in the skeletal muscle of a dystrophic mdx mouse model (Hori et al., 2011), which suggests that DUOX1 can affect the commitment of satellites cells in myogenesis, a process that is highly dependent on coordinated ROS-mediated signalling (right panel, Figure 3.3).

EXERCISE CREATES ROS-RICH ENVIRONMENTS BY INDUCING BOTH LOCAL AND SYSTEMIC ROS WAVES

Some emerging models for explaining how exercise-induced ROS leads to widespread health adaptations are an ongoing line of investigation and are summarized in Figure 3.3. Synchronized ROS-mediated signalling within contracting muscle ensures the activation of specific signalling pathways through both genomic and non-genomic mechanisms that are summarized in Figure 3.3. Non-genomic control of kinases and phosphatases through modification of cysteines drives the balance of kinase over phosphatase activity, thereby favouring the phosphorylated state of some of these effectors (Bouviere et al., 2021). Importantly, the accumulation of mRNA and consequently new-born protein in each exercise is certainly a crucial event to explain why the adaptations are achieved by a permanent stimulus provided by regular exercise training. It is hypothesized that the peaks of ROS occur prior to the increase of some mRNA at each stimulus. When exogenous antioxidants are administered, both increases of ROS and mRNA are prevented and thereby a blunted exercise response is observed (Figure 3.3b).

Fascinatingly, exercise also generates regular and synchronized ROS waves in remote tissues, such as liver, heart, adipose, among others (Davies et al., 1982; Matta et al., 2021; Muthusamy et al., 2012). It creates a transient pro-oxidative environment in those tissues that might be supported by releasable factors from contracting muscle. As a result, coordinated communication with different tissues is promoted (Louzada et al., 2020). It has recently been suggested that the first phase of the exercise-induced ROS wave occurs within contracting muscle and promotes multiple post-translational modifications in many kinases and phosphatases (Figure 3.4, panel on bottom). These are crucial for the local stimulus of muscular glucose uptake, mitochondrial biogenesis, and antioxidant capacity. Additionally, skeletal muscles also produce ROS, which could lead to the oxidation of macromolecules, such as lipid peroxidation. Interestingly, products of lipid peroxidation (13-hydroxyoctadecadienoic acid [HODE] and 4-hydroxynonenal [HNE]) are capable of binding to nuclear receptors in remote cells. In addition, myokines and exosomes might be released from skeletal muscle in response to intracellular ROS stimulated by physical exercise. In addition, lactate is a metabolite that was also referred as "lactormone" due to its properties to modulate the redox state (Brooks, 2018) and control PGC-1α expression in a redox-mediated manner

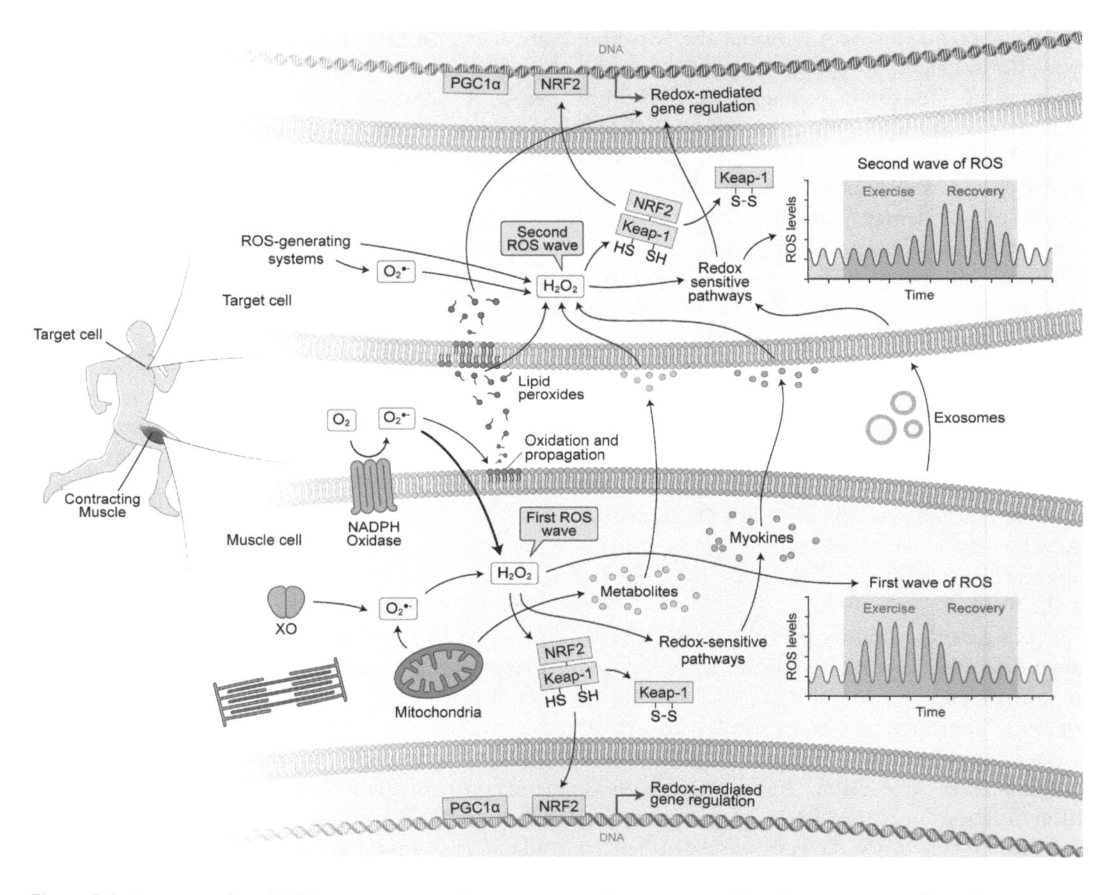

Figure 3.4 Exercise-induced ROS waves concept. Contracting muscle creates an ROS-rich environment through consecutive (ROS) spikes. The first ROS wave is produced mainly by skeletal muscle NADPH oxidase (NOX2) to induce the activation of ROS-mediated signalling pathways. Skeletal muscle releases signals to remote tissues in many ways during muscle contraction. For example, metabolites (e.g., lactate) released by skeletal muscle can be taken up and metabolized in remote tissues, an event that can stimulate ROS production. Moreover, myokines can also control the production of ROS in remote tissues. Products of lipid peroxidation (13-hydroxyoctadecadienoic acid [HODE] and 4-hydroxynonenal [HNE]) produced during skeletal muscle contraction are capable of binding to nuclear receptors in remote cells. Finally, exosomes released by the muscles undergoing contraction might stimulate signalling pathways in remote tissues. ROS-rich environments in remote tissues following physical exercise are proposed as a second ROS wave that induces ROS-mediated signalling in non-contracting cells. (Adapted from Louzada et al., 2020.)

(Nalbandian et al., 2019). Once in the blood circulation, all these mediators communicate to remote tissues during and after exercise periods (Figure 3.4 panel on top). Consequently, the second phase of ROS waves might stimulate some redox-sensitive signalling pathways in non-contracting tissues, leading to widespread responses to physical exercise. For example, activation of the nuclear factor erythroid 2-related factor 2 (Nrf2) controls the expression of more than 200 cytoprotective genes (Done and Traustadóttir, 2016). Following exercise, it is established that Nrf2 activation occurs within contracting muscle and also in several remote tissues, including kidney (Pala et al., 2016), brain (Wafi et al., 2019), liver (Rojo de la Vega and Zhang, 2018), myocardium (Shanmugam et al., 2019), and adipose tissues (Matta et al., 2021). This leads to the upregulation of endogenous antioxidant defences and an overall greater ability to counteract the damaging effects of oxidative stress (Louzada et al., 2020).

Finally, the importance of modulating the redox environment status was recently confirmed by the discovery of an endogenous factor capable of promoting "reductive stress" that counteract the physiological ROS bursts induced by exercise (Takamura, 2020). Elegantly, Musi et al. demonstrated that selenoprotein P (SEPP1) enables communication between the liver and muscle to upregulate GPX1, which maintains a reduced intracellular environment that is related to a lack of ROS-induced AMPK/ PGC-1α activation following physical exercise. As a result, the beneficial effects of exercise are hampered in some obese patients with high circulating levels of SEPP1 (Misu et al., 2017).

CONCLUSION

Understanding the type, magnitude, duration, and location of ROS production is a trending topic in exercise biology. Although the main source of ROS is still debated, NOX2 is emerging as an important source of ROS in contracting muscle. However, only a small piece of this complex puzzle has been placed to uncover the fine-tuned redox-mediated signalling that occurs in contracting muscle. In addition, the ROS-rich environment that is created after exercise is not exclusively restricted to contracting muscle but is also spread to some non-contracting tissues during and after physical exercise. Myokines, exosomes, metabolites (e.g., lactate), and lipid oxidation mediators may help the contracting muscle communicate with the whole body, mainly providing its beneficial effects via PGC-1α and Nrf2, two redox-sensitive pathways. Finally, the identification of new mediators that disrupt exercise-stimulated ROS signalling and hamper the beneficial effects of exercise will pave new ways in the exercise performance field and create new specific targets for therapeutic interventions to treat disorders caused by defective redox-mediated signalling.

ACKNOWLEDGEMENTS

We also would like to thank Sapiens Scientific Illustrations for performing artworks.

CONFLICTS OF INTEREST

The authors declare no conflict of interest.

REFERENCES

Abraham, A.G., O'Neill, E., 2014. PI3K/Akt-mediated regulation of p53 in cancer. *Biochemical Society Transactions* 42, 798–803. https://doi.org/10.1042/BST20140070.

Acharya, S., Peters, A.M., Norton, A.S., Murdoch, G.K., Hill, R.A., 2013. Change in Nox4 expression is accompanied by changes in myogenic marker expression in differentiating C2C12 myoblasts. *Pflügers Archiv: European Journal of Physiology* 465, 1181–1196. https://doi.org/10.1007/s00424-013-1241-0.

Ambasta, R.K., Kumar, P., Griendling, K.K., Schmidt, H.H.H.W., Busse, R., Brandes, R.P., 2004. Direct interaction of the novel Nox proteins with p22phox is required for the formation of a functionally active NADPH oxidase. *Journal of Biological Chemistry* 279, 45935–45941. https://doi.org/10.1074/jbc.M406486200.

Ameziane-El-Hassani, R., Schlumberger, M., Dupuy, C., 2016. NADPH oxidases: New actors in thyroid cancer? *Nature Reviews Endocrinology* 12, 485–494. https://doi.org/10.1038/nrendo.2016.64.

Atherton, P.J., Babraj, J., Smith, K., Singh, J., Rennie, M.J., Wackerhage, H., 2005. Selective activation of AMPK-PGC-1α or PKB-TSC2-mTOR signaling can explain specific adaptive responses to endurance or resistance training-like electrical muscle stimulation. *FASEB Journal* 19, 786–788. https://doi.org/10.1096/04-2179fje.

Barbieri, E., Sestili, P., 2012. Reactive oxygen species in skeletal muscle signaling. *Journal of Signal Transduction* 2012, 1–17. https://doi.org/10.1155/2012/982794.

Bellamy, L.M., Joanisse, S., Grubb, A., Mitchell, C.J., McKay, B.R., Phillips, S.M., Baker, S., Parise, G., 2014. The acute satellite cell response and skeletal muscle hypertrophy following resistance training. *PLoS ONE* 9, 17–21. https://doi.org/10.1371/journal.pone.0109739.

Bjørnsen, T., Salvesen, S., Berntsen, S., Hetlelid, K.J., Stea, T.H., Lohne-Seiler, H., Rohde, G., Haraldstad, K., Raastad, T., Køpp, U., Haugeberg, G., Mansoor, M.A., Bastani, N.E., Blomhoff, R., Stølevik, S.B., Seynnes, O.R., Paulsen, G., 2016. Vitamin C and E supplementation blunts increases in total lean body mass in elderly men after strength training. *Scandinavian Journal of Medicine & Science in Sports* 26, 755–763. https://doi.org/10.1111/sms.12506.

Bouviere, J., Fortunato, R.S., Dupuy, C., Werneck-de-Castro, J.P., Carvalho, D.P., Louzada, R.A., 2021. Exercise-stimulated ROS sensitive signaling pathways in skeletal muscle. *Antioxidants (Basel)* 10, 537. https://doi.org/10.3390/antiox10040537.

Braakhuis, A.J., Hopkins, W.G., Lowe, T.E., 2014. Effects of dietary antioxidants on training and performance in female runners. *European Journal of Sport Science* 14, 160–168. https://doi.org/10.1080/17461391.2013.785597.

Brinkmeier, H., 2011. TRP channels in skeletal muscle: Gene expression, function and implications for disease. *Advances in Experimental Medicine and Biology* 704, 749–758. https://doi.org/10.1007/978-94-007-0265-3_39.

Brooks, G.A., 2018. The science and translation of lactate shuttle theory. *Cell Metabolism* 27, 757–785. https://doi.org/10.1016/j.cmet.2018.03.008.

Carré, A., Louzada, R.A.N., Fortunato, R.S., Ameziane-El-Hassani, R., Morand, S., Ogryzko, V., de Carvalho, D.P., Grasberger, H., Leto, T.L., Dupuy, C., 2015. When an intramolecular disulfide bridge governs the interaction of DUOX2 with its partner DUOXA2. *Antioxid. Redox Signal.* 23, 724–733. https://doi.org/10.1089/ars.2015.6265.

Chen, H.-F., Chuang, H.-C., Tan, T.-H., 2019. Regulation of dual-specificity phosphatase (DUSP) ubiquitination and protein stability. *International Journal of Molecular Sciences* 20. https://doi.org/10.3390/ijms20112668.

da Justa Pinheiro, C.H., Silveira, L.R., Nachbar, R.T., Vitzel, K.F., Curi, R., 2010. Regulation of glycolysis and expression of glucose metabolism-related genes by reactive oxygen species in contracting skeletal muscle cells. *Free Radical Biology and Medicine* 48, 953–960. https://doi.org/10.1016/j.freeradbiomed.2010.01.016.

Davies, K.J., Quintanilha, A.T., Brooks, G.A., Packer, L., 1982. Free radicals and tissue damage produced by exercise. *Biochemical and Biophysical Research Communications* 107, 1198–1205. https://doi.org/10.1016/s0006-291x(82)80124-1.

Done, A.J., Traustadóttir, T., 2016. Nrf2 mediates redox adaptations to exercise. *Redox Biology* 10, 191–199. https://doi.org/10.1016/j.redox.2016.10.003.

Egan, B., Carson, B.P., Garcia-Roves, P.M., Chibalin, A.V., Sarsfield, F.M., Barron, N., McCaffrey, N., Moyna, N.M., Zierath, J.R., O'Gorman, D.J., 2010. Exercise intensity-dependent regulation of peroxisome proliferator-activated receptor γ coactivator-1α mRNA abundance is associated with differential activation of upstream signalling kinases in human skeletal muscle. *Journal of Physiology* 588, 1779–1790. https://doi.org/10.1113/jphysiol.2010.188011.

Erickson, J.R., Joiner, M.A., Guan, X., Kutschke, W., Yang, J., Oddis, C.V., Bartlett, R.K., Lowe, J.S., O'Donnell, S.E., Aykin-Burns, N., Zimmerman, M.C., Zimmerman, K., Ham, A.J.L., Weiss, R.M., Spitz, D.R., Shea, M.A., Colbran, R.J., Mohler, P.J., Anderson, M.E., 2008. A dynamic pathway for calcium-independent activation of CaMKII by methionine oxidation. *Cell* 133, 462–474. https://doi.org/10.1016/j.cell.2008.02.048.

Erickson, J.R., Julie He, B., Grumbach, I.M., Anderson, M.E., 2011. CaMKII in the cardiovascular system: Sensing redox states. *Physiological Reviews* 91, 889–915. https://doi.org/10.1152/physrev.00018.2010.

Espinosa, A., Leiva, A., Peña, M., Müller, M., Debandi, A., Hidalgo, C., Angélica Carrasco, M., Jaimovich, E., 2006. Myotube depolarization generates reactive oxygen species through NAD(P)H oxidase; ROS-elicited Ca^{2+} stimulates ERK, CREB, early genes. *Journal of Cellular Physiology* 209, 379–388. https://doi.org/10.1002/jcp.20745.

Gomez-Cabrera, M.C., Arc-Chagnaud, C., Salvador-Pascual, A., Brioche, T., Chopard, A., Olaso-Gonzalez, G., Viña, J., 2020. Redox modulation of muscle mass and function. *Redox Biology* 35, 101531. https://doi.org/10.1016/j.redox.2020.101531.

Gomez-Cabrera, M.C., Borrás, C., Pallardo, F.V., Sastre, J., Ji, L.L., Viña, J., 2005. Decreasing xanthine oxidase-mediated oxidative stress prevents useful cellular adaptations to exercise in rats. *Journal of Physiology* 567, 113–120. https://doi.org/10.1113/jphysiol.2004.080564.

Gomez-Cabrera, M.C., Close, G.L., Kayani, A., McArdle, A., Viña, J., Jackson, M.J., 2010. Effect of xanthine oxidase-generated extracellular superoxide on skeletal muscle force generation. *American Journal of Physiology - Regulatory Integrative and Comparative Physiology* 298, 2–8. https://doi.org/10.1152/ajpregu.00142.2009.

Gomez-Cabrera, M.C., Domenech, E., Romagnoli, M., Arduini, A., Borras, C., Pallardo, F.V., Sastre, J., Viña, J., 2008. Oral administration of vitamin C decreases muscle mitochondrial biogenesis and hampers training-induced adaptations in endurance performance. *American Journal of Clinical Nutrition* 87, 142–149.

Gomez-Cabrera, M.C., Ristow, M., Viña, J., 2012. Antioxidant supplements in exercise: worse than useless? *American Journal of Physiology-Endocrinology and Metabolism* 302, E476–477; author reply E478–479. https://doi.org/10.1152/ajpendo.00567.2011.

Goncalves, R.L.S., Quinlan, C.L., Perevoshchikova, I.V., Hey-Mogensen, M., Brand, M.D., 2015. Sites of superoxide and hydrogen peroxide production by muscle mitochondria assessed ex vivo under conditions mimicking rest and exercise. *Journal of Biological Chemistry* 290, 209–227. https://doi.org/10.1074/jbc.M114.619072.

Handayaningsih, A.E., Iguchi, G., Fukuoka, H., Nishizawa, H., Takahashi, M., Yamamoto, M., Herningtyas, E.H., Okimura, Y., Kaji, H., Chihara, K., Seino, S., Takahashi, Y., 2011. Reactive oxygen species play an essential role in IGF-I signaling and IGF-I-induced myocyte hypertrophy in C2C12 myocytes. *Endocrinology* 152, 912–921. https://doi.org/10.1210/en.2010-0981.

Handschin, C., Spiegelman, B., 2008. The role of exercise and PGC1alpha in inflammation and chronic disease. *Nature* 454, 463–469.

Handschin, C., Spiegelman, B.M., 2011. PGC-1 coactivators and the regulation of skeletal muscle fiber-type determination. *Cell Metabolism* 13, 351. https://doi.org/10.1016/j.cmet.2011.03.008.

Hawley, J.A., Hargreaves, M., Joyner, M.J., Zierath, J.R., 2014. Integrative biology of exercise. *Cell* 159, 738–749. https://doi.org/10.1016/j.cell.2014.10.029.

Hellsten-Westing, Y., 1993. Immunohistochemical localization of xanthine oxidase in human cardiac and skeletal muscle. *Histochemistry* 100, 215–222. https://doi.org/10.1007/BF00269094.

Henríquez-Olguín, C., Boronat, S., Cabello-Verrugio, C., Jaimovich, E., Hidalgo, E., Jensen, T.E., 2019a. The emerging roles of nicotinamide adenine dinucleotide phosphate oxidase 2 in skeletal muscle redox signaling and metabolism. *Antioxidants & Redox Signaling* 31, 1371–1410. https://doi.org/10.1089/ars.2018.7678.

Henríquez-Olguín, C., Díaz-Vegas, A., Utreras-Mendoza, Y., Campos, C., Arias-Calderón, M., Llanos, P., Contreras-Ferrat, A., Espinosa, A., Altamirano, F., Jaimovich, E., Valladares, D.M., 2016. NOX2 inhibition impairs early muscle gene expression induced by a single exercise bout. *Frontiers in Physiology* 7. https://doi.org/10.3389/fphys.2016.00282.

Henríquez-Olguin, C., Knudsen, J.R., Raun, S.H., Li, Z., Dalbram, E., Treebak, J.T., Sylow, L., Holmdahl, R., Richter, E.A., Jaimovich, E., Jensen, T.E., 2019. Cytosolic ROS production by NADPH oxidase 2 regulates muscle glucose uptake during exercise. *Nature Communications* 10, 4623. https://doi.org/10.1038/s41467-019-12523-9.

Henriquez-Olguin, C., Meneses-Valdes, R., Jensen, T.E., 2020. Compartmentalized muscle redox signals controlling exercise metabolism - Current state, future challenges. *Redox Biology* 35, 101473. https://doi.org/10.1016/j.redox.2020.101473.

Henríquez-Olguín, C., Renani, L.B., Arab-Ceschia, L., Raun, S.H., Bhatia, A., Li, Z., Knudsen, J.R., Holmdahl, R., Jensen, T.E., 2019b. Adaptations to high-intensity interval training in skeletal muscle require NADPH oxidase 2. *Redox Biology* 24. https://doi.org/10.1016/j.redox.2019.101188.

Hidalgo, C., Sánchez, G., Barrientos, G., Aracena-Parks, P., 2006. A transverse tubule NADPH oxidase activity stimulates calcium release from isolated triads via ryanodine receptor type 1 S-glutathionylation. *Journal of Biological Chemistry* 281, 26473–26482. https://doi.org/10.1074/jbc.M600451200.

Higaki, Y., Mikami, T., Fujii, N., Hirshman, M.F., Koyama, K., Seino, T., Tanaka, K., Goodyear, L.J., 2008. Oxidative stress stimulates skeletal muscle glucose uptake through a phosphatidylinositol 3-kinase-dependent pathway. *American Journal of Physiology - Endocrinology and Metabolism* 294. https://doi.org/10.1152/ajpendo.00150.2007.

Hori, Y.S., Kuno, A., Hosoda, R., Tanno, M., Miura, T., Shimamoto, K., Horio, Y., 2011. Resveratrol ameliorates muscular pathology in the dystrophic mdx mouse, a model for Duchenne muscular dystrophy. *Journal of Pharmacology and Experimental Therapeutics* 338, 784–794. https://doi.org/10.1124/jpet.111.183210.

Jackson, M.J., 2011. Control of reactive oxygen species production in contracting skeletal muscle. *Antioxidants and Redox Signaling* 15, 2477–2486. https://doi.org/10.1089/ars.2011.3976.

Jackson, M.J., Vasilaki, A., McArdle, A., 2016. Cellular mechanisms underlying oxidative stress in human exercise. *Free Radical Biology and Medicine* 98, 13–17. https://doi.org/10.1016/j.freeradbiomed.2016.02.023.

Ji, L.L., 2015. Redox signaling in skeletal muscle: Role of aging and exercise. *Advances in Physiology Education* 39, 352–359. https://doi.org/10.1152/advan.00106.2014.

Ji, L.L., 2008. Modulation of skeletal muscle antioxidant defense by exercise: Role of redox signaling. *Free Radical Biology and Medicine* 44, 142–152. https://doi.org/10.1016/j.freeradbiomed.2007.02.031.

Kamata, H., Honda, S.-I., Maeda, S., Chang, L., Hirata, H., Karin, M., 2005. Reactive oxygen species promote TNFalpha-induced death and sustained JNK activation by inhibiting MAP kinase phosphatases. *Cell* 120, 649–661. https://doi.org/10.1016/j.cell.2004.12.041.

Katz, A., 2007. Modulation of glucose transport in skeletal muscle by reactive oxygen species. *Journal of Applied Physiology* 102, 1671–1676. https://doi.org/10.1152/japplphysiol.01066.2006

Kjøbsted, R., Roll, J.L.W., Jørgensen, N.O., Birk, J.B., Foretz, M., Viollet, B., Chadt, A., Al-Hasani, H., Wojtaszewski, J.F.P., 2019. AMPK and TBC1D1 regulate muscle glucose uptake after, but not during, exercise and contraction. *Diabetes* 68, 1427–1440. https://doi.org/10.2337/db19-0050.

Kozakowska, M., Pietraszek-Gremplewicz, K., Jozkowicz, A., Dulak, J., 2015. The role of oxidative stress in skeletal muscle injury and regeneration: Focus on antioxidant enzymes. *Journal of Muscle Research and Cell Motility* 36, 377–393. https://doi.org/10.1007/s10974-015-9438-9.

Le Moal, E., Pialoux, V., Juban, G., Groussard, C., Zouhal, H., Chazaud, B., Mounier, R., 2017. Redox control of skeletal muscle regeneration. *Antioxidants and Redox Signaling* 27, 276–310. https://doi.org/10.1089/ars.2016.6782.

Lee, S.-R., Yang, K.-S., Kwon, J., Lee, C., Jeong, W., Rhee, S.G., 2002. Reversible inactivation of the tumor suppressor PTEN by H_2O_2. *Journal of Biological Chemistry* 277, 20336–20342. https://doi.org/10.1074/jbc.M111899200.

Little, J.P., Safdar, A., Cermak, N., Tarnopolsky, M.A., Gibala, M.J., 2010. Acute endurance exercise increases the nuclear abundance of PGC-1α in trained human skeletal muscle. *American Journal of Physiology - Regulatory Integrative and Comparative Physiology* 298. https://doi.org/10.1152/ajpregu.00409.2009.

Loureiro, A.C.C., Rêgo-Monteiro, I.C.D., Louzada, R.A., Ortenzi, V.H., Aguiar, A.P.D., Abreu, E.S.D., Cavalcanti-De-Albuquerque, J.P.A., Hecht, F., Oliveira, A.C.D., Ceccatto, V.M., Fortunato, R.S., Carvalho, D.P., 2016. Differential expression of NADPH oxidases depends on skeletal muscle fiber type in rats. *Oxidative Medicine and Cellular Longevity* 2016. https://doi.org/10.1155/2016/6738701.

Louzada, R.A., Bouviere, J., Matta, L.P., Werneck-de-Castro, J.P., Dupuy, C., Carvalho, D.P., Fortunato, R.S., 2020. Redox signaling in widespread health benefits of exercise. *Antioxidants & Redox Signaling*. https://doi.org/10.1089/ars.2019.7949.

Louzada, R.A., Corre, R., Ameziane-El-Hassani, R., Hecht, F., Cazarin, J., Buffet, C., Carvalho, D.P., Dupuy, C., 2018. Conformation of the N-terminal ectodomain elicits different effects on duox function: a potential impact on congenital hypothyroidism caused by a H_2O_2 production defect. *Thyroid* 28, 1052–1062. https://doi.org/10.1089/thy.2017.0596.

Low, I.C.C., Loh, T., Huang, Y., Virshup, D.M., Pervaiz, S., 2014. Ser70 phosphorylation of Bcl-2 by selective tyrosine nitration of PP2A-B56δ stabilizes its antiapoptotic activity. *Blood* 124, 2223–2234. https://doi.org/10.1182/blood-2014-03-563296.

Matta, L., Fonseca, T.S., Faria, C.C., Lima-Junior, N.C., De Oliveira, D.F., Maciel, L., Boa, L.F., Pierucci, A.P.T.R., Ferreira, A.C.F., Nascimento, J.H.M., Carvalho, D.P., Fortunato, R.S., 2021. The effect of acute aerobic exercise on redox homeostasis and mitochondrial function of rat white adipose tissue [WWW Document]. *Oxidative Medicine and Cellular Longevity*. https://doi.org/10.1155/2021/4593496.

McCormick, R., Pearson, T., Vasilaki, A., 2016. Manipulation of environmental oxygen modifies reactive oxygen and nitrogen species generation during myogenesis. *Redox Biology* 8, 243–251. https://doi.org/10.1016/j.redox.2016.01.011.

Michaelson, L.P., Shi, G., Ward, C.W., Rodney, G.G., 2010. Mitochondrial redox potential during contraction in single intact muscle fibers. *Muscle Nerve* 42, 522–529. https://doi.org/10.1002/mus.21724.

Misu, H., Takayama, H., Saito, Y., Mita, Y., Kikuchi, A., Ishii, K.A., Chikamoto, K., Kanamori, T., Tajima, N., Lan, F., Takeshita, Y., Honda, M., Tanaka, M., Kato, S., Matsuyama, N., Yoshioka, Y., Iwayama, K., Tokuyama, K., Akazawa, N., Maeda, S., Takekoshi, K., Matsugo, S., Noguchi, N., Kaneko, S., Takamura, T., 2017. Deficiency of the hepatokine selenoprotein P increases responsiveness to exercise in mice through upregulation of reactive oxygen species and AMP-activated protein kinase in muscle. *Nature Medicine* 23, 508–516. https://doi.org/10.1038/nm.4295.

Miura, S., Kawanaka, K., Kai, Y., Tamura, M., Goto, M., Shiuchi, T., Minokoshi, Y., Ezaki, O., 2007. An increase in murine skeletal muscle peroxisome proliferator-activated receptor-γ coactivator-1α (PGC-1α) mRNA in response to exercise is mediated by β-adrenergic receptor activation. *Endocrinology* 148, 3441–3448. https://doi.org/10.1210/en.2006-1646.

Mofarrahi, M., Brandes, R.P., Gorlach, A., Hanze, J., Terada, L.S., Quinn, M.T., Mayaki, D., Petrof, B., Hussain, S.N.A., 2008. Regulation of proliferation of skeletal muscle precursor cells by NADPH oxidase. *Antioxidants and Redox Signaling* 10, 559–574. https://doi.org/10.1089/ars.2007.1792.

Mollica, J.P., Dutka, T.L., Merry, T.L., Lamboley, C.R., Mcconell, G.K., Mckenna, M.J., Murphy, R.M., Lamb, G.D., 2012. S-Glutathionylation of troponin I (fast) increases contractile apparatus Ca^{2+} sensitivity in fast-twitch muscle fibres of rats and humans. *Journal of Physiology* 590, 1443–1463. https://doi.org/10.1113/jphysiol.2011.224535.

Murata, H., Ihara, Y., Nakamura, H., Yodoi, J., Sumikawa, K., Kondo, T., 2003. Glutaredoxin exerts an antiapoptotic effect by regulating the redox state of Akt. *Journal of Biological Chemistry* 278, 50226–50233. https://doi.org/10.1074/jbc.M310171200.

Musaró, A., Giacinti, C., Borsellino, G., Dobrowolny, G., Pelosi, L., Cairns, L., Ottolenghi, S., Cossu, G., Bernardi, G., Battistini, L., Molinaro, M., Rosenthal, N., 2004. Stem cell-mediated muscle regeneration is enhanced by local isoform of insulin-like growth factor 1. *Proceedings of the National Academy of Sciences of the United States of America* 101, 1206–1210. https://doi.org/10.1073/pnas.0303792101.

Muthusamy, V.R., Kannan, S., Sadhaasivam, K., Gounder, S.S., Davidson, C.J., Boeheme, C., Hoidal, J.R., Wang, L., Rajasekaran, N.S., 2012. Acute exercise stress activates Nrf2/ARE signaling and promotes antioxidant mechanisms in the myocardium. *Free Radical Biology and Medicine* 52, 366–376. https://doi.org/10.1016/j.freeradbiomed.2011.10.440.

Nalbandian, M., Radak, Z., Takeda, M., 2019. N-acetyl-L-cysteine prevents lactate-mediated PGC1-alpha expression in C2C12 myotubes. *Biology* 8, 44. https://doi.org/10.3390/biology8020044.

Narkar, V.A., Downes, M., Yu, R.T., Embler, E., Wang, Y.X., Banayo, E., Mihaylova, M.M., Nelson, M.C., Zou, Y., Juguilon, H., Kang, H., Shaw, R.J., Evans, R.M., 2008. AMPK and PPARδ agonists are exercise mimetics. *Cell* 135, 189. https://doi.org/10.1016/j.cell.2008.06.051.

Osório Alves, J., Matta Pereira, L., Cabral Coutinho do Rêgo Monteiro, I., Pontes Dos Santos, L.H., Soares Marreiros Ferraz, V., Carneiro Loureiro, A.C., Calado Lima, C., Leal-Cardoso, J.H., Pires Carvalho, D., Soares Fortunato, R., Marilande Ceccatto, V., 2020. Strenuous acute exercise induces slow and fast twitch-dependent NADPH oxidase expression in rat skeletal muscle. *Antioxidants (Basel)* 9. https://doi.org/10.3390/antiox9010057.

Pala, R., Orhan, C., Tuzcu, M., Sahin, N., Ali, S., Cinar, V., Atalay, M., Sahin, K., 2016. Coenzyme Q10 supplementation modulates NFκB and Nrf2 pathways in exercise training. *Journal of Sports Science and Medicine* 15, 196–203.

Pedersen, B.K., Saltin, B., 2015. Exercise as medicine - Evidence for prescribing exercise as therapy in 26 different chronic diseases. *Scandinavian Journal of Medicine and Science in Sports* 25, 1–72. https://doi.org/10.1111/sms.12581.

Piao, Y.J., Seo, Y.H., Hong, F., Kim, J.H., Kim, Y.J., Kang, M.H., Kim, B.S., Jo, S.A., Jo, I., Jue, D.M., Kang, I., Ha, J., Kim, S.S., 2005. Nox 2 stimulates muscle differentiation via NF-κB/iNOS pathway. *Free Radical Biology and Medicine* 38, 989–1001. https://doi.org/10.1016/j.freeradbiomed.2004.11.011.

Powers, S.K., Jackson, M.J., 2008. Exercise-induced oxidative stress: Cellular mechanisms and impact on muscle force production. *Physiological Reviews* 88, 1243–1276. https://doi.org/10.1152/physrev.00031.2007.

Powers, S.K., Radak, Z., Ji, L.L., 2016. Exercise-induced oxidative stress: Past, present and future. *Journal of Physiology* 594, 5081–5092. https://doi.org/10.1113/JP270646.

Qaisar, R., Bhaskaran, S., Van Remmen, H., 2016. Muscle fiber type diversification during exercise and regeneration. *Free Radical Biology and Medicine* 98, 56–67. https://doi.org/10.1016/j.freeradbiomed.2016.03.025.

Ristow, M., Zarse, K., Oberbach, A., Klöting, N., Birringer, M., Kiehntopf, M., Stumvoll, M., Kahn, C.R., Blüher, M., 2009. Antioxidants prevent health-promoting effects of physical exercise in humans. *Proceedings of the National Academy of Sciences of the United States of America* 106, 8665–8670. https://doi.org/10.1073/pnas.0903485106.

Rodriguez, M.C., Tarnopolsky, M.A., 2003. Patients with dystrophinopathy show evidence of increased oxidative stress. *Free Radical Biology and Medicine* 34, 1217–1220. https://doi.org/10.1016/S0891-5849(03)00141-2.

Rojo de la Vega, M., Zhang, D.D., 2018. NRF2 induction for NASH treatment: A new hope rises. *Cellular and Molecular Gastroenterology and Hepatology* 5, 422–423. https://doi.org/10.1016/j.jcmgh.2017.12.009.

Russell, A.P., Feilchenfeldt, J., Schreiber, S., Praz, M., Crettenand, A., Gobelet, C., Meier, C.A., Bell, D.R., Kralli, A., Giacobino, J.P., Dériaz, O., 2003. Endurance training in humans leads to fiber type-specific increases in levels of peroxisome proliferator-activated receptor-γ coactivator-1 and peroxisome proliferator-activated receptor-α in skeletal muscle. *Diabetes* 52, 2874–2881. https://doi.org/10.2337/diabetes.52.12.2874.

Ryan, M.J., Jackson, J.R., Hao, Y., Leonard, S.S., Alway, S.E., 2011. Inhibition of xanthine oxidase reduces oxidative stress and improves skeletal muscle function in response to electrically stimulated isometric contractions in aged mice. *Free Radical Biology and Medicine* 51, 38–52. https://doi.org/10.1016/j.freeradbiomed.2011.04.002.

Sakamoto, K., Arnolds, D.E., Fujii, N., Kramer, H.F., Hirshman, M.F., Goodyear, L.J., 2006. Role of Akt2 in contraction-stimulated cell signaling and glucose uptake in skeletal muscle. *American Journal of Physiology-Endocrinology and Metabolism* 291, E1031–1037. https://doi.org/10.1152/ajpendo.00204.2006.

Sakellariou, G.K., Vasilaki, A., Palomero, J., Kayani, A., Zibrik, L., McArdle, A., Jackson, M.J., 2013. Studies of mitochondrial and nonmitochondrial sources implicate nicotinamide adenine dinucleotide phosphate oxidase(s) in the increased skeletal muscle superoxide generation that occurs during contractile activity. *Antioxidants and Redox Signaling* 18, 603–621. https://doi.org/10.1089/ars.2012.4623.

Sambasivan, R., Yao, R., Kissenpfennig, A., van Wittenberghe, L., Paldi, A., Gayraud-Morel, B., Guenou, H., Malissen, B., Tajbakhsh, S., Galy, A., 2011. Pax7-expressing satellite cells are indispensable for adult skeletal muscle regeneration. *Development* 138, 3647–3656. https://doi.org/10.1242/dev.067587.

Sandiford, S.D., Kennedy, K.A., Xie, X., Pickering, J.G., Li, S.S., 2014. Dual Oxidase Maturation factor 1 (DUOXA1) overexpression increases reactive oxygen species production and inhibits murine muscle satellite cell differentiation. *Cell Communication and Signaling* 12, 1–15. https://doi.org/10.1186/1478-811X-12-5.

Schiaffino, S., Reggiani, C., 2011. Fiber types in mammalian skeletal muscles. *Physiological Reviews* 91, 1447–1531. https://doi.org/10.1152/physrev.00031.2010.

Shanmugam, G., Challa, A.K., Devarajan, A., Athmanathan, B., Litovsky, S.H., Krishnamurthy, P., Davidson, C.J., Rajasekaran, N.S., 2019. Exercise mediated Nrf2 signaling protects the myocardium from isoproterenol-induced pathological remodeling. *Frontiers in Cardiovascular Medicine* 6, 68. https://doi.org/10.3389/fcvm.2019.00068.

Shanmugasundaram, K., Nayak, B.K., Friedrichs, W.E., Kaushik, D., Rodriguez, R., Block, K., 2017. NOX4 functions as a mitochondrial energetic sensor coupling cancer metabolic reprogramming to drug resistance. *Nature Communications* 8, 997. https://doi.org/10.1038/s41467-017-01106-1.

Shao, D., Oka, S.I., Liu, T., Zhai, P., Ago, T., Sciarretta, S., Li, H., Sadoshima, J., 2014. A redox-dependent mechanism for regulation of AMPK activation by thioredoxin1 during energy starvation. *Cell Metabolism* 19, 232–245. https://doi.org/10.1016/j.cmet.2013.12.013.

Silveira, L.R., Pilegaard, H., Kusuhara, K., Curi, R., Hellsten, Y., 2006. The contraction induced increase in gene expression of peroxisome proliferator-activated receptor (PPAR)-γ coactivator 1α (PGC-1α), mitochondrial uncoupling protein 3 (UCP3) and hexokinase II (HKII) in primary rat skeletal muscle cells is dependent on rea. *Biochimica et Biophysica Acta - Molecular Cell Research* 1763, 969–976. https://doi.org/10.1016/j.bbamcr.2006.06.010.

Smith, M.A., Reid, M.B., 2006. Redox modulation of contractile function in respiratory and limb skeletal muscle. *Respiratory Physiology and Neurobiology* 151, 229–241. https://doi.org/10.1016/j.resp.2005.12.011.

Spangenburg, E.E., Le Roith, D., Ward, C.W., Bodine, S.C., 2008. A functional insulin-like growth factor receptor is not necessary for load-induced skeletal muscle hypertrophy. *The Journal of Physiology (London)* 586, 283–291. https://doi.org/10.1113/jphysiol.2007.141507.

Specht, K.S., Kant, S., Addington, A.K., McMillan, R., Hulver, M.W., Learnard, H., Campbell, M., Donnelly, S., Caliz, A.D., Pei, Y., Reif, M., Bond, J., DeMarco, A., Craige, B., Keaney, J.F., Craige, S.M., 2021. Nox4 mediates skeletal muscle metabolic responses to exercise. *Molecular Metabolism* 45:101160. https://doi.org/10.1016/j.molmet.2020.101160.

St-Pierre, J., Drori, S., Uldry, M., Silvaggi, J.M., Rhee, J., Jäger, S., Handschin, C., Zheng, K., Lin, J., Yang, W., Simon, D.K., Bachoo, R., Spiegelman, B.M., 2006. Suppression of reactive oxygen species and neurodegeneration by the PGC-1 transcriptional coactivators. *Cell* 127, 397–408. https://doi.org/10.1016/j.cell.2006.09.024.

Sun, J., Xin, C., Eu, J.P., Stamler, J.S., Meissner, G., 2001. Cysteine-3635 is responsible for skeletal muscle ryanodine receptor modulation by NO. *Proceedings of the National Academy of Sciences of the United States of America* 98, 11158–11162. https://doi.org/10.1073/pnas.201289098.

Sun, Q.A., Hess, D.T., Nogueira, L., Yong, S., Bowles, D.E., Eu, J., Laurita, K.R., Meissner, G., Stamler, J.S., 2011. Oxygen-coupled redox regulation of the skeletal muscle ryanodine receptor-Ca^{2+} release channel by NADPH oxidase 4. *Proceedings of the National Academy of Sciences of the United States of America* 108, 16098–16103. https://doi.org/10.1073/pnas.1109546108.

Sun, Q.A., Wang, B., Miyagi, M., Hess, D.T., Stamler, J.S., 2013. Oxygen-coupled redox regulation of the skeletal muscle ryanodine receptor/Ca^{2+} release channel (RyR1): Sites and nature of oxidative modification. *Journal of Biological Chemistry* 288, 22961–22971. https://doi.org/10.1074/jbc.M113.480228.

Takamura, T., 2020. Hepatokine selenoprotein P-mediated reductive stress causes resistance to intracellular signal transduction. *Antioxid. Redox Signal.* https://doi.org/10.1089/ars.2020.8087.

Wadley, G.D., Nicolas, M.A., Hiam, D.S., McConell, G.K., 2013. Xanthine oxidase inhibition attenuates skeletal muscle signaling following acute exercise but does not impair mitochondrial adaptations to endurance training. *American Journal of Physiology - Endocrinology and Metabolism* 304, 853–862. https://doi.org/10.1152/ajpendo.00568.2012.

Wafi, A.M., Yu, L., Gao, L., Zucker, I.H., 2019. Exercise training upregulates Nrf2 protein in the rostral ventrolateral medulla of mice with heart failure. *Journal of Applied Physiology.* https://doi.org/10.1152/japplphysiol.00469.2019.

Wong, H.S., Dighe, P.A., Mezera, V., Monternier, P.A., Brand, M.D., 2017. Production of superoxide and hydrogen peroxide from specific mitochondrial sites under different bioenergetic conditions. *Journal of Biological Chemistry* 292, 16804–16809. https://doi.org/10.1074/jbc.R117.789271.

Xia, R., Webb, J.A., Gnall, L.L.M., Cutler, K., Abramson, J.J., 2003. Skeletal muscle sarcoplasmic reticulum contains a NADH-dependent oxidase that generates superoxide. *American Journal of Physiology - Cell Physiology* 285, C215.

Yan, Z., 2009. Exercise, PGC-1α, and metabolic adaptation in skeletal muscle. *Applied Physiology, Nutrition and Metabolism* 34, 424–427. https://doi.org/10.1139/H09-030.

Zmijewski, J.W., Banerjee, S., Bae, H., Friggeri, A., Lazarowski, E.R., Abraham, E., 2010. Exposure to hydrogen peroxide induces oxidation and activation of AMP-activated protein kinase. *Journal of Biological Chemistry* 285, 33154–33164. https://doi.org/10.1074/jbc.M110.143685.

CHAPTER FOUR

Oxygen Transport

A REDOX O_2DYSSEY

P.N. Chatzinikolaou, N.V. Margaritelis, A.N. Chatzinikolaou, V. Paschalis, A.A. Theodorou, I.S. Vrabas, A. Kyparos and M.G. Nikolaidis

CONTENTS

Introduction / 41
A Quantitative Snapshot of Oxygen Transport / 42
Lungs / 43
Erythrocytes / 45
  Oxygen / 45
  Redox Network / 45
  Energetics / 46
  A Computational Model / 48
Microcirculation / 48
  Structure / 48
  Regulation / 49
Muscle / 49
Mitochondria / 51
  Oxygen Flow and Consumption / 51
  Oxygen Consumption in Cellular (Redox) Processes Beyond Mitochondrial Respiration / 51
Conclusion / 53
References / 53

INTRODUCTION

Oxygen (O_2), a stable di-radical, is central for almost all physiological systems (e.g., respiratory, cardiovascular and muscular), which coordinate in an integrated manner to match the metabolic demands of active tissues. During exercise, a mismatch between oxygen supply and demand decreases performance and increases the rate of fatigue development (Amann and Calbet, 2008). Hence, several exercise and nutritional interventions have been prescribed with the aim to decrease fatigue and improve performance by modulating oxygen metabolism.

Reactive oxygen and nitrogen species (RONS) are involved in many biological processes (e.g., ageing, stress adaptation) and in human disease (e.g., cancer, cardiovascular disease). Of note, specific RONS

DOI: 10.1201/9781003051619-4

[e.g., hydrogen peroxide (H_2O_2) and nitric oxide ($NO^{\bullet}$)] are identified as pleiotropic signaling molecules (Cobley et al., 2017; Sies & Jones, 2020). In the same context, redox biology is an integral part of exercise metabolism, since targeted RONS regulate diverse exercise responses (e.g., vasodilation, cell bioenergetics) and adaptations (e.g., angiogenesis, mitochondrial biogenesis) (Margaritelis et al., 2020a), while complex redox modifications (e.g., S-glutathionylation of proteins) are implicated in the development of exercise-induced muscle fatigue (Debold, 2015; Cheng et al., 2016). With respect to oxygen metabolism, $NO^{\bullet}$ is implicated in oxygen transport and utilization, and, therefore, its precursors (e.g., dietary nitrates) are considered ergogenic aids (Jones et al., 2018). Although accumulating evidence indicates that redox processes regulate oxygen fate, the literature may be perceived as fragmented especially under physiological in vivo conditions (e.g., exercise). This is exemplified by the fact that the redox components (i.e., signaling as well as structural and functional modifications mediated by RONS) at each step of the oxygen pathway are still unidentified.

Therefore, the aim of the present work is to shed light on the most critical redox components at each step of the oxygen pathway, from atmospheric air to muscle mitochondria. We will follow a quantitative and figure-based approach by presenting, as realistically as possible, what is happening from the molecular to the physiological level. The starting point for modeling biological processes is to visualize them by including the most important physiological and structural characteristics, and excluding dispensable ones from the model (Phillips et al., 2013). This is the first domain highlighting the link between oxygen transport, redox regulation and fatigue development during exercise. Identifying sources of reactive species, oxidative targets and redox mechanisms that modulate oxygen delivery and utilization at each step may reveal novel pathways for nutritional interventions with ergogenic potential. Moreover, with the numerical depictions of oxygen transport, we aspire to provide experimentally testable quantitative predictions in exercise physiology and sports nutrition.

A QUANTITATIVE SNAPSHOT OF OXYGEN TRANSPORT

The pathway of oxygen from atmosphere to mitochondria involves a series of transfer steps, which are depicted in Figure 4.1. The resting whole-body oxygen consumption (VO_2) of a healthy young adult with a cardiac output of 5 L/min is ≈3.5 mL/kg/min (Joyner and Casey, 2015). Maximal exercise evokes large changes in oxygen transport and utilization as indicated by the 50- to 100-fold increase in skeletal muscle blood flow and the 10-fold increase in VO_2max (Joyner and Dempsey, 2018). VO_2 max can be calculated from the Fick equation as the product of cardiac output (Q) and arterio-venous difference of oxygen (a – vO_2 diff) (equation 4.1) (Opondo, Sarma and Levine, 2015).

$$VO_2 = \dot{Q} \times (a - vO_2\text{diff}) \qquad (4.1)$$

Oxygen delivery (DO_2) can be defined as the product of cardiac output and arterial oxygen content. Oxygen content per liter of arterial blood can be calculated from the concentration of hemoglobin (cHb), its binding capacity and saturation level (%), plus the partial pressure of arterial oxygen (PaO_2= 100 mm Hg) and the 0.03 mL of oxygen dissolved in plasma (eq. 4.2) (Pittman, 2016). The binding capacity of oxygen is reported to be ≈1.31–1.39 mL O_2/g Hb, with the most common value reported in the region of 1.34 mL O_2/g Hb (Pittman, 2016). Therefore, for a young male with a 70 kg body mass, cardiac output of 5 L/min and hemoglobin concentration 150 g per liter of blood, the resulting rate of oxygen delivery at rest is 1020 mL/min or 14.57 mL/min/kg. At maximal exercise, and with a cardiac output of 20 L/min, the maximum capacity of oxygen delivery would increase to 4,080 mL/min or 58.3 mL/kg/min [assuming oxygen saturation levels remain constant (100%)]. Yet, the various dyshemoglobin species (e.g., methemoglobin, carbaminohemoglobin), which impair the binding capacity of hemoglobin, must be considered when calculating total oxygen delivery. For instance, increased oxidation of hemoglobin to methemoglobin (which cannot carry oxygen) leading to a 10% increase in dyshemoglobin species could reduce oxygen delivery to 13.14 and 52.54 mL/min/kg at rest and maximal exercise, respectively. At rest or low intensities, these effects (i.e., ≈1.43 mL/min/kg decrease) may be compensated by hemodynamic changes (e.g., increased heart rate); yet at maximal exercise, the difference in oxygen delivery (i.e., ≈5.76 mL/min/kg decrease) could

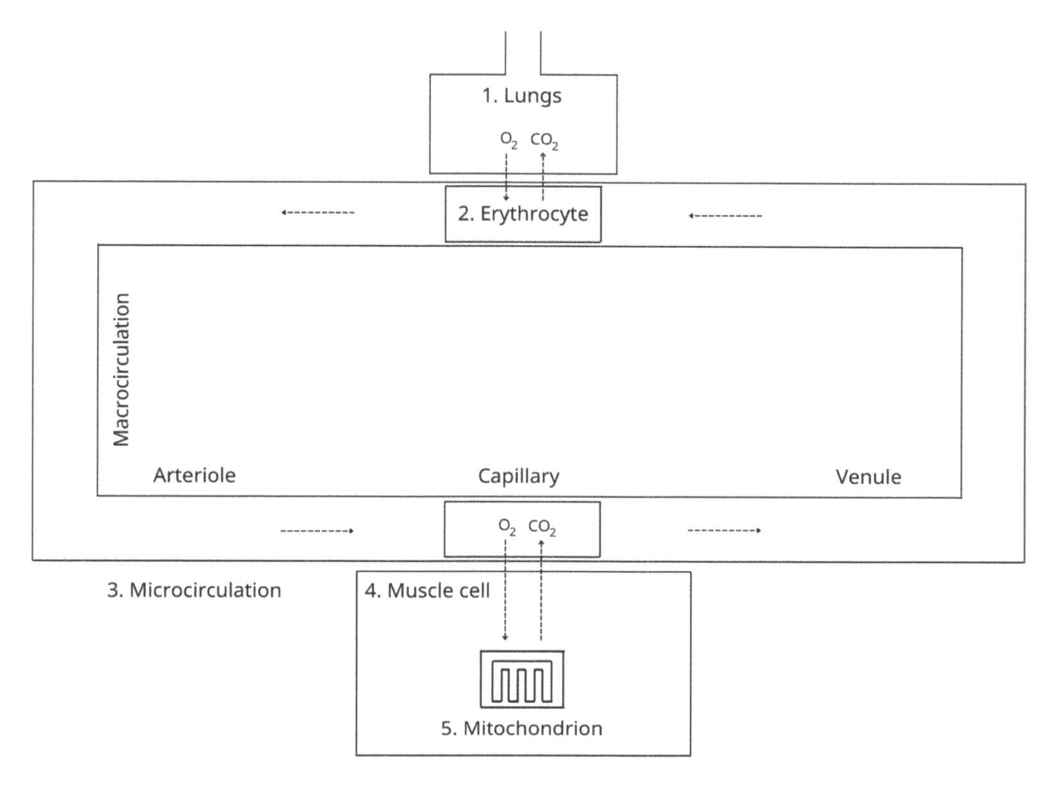

Figure 4.1 Overview of the oxygen (O_2) pathway and the critical transfer steps from atmosphere to muscle mitochondria. First, ventilation moves air into the lungs (1), where pulmonary blood is oxygenated, and carbon dioxide (CO_2) is removed from the body. Then, oxygen, bound to erythrocytes (2) and dissolved in plasma, is transported within the macrocirculation to the microcirculation (3), with the latter regulating the distribution of blood and oxygen to the tissues. Finally, oxygen is diffused from the capillaries into the muscle cell (4) and ultimately to muscle mitochondria (5), where it is utilized for energy production.

increase fatigue and impair exercise performance. On this basis, conditions characterized by very high or chronic erythrocyte oxidative stress (e.g., an unaccustomed eccentric exercise bout eliciting excessive muscle damage or during a clinical condition such as type-II diabetes; (Nikolaidis et al., 2012; Parker et al., 2017)) can increase dyshemoglobin (e.g., methemoglobin) species and reduce oxygen delivery.

$$DO_2 = CO \times \left[(1.34 \times cHb \times SaO_2) + (0.03 \times PaO_2) \right] \quad (4.2)$$

LUNGS

At the first step of oxygen transport, the atmospheric air is channeled to the lungs through ventilation at the alveolar-capillary region, to replace oxygen and remove carbon dioxide (Figure 4.2). The gas exchange between the alveoli and pulmonary capillaries is facilitated via passive gas diffusion, from a high to low-pressure gradient. In human lungs, the number of alveoli is $\approx 4.8 \times 10^8$ (Ochs et al., 2004) and is estimated to occupy a large surface area of $\approx 60–80\,m^2$ (Pittman, 2016). The alveoli are in close contact with the pulmonary capillaries, separated only by a thin air-blood barrier $\approx 0.2–0.6\,\mu m$ thick (Hall, 2016). Each alveolus has a wide diameter of $\approx 200\,\mu m$ (Ochs et al., 2004), whereas a pulmonary capillary could be as small as $\approx 3–5\,\mu m$ (Hall, 2016; Pittman, 2016; Kuck, Peart and Simmonds, 2020). The erythrocytes are evidently larger cells (i.e., $\approx 7–8\,\mu m$ diameter) than capillaries and have to change their shape and mechanical properties

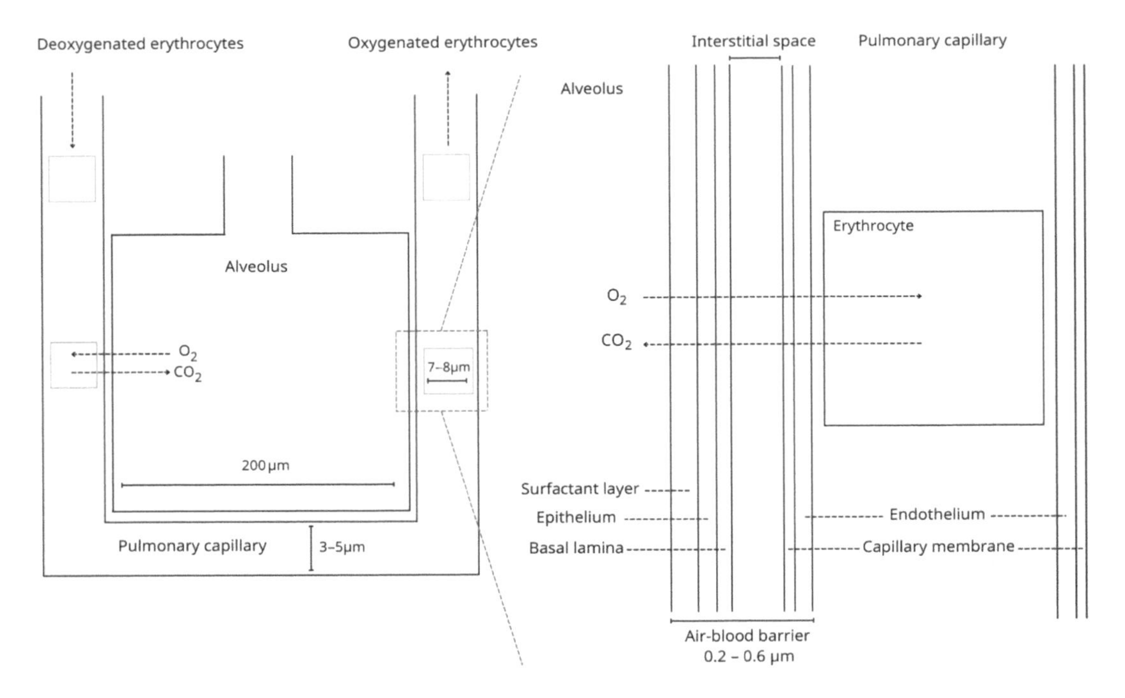

Figure 4.2 Gas exchange at the alveolar-capillary level in the lungs. In the left panel, the transfer of oxygen and carbon dioxide between an alveolus and erythrocytes is depicted. The erythrocytes change their shape (deformability) to travel through the pulmonary capillaries, which are much smaller in diameter (≈3–5 μm) compared to erythrocytes (≈7–8 μm). In the right panel, the various membrane layers that oxygen must traverse to reach the erythrocyte are presented. The air-blood barrier, consisting of the membrane layers of alveolus, the interstitial space and capillary wall, is very thin (≈0.2–0.6 μm) to facilitate oxygen diffusion. This thin air-blood barrier, the aggregate number of erythrocytes adjacent to the pulmonary capillaries (see also Figure 4.3) and the close contact of the alveoli with the capillaries ensure the diffusion of oxygen from lungs to the blood, which is the first transfer step in the oxygen pathway.

to traverse through the pulmonary capillaries and bind oxygen (Kuck, Peart and Simmonds, 2020) (Figure 4.3). Overall, the large surface area of the alveolar region, the thin air-blood barrier, the difference in gas pressure gradients and the erythrocyte's ability to deform greatly enhance gas diffusion at this step.

Lungs are characterized by a rich redox network; however, human studies directly examining lung tissue are scarce and mostly focused on diseased populations due to ethical restrictions (i.e., biopsies). Under pathophysiological conditions characterized by increased inflammation and oxidative stress, the close contact between erythrocytes and alveoli facilitates the diffusion of RONS between them. Specifically, superoxide $\left(O_2^{\cdot -}\right)$ and H_2O_2 from alveolar macrophages and NADPH oxidases can damage the alveolar cells and impair physiological respiration (Boots, Haenen and Bast, 2003; Kato and Hecker, 2020). Moreover, oxidative stress- induced erythrocyte lysis may lead to oxidized hemoglobin release to the alveoli, which can damage alveolar cells (Chintagari, Jana and Alayash, 2016). We speculate that this may be transiently the case also during high-intensity exercise, which has been reported to impair the air-blood barrier and promote erythrocyte movement into the bronchoalveolar fluid (Hopkins et al., 1997).

The respiratory system per se affects oxygen transport and contributes to fatigue (Dempsey, La Gerche and Hull, 2020). During exercise-induced hypoxemia (a frequent phenomenon in trained athletes), arterial oxygen saturation (SaO_2) can reach as low as 88%. For every 1% drop in SaO_2, VO_2max decreases by 2% (Dempsey, La Gerche and Hull, 2020). Redox-related processes could be involved in this exercise-induced

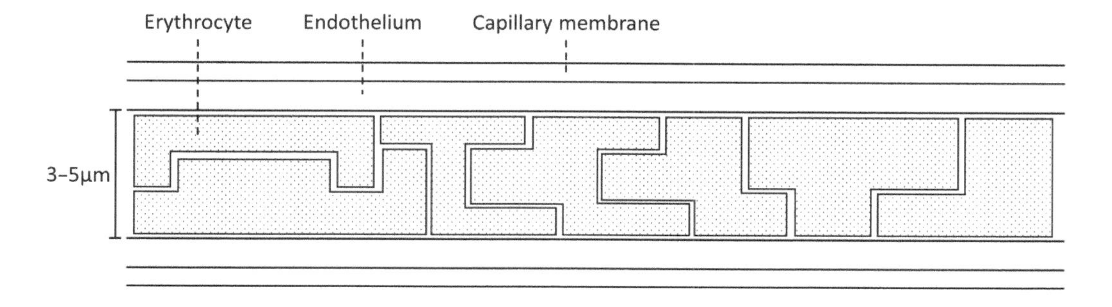

Figure 4.3 Erythrocyte "Tetris". Erythrocytes are amazingly deformable structures able to recover their initial shape even after large deformations as when passing through tight blood capillaries (≈3–5 μm). The erythrocyte's deformability depends on its physical (i.e., membrane, cytoskeleton) and metabolic (i.e., redox network and energetics) characteristics. Inability to change their shape (e.g., due to oxidative modifications on their membrane) could lead to cell lysis or increased shear stress on the capillary walls that damages the endothelium, leading to impaired blood flow and oxygen delivery. The figure is constructed based on transmission electron microscopy images.

SaO_2 drop. Exercise has been reported to induce oxidative stress in the lungs of healthy adults, predominantly by enzymes (e.g., NADPH oxidases) residing at the outer layers of lung cells and in various leukocytes (Araneda, Carbonell and Tuesta, 2016). Moreover, hemoglobin oxidation during exercise decreases its oxygen-carrying capacity. Noteworthy, both these redox responses are amplified under certain environmental conditions frequently faced by athletes (e.g., cold and altitude). Collectively, it becomes reasonable to argue that redox processes (e.g., hemoglobin oxidation, oxidative damage to alveolar cells) are involved in the first step of oxygen pathway, yet it is still unknown to what extent these processes limit oxygen transport.

ERYTHROCYTES

Oxygen

Oxygen diffusing to the pulmonary blood binds to erythrocytes (99%) and about 1% remains dissolved in plasma. An average male adult, with a total blood volume of 5 L, has 2.2 L of erythrocytes. Furthermore, 1 L of blood contains approximately 150 and 130 g of hemoglobin in men and women, respectively. As one gram of hemoglobin can bind ≈1.34 mL of oxygen, a liter of blood contains ≈201 mL of oxygen in men and ≈177 mL in women (Mairbäurl, 2013). In addition, each erythrocyte has $\approx 2.74 \times 10^8$ hemoglobin molecules (Milo and Rob, 2016). The erythrocyte, as the most abundant cell in the body ($\approx 3 \times 10^{13}$) (Milo and Rob, 2016), serves as a main gas transporter from lung to tissue and vice versa.

Chronic exercise increases plasma volume, erythrocyte count and total hemoglobin content, which are fundamental determinants for VO_2max improvements (Sawka et al., 2000; Mairbäurl, 2013; Saunders et al., 2013). In particular, exercise training can increase total blood volume by ≈10%–15%, by inducing blood volume expansion and increasing erythrocyte count (Sawka et al., 2000). Blood volume expansion augments cardiac output, while the increase in erythrocyte count and hemoglobin content enhances the oxygen-carrying capacity and subsequently VO_2max (Lundby et al., 2017; Mairbäurl, 2013). The central role of blood is also exemplified by the fact that an acute loss of 450 mL reduces VO_2max by 7% probably due to reduced stroke volume (Skattebo et al., 2021). These effects explain why the World Anti-Doping Agency prohibits blood doping (e.g., transfusion of blood or administration of erythropoietin).

Redox Network

To preserve its mechanical properties (e.g., deformability) and maintain hemoglobin in a reduced state, the erythrocyte is equipped with a rich and well-orchestrated redox network consisting of (1) enzymatic antioxidants such as superoxide dismutase (SOD), catalase (Cat), glutathione peroxidase (GPx) and peroxiredoxins (Prx); (2) non-enzymatic antioxidants such as glutathione (GSH), α-tocopherol (vitamin E) and ascorbate (vitamin C); and (3) reducing equivalents, namely, NADH and NADPH (Kuhn et al., 2017) (Figure 4.4).

Hemoglobin autooxidation, free Fe^{3+} and NADPH oxidases are the major sources of RONS in erythrocytes (George et al., 2013; Kuhn et al., 2017). In addition, an erythrocyte membrane is rich in polyunsaturated fatty acids (Poppitt et al., 2005), which are more prone to oxidation compared to other lipids due to their double-bond structure. This is important because excessive peroxidation of the lipid membrane can lead to erythrocyte lysis (Kuhn et al., 2017). Hemoglobin autooxidation (Hb-Fe^{2+} oxidation to Hb-Fe^{3+}, in which the sixth coordination position of the heme iron is occupied by hydroxide or water) produces methemoglobin and $O_2^{\bullet-}$ that is rapidly dismutated to H_2O_2. H_2O_2 can subsequently participate in the Fenton reaction and produce hydroxyl radical (OH, the most reactive RONS). Additionally, the most frequently encountered redox modifications in hemoglobin are tyrosine nitration (Koppenol, 2012; Barbieri et al., 2013), carbonylation (Subudhi et al., 2001) and thiol oxidation (Gwozdzinski et al., 2017). Some of these redox modifications have also been detected during exercise (Petibois and Déléris, 2005; Gwozdzinski et al., 2017). Once oxidized, hemoglobin can lead to the formation of Heinz bodies (aggregated hemoglobin molecules), which also damage the erythrocyte's membrane, decrease its deformability and impair its oxygen-carrying capacity (Barodka et al., 2014; Mohanty, Nagababu and Rifkind, 2014; Kuhn et al., 2017).

Exercise also increases erythrocyte oxidative stress and promotes cell lysis (Sentürk et al., 2001; Theodorou et al., 2010). Hemolysis can induce vasoconstriction by scavenging $NO^{\bullet}$ from free hemoglobin (Ferguson et al., 2018) and may lead to short-term impairment in endothelial function (Rakobowchuk et al., 2017). In addition, hemolysis-derived hemoglobin can decrease $NO^{\bullet}$ bioavailability in muscle, impairing microvascular control and muscle contraction (Ferguson et al., 2018). Collectively, the central task of the antioxidant system in the erythrocyte is to preserve hemoglobin in a reduced state and prevent (per) oxidation of cellular biomolecules.

Energetics

As erythrocytes lack mitochondria, they rely solely on glucose to produce ATP. In a reference man, erythrocytes consume ≈20 g of glucose per day, which is ≈10% of total body glucose metabolism and have a rate of glucose utilization of ≈10 g glucose/kg of erythrocyte per day (Baynes and Dominiczak, 2019). An approximate 90% of the glucose is metabolized via glycolysis and the rest 10% is used in the pentose phosphate pathway (also referred to as pentose shunt). Overall, glucose metabolism in erythrocytes is essential for (1) ATP production, which is relevant for the membrane and cytoskeleton's properties (e.g., deformability); (2) maintenance of redox network (e.g., NADPH and glutathione); (3) reduction of methemoglobin to hemoglobin by NADH-cytochrome b5 reductase; and (4) production of 2,3-bisphosphoglycerate, which binds to hemoglobin and releases oxygen (Van Wijk & Van Solinge, 2005) (Figure 4.4).

Clearly, erythrocyte metabolism and redox features are tightly intertwined and strongly dependent on adequate glucose availability. For instance, in glycolysis-associated enzymopathies (e.g., G6PD-deficiency), conditions of decreased glucose uptake (e.g., excessive fasting) and environmental settings that stimulate rerouting of the carbohydrate flux (e.g., hypoxia). All these cases are characterized by an impaired antioxidant system, where the erythrocyte becomes susceptible to lysis dropping its oxygen-carrying capacity (Rogers et al., 2009; Rifkind and Nagababu, 2013; Kuhn et al., 2017; Georgakouli et al., 2019; Wang et al., 2020). Regarding sports nutrition, the ergogenic effect of carbohydrates during exercise can be reasonably attributed not only to their energetic potential as substrate but also to their redox-related reductive power (i.e., NADPH regeneration). Moreover, training sometimes in a carbohydrate-depleted state has been proposed to stimulate cell signaling processes for long-term adaptations (Impey et al., 2018). In the same vein, we speculate that this may be also the case for redox adaptations after exercise training, such as the enhanced gene expression of antioxidant enzymes. Of course, this hypothesis rests on tentative ground without experimental evidence.

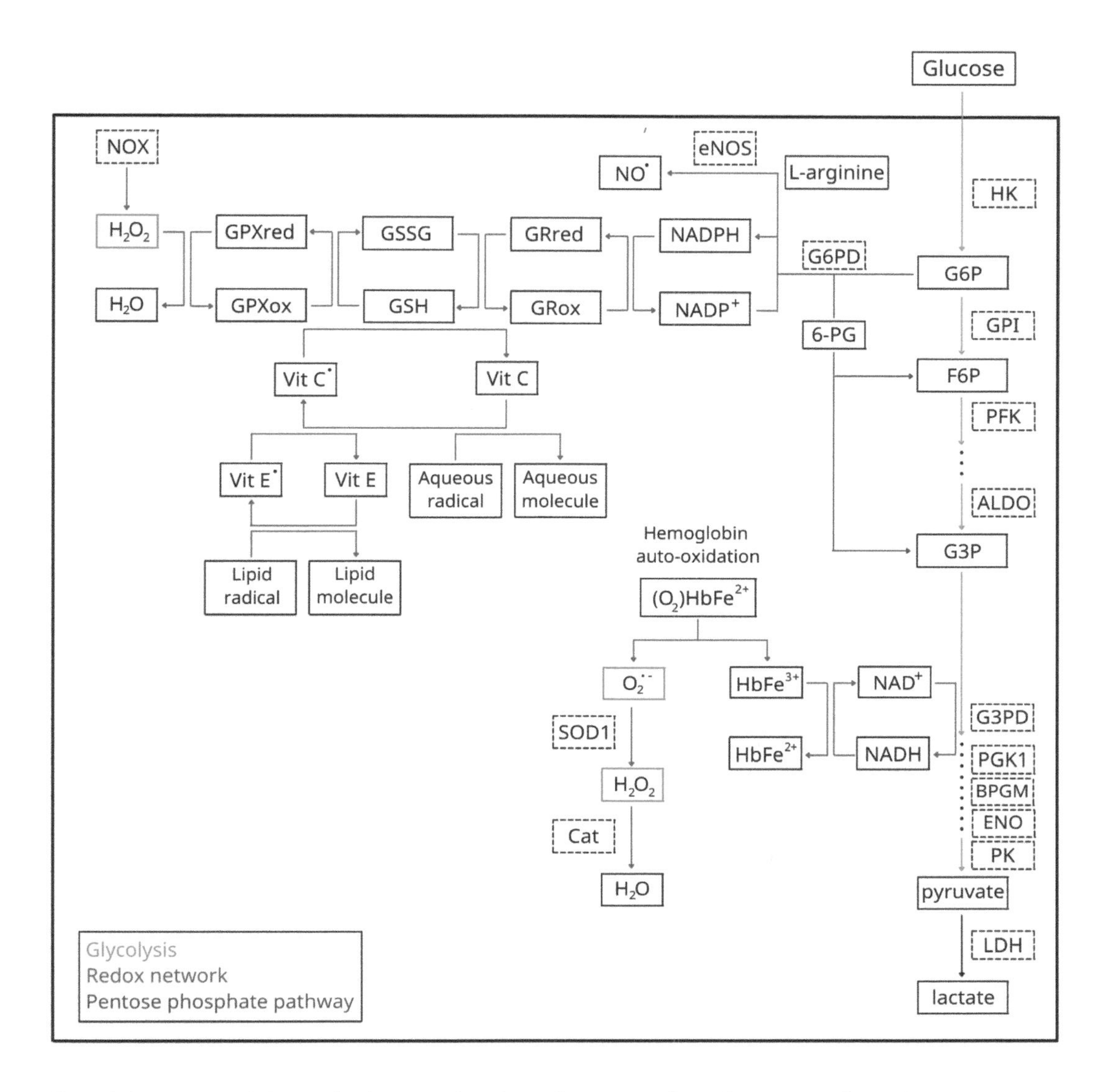

Figure 4.4 Energy metabolism and redox network in the erythrocyte. Briefly, as glucose enters the erythrocyte, it is utilized simultaneously by two pathways, namely the pentose phosphate pathway (also referred to as pentose phosphate shunt) and glycolysis. Under physiological conditions, ≈90% of the glucose is utilized (orange color) by glycolysis. This pathway also ensures the recycle of NADH which reduces methemoglobin (it cannot carry oxygen) back to hemoglobin. The rest ≈10% of glucose is utilized by the pentose phosphate pathway (green color), which is strongly associated with the erythrocyte's redox network. This pathway is responsible for maintaining the concentration of NADPH, which in turn recycles glutathione (GSH). In red color is depicted the production of superoxide $(O_2^{\cdot-})$ from hemoglobin auto-oxidation and of hydrogen peroxide (H_2O_2) from NADPH oxidases (NOX) after dismutation of $O_2^{\cdot-}$ by SOD. Downregulation of these pathways, due to reduced glucose availability, could increase the concentration of methemoglobin species and decrease erythrocyte's oxygen-carrying capacity, leading to fatigue. ALDO: aldolase; BPGM: 3-phosphoclycerate mutase; ENO: enolase; F6P: fructose-6-phosphate; FBP: fructose-1,6-bisphosphate; G3P: glyceraldehyde-3-phosphate; G3PD: glyceraldehyde-3-phosphate dehydrogenase; G6P: glucose-6-phosphate; G6PD: glucose-6-phosphate dehydrogenase; GPI: glucose-2-phosphate isomerase; LDH: lactate dehydrogenase; PFK: phosphofructokinase; PGK1: phosphoglycerate kinase; PK: pyruvate kinase; 6-PG: 6-phosphogluconolactone. All enzymes of erythrocyte glycolysis are presented, but some steps of the pathway not relevant to our discussion were omitted and are shown with dots.

A Computational Model

We have recently argued that even basic mathematical equations may refine the broad understanding of exercise redox biology and define the borders between cellular signaling and oxidative stress (Nikolaidis, Margaritelis and Matsakas, 2020). In this context, the role of erythrocyte glutathione on lipid peroxidation can be partially predicted with simple computational calculations. The biochemical reactions [reaction I] and [reaction II], showing the reduction of lipid hydroperoxides (R-OOH) from glutathione and the recycling of glutathione from NADPH, were used to create a minimal "proof-of-concept" computational model. Glutathione (Margaritelis et al., 2020b), lipid hydroperoxides (Niki, 2014), alcohols (R-OH) (Tsalouhidou, Petridou and Mougios, 2009) and NADP(H) (Zerez et al., 1988) data were collected from the literature. All biochemical kinetic rates (Benfeitas et al., 2014) were computed in R (Version 4.0.3; R Core Team, 2020). The model was subsequently developed in COPASI (Hoops et al., 2006) to simulate the effect of low (2.1 μmol/g Hb) and normal (3.4 μmol/g Hb) glutathione levels on lipid peroxidation. The model showed that low glutathione (dashed line) resulted in an average 40% increase (mean of three time points until reaction plateau) in R-OOH concentration compared to normal glutathione levels (solid line) (Figure 4.5). These results corroborate experimental data from our group, where glutathione deficient individuals showed on average 25% increased (mean of three studies) F2-isoprostanes levels compared to individuals with normal glutathione levels (a situation resembling ferroptosis (erythrocyte apoptosis) which is also characterized by low levels of glutathione and high levels of lipid peroxidation (Li et al., 2020)). Although our model is simple, the experimental verification of the model's results highlights its feasibility.

$$2GSH + R\text{-}OOH \rightarrow GSSG + R\text{-}OH + H_2O$$
[reaction I]

$$GSSG + NADPH + H^+ \rightarrow 2GSH + NADP^+$$
[reaction II]

MICROCIRCULATION

Structure

The intermediate step of the oxygen pathway is its transport from the macro- to the microcirculation. The microcirculation, as coordinated by resistance arteries, arterioles and capillaries, regulates the delivery of oxygen to muscles. Since capillaries lack vascular smooth muscle cells, the delivery of oxygen is controlled by the vasodilatory state

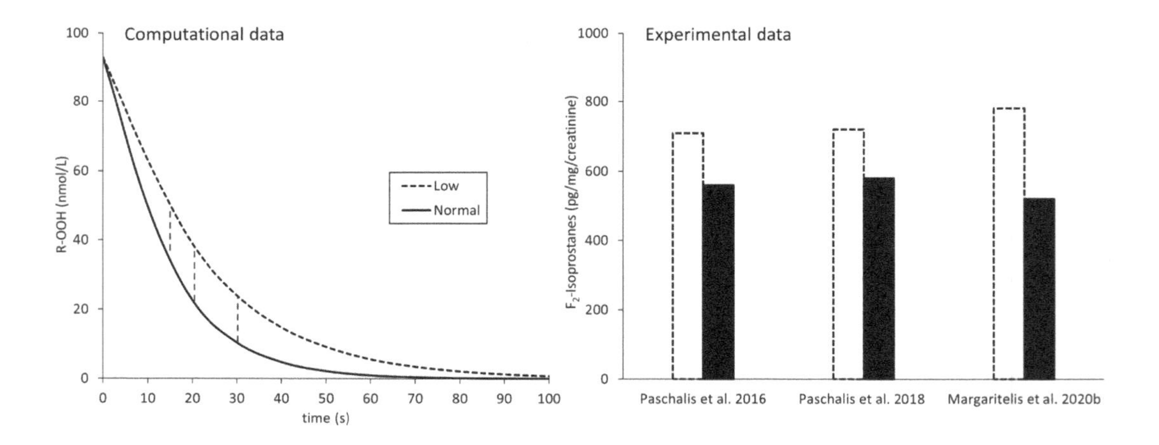

Figure 4.5 Results of the computational model describing lipid peroxidation under two glutathione (GSH) levels (low and normal). In the left panel, the computational data from the model formed in the complex pathway simulator (COPASI) are presented. The simulation of low erythrocyte glutathione levels yielded increased lipid hydroperoxides (R-OOH) compared to normal glutathione. In the right panel, the experimental data from three independent studies of our group are presented. In these studies, participants with low erythrocyte glutathione exhibited increased F_2-isoprostanes compared to participants with normal glutathione levels. The agreement between the computational and experimental data highlights the feasibility of integrated, computational and experimental, strategies.

of the terminal arterioles (Murrant et al., 2021). Each capillary unit (length 850 μm) delivers blood and oxygen to only a small region of ≈12 muscle fibers, originating from different motor units. Myocytes require many microvascular units to be perfused to deliver oxygen along their entire length (4–20 cm in human sartorius muscle) (Murrant et al., 2021).

Increased capillary density following chronic endurance exercise increases the surface area of capillary per muscle cell and shortens the oxygen diffusion distances (Lundby et al., 2017). On the other, the structure of microcirculation can be disrupted in conditions characterized by inflammation and edema leading to impaired oxygen delivery to the muscle via two ways. First, erythrocyte's membrane integrity is compromised in sites of inflammation due to excessive lipid peroxidation (Benfeitas et al., 2014; Jelcic et al., 2017). Second, edema increases inter-capillary distances reducing the oxygen pressure gradient from capillary to muscle (Leach and Treacher, 2002). In a physiological setting, this may be relevant after unaccustomed exercise, which induces inflammation (Margaritelis et al., 2020c) and edema (Fatouros and Jamurtas, 2016).

Regulation

Several redox processes have been reported to regulate microvascular blood flow and, consequently, influence oxygen transport to muscle (Trinity, Broxterman and Richardson, 2016; Costa et al., 2020; Kadlec and Gutterman, 2020). Blood flow induces shear stress on the endothelium, increasing the activity of $NO^{•}$ synthases (eNOS), which utilize oxygen and L-arginine to produce $NO^{•}$ and L-citrulline (Panday, Kar and Kavdia, 2021). Nitric oxide subsequently activates the soluble guanylate cyclase pathway in the vascular smooth muscle cells leading to vascular wall expansion and increased blood flow (Simmonds, Detterich and Connes, 2014) (Figure 4.6). In the same context, chronic supplementation with L-citrulline has been reported to increase $NO^{•}$ levels, peripheral blood flow, muscle oxygenation and exercise performance (Figueroa et al., 2017).

Along with shear stress, erythrocyte-derived RONS (e.g., $NO^{•}$, H_2O_2) have also been reported to be involved in microvascular blood flow regulation (Simmonds, Detterich and Connes, 2014). In particular, during a mismatch between oxygen supply and demand, erythrocytes "sense the state" of local tissue oxygenation and regulate the local blood flow via three possible $NO^{•}$-mediated mechanisms: (1) release of ATP to stimulate the production of $NO^{•}$ by the endothelial cells, (2) $NO^{•}$ release from nitrosohemoglobin upon deoxygenation of hemoglobin and (3) the nitrate-nitrite-$NO^{•}$ reduction pathway (Simmonds, Detterich and Connes, 2014; Helms, Gladwin and Kim-Shapiro, 2018; Richardson, Kuck and Simmonds, 2020). In addition, nitrosothiols (Koppenol, 2012) and peroxynitrite (Nossaman et al., 2007), which can be formed in the erythrocyte and in microvascular cells, have also been reported to induce vasodilatory responses, yet in non-exercising conditions. Beyond $NO^{•}$, H_2O_2 produced mainly by endothelial NADPH oxidases has also been reported to regulate microvascular blood flow (Costa et al., 2020); yet, the available mechanistic data about its involvement are still limited and there is also substantial H_2O_2 levels (≈1–5 μM) in plasma (Forman, Bernardo and Davies, 2016).

Regarding the effect of antioxidants, an interesting computational study examined the effects of various ascorbate concentrations on $NO^{•}$ metabolism and endothelial function (Panday, Kar and Kavdia, 2021). The authors emphasized that under oxidative stress, ascorbate improved $NO^{•}$ production and maintained endothelial function by increasing tetrahydrobiopterin availability. This finding corroborated previous experiments that reported endothelial dysfunction in smokers, autoimmune and neurovascular patients, which was reversed after tetrahydrobiopterin supplementation (Heitzer et al., 2000a, b; Machin et al., 2016; Rodriguez-Miguelez et al., 2018).

MUSCLE

The final transfer step of oxygen is the pathway from capillaries to muscle mitochondria. The delivery of oxygen from capillaries to sarcolemma is proportional to oxygen diffusive capacity and the oxygen pressure difference between the two compartments. The diffusive capacity of oxygen depends on erythrocyte flux, viscosity and the aggregate number of erythrocytes within capillaries adjacent to the muscle cell (Hirai et al., 2019). The pathway from the erythrocyte to the capillary wall, interstitium and sarcolemma is known as the carrier-free area and is relatively short (1.5 μm in rat spinotrapezius muscle) (Hirai et al., 2018).

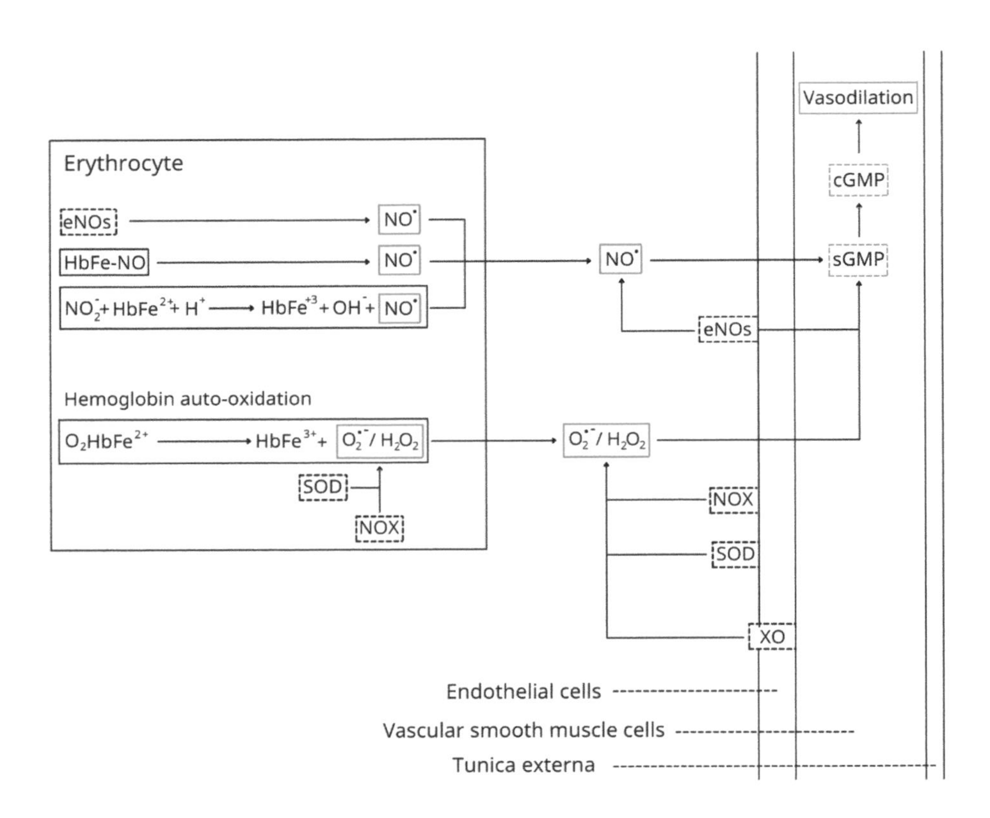

Figure 4.6 Redox regulation of microcirculation. Arterioles consist of three layers, the tunica externa, the vascular smooth muscles cells and the endothelial cells. Nitric oxide (NO•) is the major reactive nitrogen species controlling the vasodilation of the vascular muscle cells, by activating the soluble guanosine monophosphate (sGMP), inducing thereby vasodilation and increasing microvascular blood flow. In addition, erythrocyte-derived nitric oxide has also been reported to regulate vasodilation during certain conditions (e.g., low oxygen availability). Furthermore, superoxide $(O_2^{•-})$ and hydrogen peroxide (H_2O_2), produced from several vascular and erythrocyte sources, have been shown to activate the soluble guanosine monophosphate pathway and increase vasodilation during exercise. Abbreviations: cGMP: cyclic guanosine monophosphate; eNOS: endothelial nitric oxide synthase; H_2O_2: hydrogen peroxide; NOX: NADPH oxidase; $O_2^{•-}$: superoxide; SOD1: superoxide dismutase, XO: xanthine oxidase.

Despite its size, the carrier-free area is the first critical step in the oxygen delivery pathway, due to the small oxygen pressure difference from the interstitial to intramyocellular space (Hirai et al., 2019).

Having crossed the carrier-free area, oxygen has now two complementary routes to follow. The first one is through aquaporins, protein channels residing on cell membranes that facilitate the transport mainly of water molecules between cells. Since the mechanisms for the transfer of oxygen through aquaporins are poorly understood and the existing data is limited and controversial (Clanton, Hogan and Gladden, 2013), we will focus our discussion on the second route, namely, lateral diffusion (Pias, 2020). Lateral diffusion via lipid networks is proposed to enhance oxygen diffusion across membranes by providing a pathway of lower resistance, higher solubility and fewer membrane-associated barriers (Clanton, Hogan and Gladden, 2013; Pias, 2020).

A recent paper provided interesting quantitative data by modeling oxygen diffusion across three different pathways, including cell bodies, interstitial fluid and lipid networks (Pias, 2020). It was estimated that oxygen diffusion along the lipid network may contribute approximately 20%–30% of the oxygen delivered (Pias, 2020). From our point of view, this provides an energetic advantage, since lateral diffusion via lipid networks also ensures that oxygen will interact with oxidative enzymes residing in lipid membranes (e.g., NADPH oxidase). In support of lateral diffusion, the clustering of mitochondria below the sarcolemma could enhance oxygen diffusion by decreasing the path length between capillaries and mitochondria (Clanton, Hogan and Gladden, 2013). Another intriguing hypothesis is that lipid networks could serve as oxygen storage and enhance oxygen distribution via lateral diffusion. From an exercise perspective, the increase in intramuscular triglycerides is considered a classic beneficial adaptation (Vogt et al., 2003). Thus, apart from increasing the available metabolic substrates, the lipid droplets may also enhance oxygen diffusion toward muscle mitochondria.

MITOCHONDRIA

Oxygen Flow and Consumption

Once inside the muscle cell, oxygen flow to mitochondria is achieved by two ways: (i) as dissolved oxygen and (ii) via myoglobin-mediated delivery (Pias, 2020). Intracellular oxygen bound to myoglobin exceeds free oxygen by a ratio of 30:1, leaving an approximate sarcoplasmic free oxygen of ≈3.2 μM at 37°C (Pias, 2020). The role of myoglobin has been debated over the years, with the prevailing theories being that it serves as a (1) short-term oxygen storage (e.g., at rest), (2) oxygen transporter to mitochondria (e.g., during exercise) and (3) redox catalyst (Meyer, 2004). Oxygen binding to myoglobin during exercise could decrease the intramyocellular oxygen pressure gradient and enhance oxygen diffusion (Clanton, Hogan and Gladden, 2013). On the other, exercise-induced muscle damage increases serum myoglobin concentration (Balnave and Thompson, 1993) limiting oxygen availability intracellularly. Another molecule with the ability to store oxygen inside muscle is cytoglobin, which is found in micromolar concentrations (Fago et al., 2004). Despite the limited information in exercise conditions, cytoglobin has been reported to buffer intracellular oxygen (Clanton, Hogan and Gladden, 2013), regulate $NO^{\bullet}$ availability (Mathai et al., 2020) and serve a role in muscle repair in mice (Singh et al., 2014).

Upon reaching mitochondria, oxygen diffuses through membranes and structures. Muscle mitochondria are subdivided into two distinct subpopulations with different morphological and metabolic characteristics (Memme et al., 2021) (Figure 4.7). The first subpopulation forms clusters under the sarcolemma, proximal to the capillary and nuclei, and the second lies between the myofibrils adjacent to the Z-line of the sarcomere (Hood et al., 2019). The subsarcolemmal mitochondria provide energy for the nuclei and membrane pumps to facilitate the active transport of molecules. The intermyofibrillar mitochondria produces approximately a two-fold higher concentration of ATP compared to subsarcolemmal (Takahashi and Hood, 1996), which is directed to the muscle fibers to facilitate contraction.

Mitochondria are considered the major consumers of oxygen in the body (Pittman, 2016). Acetyl coenzyme-A is used in the Krebs cycle to produce ATP, NADH, $FADH_2$ and carbon dioxide. Electrons from the NADH and $FADH_2$ are then moved to the electron transport chain, which is the last step of aerobic respiration. The electron transport chain involves a series of redox reactions along four complexes (I-IV) on the inner mitochondrial membrane, where electrons pass rapidly from one complex to the next (Mailloux, 2015). At the end of the chain, the electrons reduce oxygen to water and produce a large amount of ATP from NADH.

Oxygen Consumption in Cellular (Redox) Processes Beyond Mitochondrial Respiration

Many researchers believe that all oxygen is used in mitochondrial respiration to produce ATP (Hill et al., 2012; Pittman, 2016). This notion is even more widely spread among exercise physiologists because skeletal muscle cells are rich in mitochondria and the energy-centric point of view prevails. However, it is now known that certain enzymes consume oxygen (e.g., oxygenases) and are localized mostly in the cytoplasm and not mitochondria (Romero et al., 2018). The most well-characterized enzymes consuming oxygen for purposes other than respiration are NADPH oxidases, NO synthases, xanthine oxidase, cyclooxygenases and lipoxygenases (Wagner, Venkataraman and Buettner, 2011). These enzymes have been found to be expressed in skeletal muscle in diverse cellular locations (Gomez-Cabrera et al., 2005; McConell et al., 2012; Henríquez-Olguin et al., 2019). As their activity increases during exercise, they become important consumers of oxygen in cases of increased contractile activity.

The quantitative data presented next refer to resting conditions since there is no data during and/or after exercise. Herst and Berridge (2007) reported that non-mitochondrial oxygen

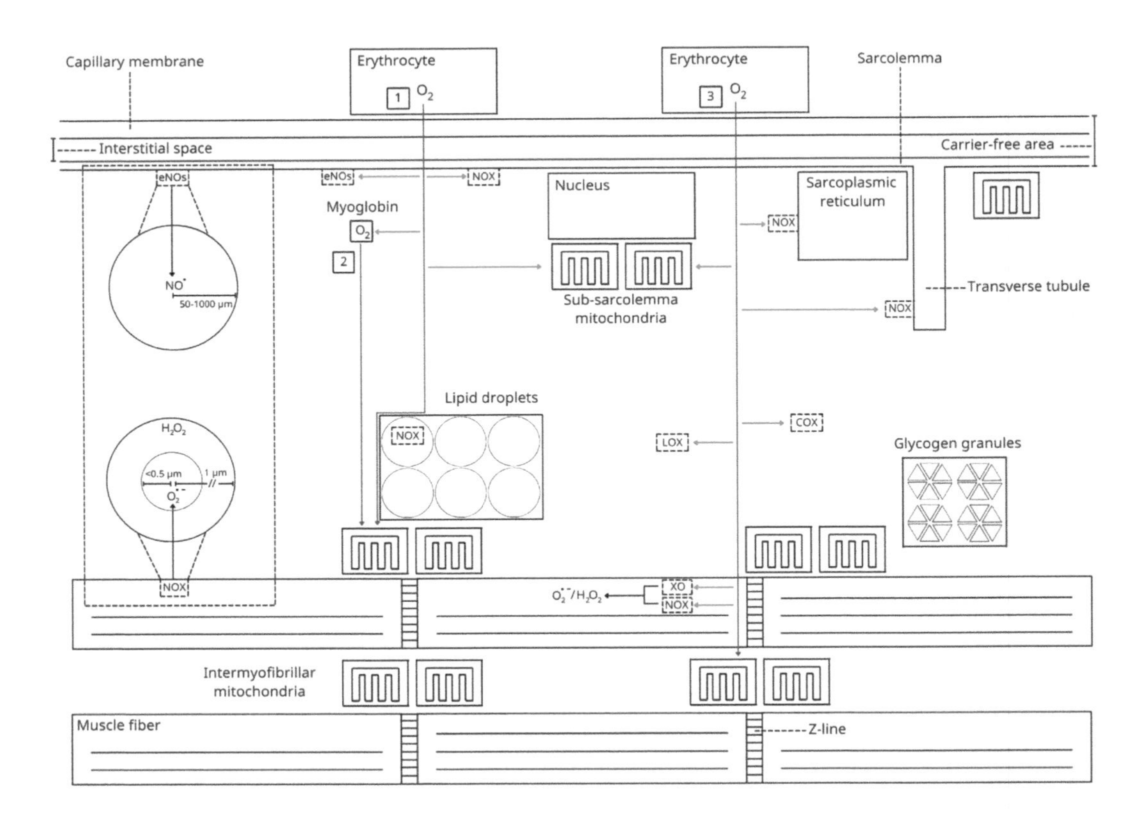

Figure 4.7 Oxygen pathways and hydrogen peroxide production and consumption sources. In this figure, the various routes, barriers and redox domains are presented. Three different pathways of oxygen from erythrocytes up to muscle mitochondria are depicted in blue. The various mitochondrial and non-mitochondrial sources of oxygen consumption are presented in red. The carrier-free area, including the capillary membrane, interstitial space and sarcolemma, is a step of substantial resistance because oxygen is no longer being transported and the oxygen pressure gradient in these regions is almost unchanged. Inside the muscle, oxygen flow to mitochondria is achieved via two additive and simultaneous ways. The myoglobin-mediated transport (pathway 2) and diffusion (pathways 1 and 3). Lateral oxygen diffusion through lipid networks (pathway 1) is proposed to enhance oxygen diffusion across membranes by providing a pathway of lower resistance, higher solubility and fewer membrane-associated barriers. Apart from energy production, oxygen is also utilized as a reductive substrate by various oxidative enzymes (e.g., NADPH oxidases [NOX], xanthine oxidases [XO], nitric oxide synthases [NOs]) and oxygenases (e.g., lipoxygenases [LOX], cyclooxygenases [COX]), which have been reported to be expressed in skeletal muscle in several cellular locations. During exercise, the non-mitochondrial sources of oxygen consumption have increased metabolic requirements and consume a significant amount of oxygen that is otherwise directed to the mitochondria.

consumption ranged from 6% to 91% (which is likely to be highly dynamic) in mitochondrial competent cancer cell lines (Herst and Berridge, 2007). Another study reported that non-mitochondrial oxygen consumption ranged from 17% to 40% in cardiomyocytes, hepatocytes and endothelial cells (Hill et al., 2012). Piccoli et al. (2005) reported that NADPH oxidase isoforms 2 and 4 consumed up to 40% of total oxygen consumption in CD34+ human hemopoietic stem cells (Piccoli et al., 2005), yet these cells are cultured at non-physiological oxygen levels limiting the translation of their findings in vivo. On the contrary, another study reported that the utilization of oxygen from NADPH oxidases accounted for less than 1% of total oxygen consumption in synaptosomes (Abdel-Rahman et al., 2016). Based on these fragmented data, it becomes clear that the widely held view that mitochondrial respiration consumes over 95% of the oxygen should be re-evaluated and always put in context (e.g., cell type, condition) (Hill et al., 2012; Pittman, 2016). Of course, the reader should keep in mind that mitochondrial respiration is highly dynamic and the major uses of oxygen vary depending on the state.

The above quantitative analysis raises a question about the molecular nature of this non-mitochondrial oxygen-consuming system. In our opinion, the most likely and most interesting (from a redox point of view) components of this system are the NADPH oxidases and NO synthases, which produce the parent reactive oxygen and nitrogen species (of course, wherever oxygen is used there is always the potential for superoxide production). These RONS serve signaling purposes controlling many important biological processes (Cobley, 2019; Margaritelis et al., 2020c). On a more theoretical note, the competition for oxygen in skeletal muscle between respirational and non-respirational sources can, at least partially, (and depending on the Km, NADPH availability and how much of each enzyme there is, in a state, competent for catalysis) determine the redox status and the energy status of the cell. Tipping the balance toward non-ATP producing routes may lead to an energy crisis and tipping toward ATP producing routes may lead to oxidative stress. Of course, both of these states can fuel the oxidative stress/energy depletion vicious circle, which could lead to chronic redox and bioenergetics adaptations. Thus, the quantitative delineation of the redox enzymes responsible for oxygen consumption should be subject to an in-depth investigation under different exercise and nutritional conditions. Moreover, it would be interesting to determine if it is possible for the non-respirational enzymes to consume a high volume of oxygen during intense exercise that could potentially limit mitochondrial respiration and, ultimately, lead to fatigue.

CONCLUSION

Oxygen is transferred from atmospheric air to mitochondria in a series of steps by means of diffusion and convection (transport). The major oxygen transfer "stages" are (1) lungs, (2) erythrocytes, (3) microcirculation, (4) muscle and (5) mitochondria. We present convincing evidence that redox mechanisms regulate oxygen transport throughout all these stages. Thus, oxygen transport can no longer be considered a redox neutral area. Following a quantitative approach, we emphasize the need to implement more spatiotemporally refined redox interventions to increase oxygen delivery and improve exercise performance. In this context, the architectural and numerical depictions of oxygen transport can provide the basis for experimentally testable predictions in redox biology, exercise physiology and sports nutrition.

REFERENCES

Abdel-Rahman, E. A. et al. (2016) 'Resolving contributions of oxygen-consuming and ROS-generating enzymes at the synapse', *Oxidative Medicine and Cellular Longevity*, 2016. doi: 10.1155/2016/1089364.

Amann, M. and Calbet, J. A. L. (2008) 'Convective oxygen transport and fatigue', *Journal of Applied Physiology*, 104, pp. 861–870.

Araneda, O. F., Carbonell, T. and Tuesta, M. (2016) 'Update on the mechanisms of pulmonary inflammation and oxidative imbalance induced by exercise', *Oxidative Medicine and Cellular Longevity*. Hindawi Publishing Corporation. doi: 10.1155/2016/4868536.

Balnave, C. D. and Thompson, M. W. (1993) 'Effect of training on eccentric exercise-induced muscle damage', *Journal of Applied Physiology*, 75(4), pp. 1545–1551.

Barbieri, M. et al. (2013) 'Nitrative stress causes nitration, oxidation, and subunit cross linking in human hemoglobin', *Zeitschrift fur Anorganische und Allgemeine Chemie*, 639(8–9), pp. 1384–1394.

Barodka, V. M. et al. (2014) 'New insights provided by a comparison of impaired deformability with erythrocyte oxidative stress for sickle cell disease', *Blood Cells, Molecules, and Diseases*, 52(4), pp. 230–235.

Baynes, J. W. and Dominiczak, M. H. (2019) *Medical Biochemistry*, Fifth Edition. Elsevier, Amsterdam.

Benfeitas, R. et al. (2014) 'Hydrogen peroxide metabolism and sensing in human erythrocytes: A validated kinetic model and reappraisal of the role of peroxiredoxin II', *Free Radical Biology and Medicine*, 74, pp. 35–49.

Boots, A. W., Haenen, G. R. M. M. and Bast, A. (2003) 'Oxidant metabolism in chronic obstructive pulmonary disease', *European Respiratory Journal*, 22(46), pp. 14–27.

Cheng, A. J. et al. (2016) 'Reactive oxygen/nitrogen species and contractile function in skeletal muscle during fatigue and recovery', *Journal of Physiology*, 594(18), pp. 5149–5160.

Chintagari, N. R., Jana, S. and Alayash, A. I. (2016) 'Oxidized ferric and ferryl forms of hemoglobin trigger mitochondrial dysfunction and injury in alveolar type i cells', *American Journal of Respiratory Cell and Molecular Biology*, 55(2), pp. 288–298.

Clanton, T. L., Hogan, M. C. and Gladden, L. B. (2013) 'Regulation of cellular gas exchange, oxygen sensing, and metabolic control', *Comprehensive Physiology*, 3(3), pp. 1135–1190.

Cobley, J. N. et al. (2017) 'Exercise redox biochemistry: Conceptual, methodological and technical recommendations', *Redox Biology*, 12, pp. 540–548.

Cobley, J. N. (2019) 'How exercise induces oxidative eustress', In Sies, H. (ed.) *Oxidative Stress: Eustress and Distress*. Elsevier, Amsterdam, pp. 447–462.

Costa, T. J. et al. (2020) 'The homeostatic role of hydrogen peroxide, superoxide anion and nitric oxide in the vasculature', *Free Radical Biology and Medicine*, 162, pp. 615–635.

Debold, E. P. (2015) 'Potential molecular mechanisms underlying muscle fatigue mediated by reactive oxygen and nitrogen species', *Frontiers in Physiology*, 6, p. 239.

Dempsey, J. A., La Gerche, A. and Hull, J. H. (2020) 'Is the healthy respiratory system built just right, overbuilt, or underbuilt to meet the demands imposed by exercise?', *Journal of Applied Physiology*, 129, pp. 1235–1256. American Physiological Society.

Fago, A. et al. (2004) 'Functional properties of neuroglobin and cytoglobin. Insights into the ancestral physiological roles of globins', *IUBMB Life*, 56(11–12), pp. 689–696.

Fatouros, I. G. and Jamurtas, A. Z. (2016) 'Insights into the molecular etiology of exercise-induced inflammation: Opportunities for optimizing performance', *Journal of Inflammation Research*, 9, pp. 175–186.

Ferguson, S. K. et al. (2018) 'Impact of cell-free hemoglobin on contracting skeletal muscle microvascular oxygen pressure dynamics', *Nitric Oxide - Biology and Chemistry*, 76, pp. 29–36.

Figueroa, A. et al. (2017) 'Influence of L-citrulline and watermelon supplementation on vascular function and exercise performance', *Current Opinion in Clinical Nutrition and Metabolic Care*, 20, pp. 92–98. Lippincott Williams and Wilkins.

Forman, H. J., Bernardo, A. and Davies, K. J. A. (2016) 'What is the concentration of hydrogen peroxide in blood and plasma?', *Archives of Biochemistry and Biophysics*, 603, pp. 48–53.

Georgakouli, K. et al. (2019) 'Exercise in glucose-6-phosphate dehydrogenase deficiency: Harmful or harmless? A narrative review', *Oxidative Medicine and Cellular Longevity*, 2019. doi: 10.1155/2019/8060193.

George, A. et al. (2013) 'Erythrocyte NADPH oxidase activity modulated by Rac GTPases, PKC, and plasma cytokines contributes to oxidative stress in sickle cell disease', *Blood*, 121(11), pp. 2099–2107.

Gomez-Cabrera, M. C. et al. (2005) 'Decreasing xanthine oxidase-mediated oxidative stress prevents useful cellular adaptations to exercise in rats', *Journal of Physiology*, 567(1), pp. 113–120.

Gwozdzinski, K. et al. (2017) 'Investigation of oxidative stress parameters in different lifespan erythrocyte fractions in young untrained men after acute exercise', *Experimental Physiology*, 102(2), pp. 190–201.

Hall, J. E. (2016) *Guyton and Hall Textbook of Medical Physiology*. Elsevier, Philadelphia, PA.

Heitzer, T., Krohn, K., et al. (2000a) 'Tetrahydrobiopterin improves endothelium-dependent vasodilation by increasing nitric oxide activity in patients with Type II diabetes mellitus', *Diabetologia*, 43(11), pp. 1435–1438.

Heitzer, T., Brockhoff, C., et al. (2000b) 'Tetrahydrobiopterin improves endothelium-dependent vasodilation in chronic smokers : Evidence for a dysfunctional nitric oxide synthase', *Circulation Research*, 86(2), pp. e36–e41.

Helms, C. C., Gladwin, M. T. and Kim-Shapiro, D. B. (2018) 'Erythrocytes and vascular function: Oxygen and nitric oxide', *Frontiers in Physiology*, 9, pp. 1–9.

Henríquez-Olguin, C. et al. (2019) 'Cytosolic ROS production by NADPH oxidase 2 regulates muscle glucose uptake during exercise', *Nature Communications*, 10(1), pp. 1–11.

Herst, P. M. and Berridge, M. V. (2007) 'Cell surface oxygen consumption: A major contributor to cellular oxygen consumption in glycolytic cancer cell lines', *Biochimica et Biophysica Acta - Bioenergetics*, 1767(2), pp. 170–177.

Hill, B. G. et al. (2012) 'Integration of cellular bioenergetics with mitochondrial quality control and autophagy', *Biological Chemistry*, 393, pp. 1485–1512.

Hirai, D. M. et al. (2018) 'Skeletal muscle microvascular and interstitial P_{O2} from rest to contractions', *Journal of Physiology*, 596(5), pp. 869–883.

Hirai, D. M. et al. (2019) 'Skeletal muscle interstitial O_2 pressures: Bridging the gap between the capillary and myocyte', *Microcirculation*, 26(5), pp. 1–25.

Hood, D. A. et al. (2019) 'Maintenance of skeletal muscle mitochondria in health, exercise, and aging', *Annual Review of Physiology*, 81, pp. 19–41.

Hoops, S. et al. (2006) 'COPASI - A COmplex PAthway SImulator', *Bioinformatics*, 22(24), pp. 3067–3074.

Hopkins, S. R. et al. (1997) 'Intense exercise impairs the integrity of the pulmonary blood-gas barrier in elite athletes', *American Journal of Respiratory and Critical Care Medicine*, 155(3), pp. 1090–1094.

Impey, S. G. et al. (2018) 'Fuel for the work required: A theoretical framework for carbohydrate periodization and the glycogen threshold hypothesis', *Sports Medicine*, 48(5), pp. 1031–1048.

Jelcic, M. et al. (2017) 'Image-based measurement of H_2O_2 reaction-diffusion in wounded zebrafish larvae', *Biophysical Journal*, 112(9), pp. 2011–2018.

Jones, A. M. et al. (2018) 'Dietary nitrate and physical performance', *Annual Review of Nutrition*, 38, pp. 303–328. Annual Reviews Inc.

Joyner, M. J. and Casey, D. P. (2015) 'Regulation of increased blood flow (Hyperemia) to muscles during exercise: A hierarchy of competing physiological needs', *Physiological Reviews*, 95(2), pp. 549–601.

Joyner, M. J. and Dempsey, J. A. (2018) 'Physiological redundancy and the integrative responses to exercise', *Cold Spring Harbor Perspectives in Medicine*, 8(5), pp. 1–12.

Kadlec, A. O. and Gutterman, D. D. (2020) 'Redox regulation of the microcirculation', *Comprehensive Physiology*, 10(1), pp. 229–260.

Kato, K. and Hecker, L. (2020) 'NADPH oxidases: Pathophysiology and therapeutic potential in age-associated pulmonary fibrosis', *Redox Biology*, 33, p. 101541.

Koppenol, W. H. (2012) 'Nitrosation, thiols, and hemoglobin: Energetics and kinetics', *Inorganic Chemistry*, 51(10), pp. 5637–5641.

Kuck, L., Peart, J. N. and Simmonds, M. J. (2020) 'Active modulation of human erythrocyte mechanics', *American Journal of Physiology - Cell Physiology*, 319(2), pp. C250–C257.

Kuhn, V. et al. (2017) 'Red blood cell function and dysfunction: Redox regulation, nitric oxide metabolism, anemia', *Antioxidants and Redox Signaling*, 26(13), pp. 718–742.

Leach, R. M. and Treacher, D. F. (2002) 'The pulmonary physician in critical care · 2: Oxygen delivery and consumption in the critically ill', *Thorax*, 57(2), pp. 170–177.

Li, J. et al. (2020) 'Ferroptosis: Past, present and future', *Cell Death and Disease*, 11(2), pp. 1–13.

Lundby, C., Montero, D. and Joyner, M. (2017) 'Biology of VO_2max: Looking under the physiology lamp', *Acta Physiologica*, 220(2), pp. 218–228.

Machin, D. R. et al. (2016) 'Impaired exercise-induced forearm blood flow in patients with systemic sclerosis (SSc) is restored after acute tetrahydrobiopterin (BH4) supplementation', *The FASEB Journal*, 30, p. 1288.1.

Mailloux, R. J. (2015) 'Teaching the fundamentals of electron transfer reactions in mitochondria and the production and detection of reactive oxygen species', *Redox Biology*, 4, pp. 381–398. Elsevier B.V.

Mairbäurl, H. (2013) 'Red blood cells in sports: Effects of exercise and training on oxygen supply by red blood cells', *Frontiers in Physiology*, 4, pp. 1–13.

Margaritelis, N. V., Paschalis, V. et al. (2020a) 'Antioxidant supplementation, redox deficiencies and exercise performance: A falsification design', *Free Radical Biology and Medicine*, 158, pp. 44–52.

Margaritelis, N. V., Theodorou, A. A. et al. (2020b) 'Eccentric exercise per se does not affect muscle damage biomarkers: Early and late phase adaptations', *European Journal of Applied Physiology*, 121, pp. 549–559.

Margaritelis, N. V. et al. (2020c) 'Redox basis of exercise physiology', *Redox Biology*, 35, p. 101499.

Mathai, C. et al. (2020) 'Emerging perspectives on cytoglobin, beyond NO dioxygenase and peroxidase', *Redox Biology*, 32, p. 101468.

McConell, G. K. et al. (2012) 'Skeletal muscle nitric oxide signaling and exercise: A focus on glucose metabolism', *American Journal of Physiology - Endocrinology and Metabolism*, 303, p. E301.

Memme, J. M. et al. (2021) 'Exercise and mitochondrial health', *Journal of Physiology*, 599(3), pp. 803–817.

Meyer, R. A. (2004) 'Aerobic performance and the function of myoglobin in human skeletal muscle', *The American Journal of Physiology-Regulatory, Integrative and Comparative Physiology*, 287(6), p. R1304.

Milo, R. and Rob, P. (2016) *Cell Biology by the Numbers*. CRC Press, Boca Raton, FL.

Mohanty, J. G., Nagababu, E. and Rifkind, J. M. (2014) 'Red blood cell oxidative stress impairs oxygen delivery and induces red blood cell aging', *Frontiers in Physiology*, 5, pp. 1–6.

Murrant, C. L. et al. (2021) 'Do skeletal muscle motor units and microvascular units align to help match blood flow to metabolic demand?', *European Journal of Applied Physiology*, 121, pp. 1241–1254.

Niki, E. (2014) 'Role of vitamin e as a lipid-soluble peroxyl radical scavenger: In vitro and in vivo evidence', *Free Radical Biology and Medicine*, 66, pp. 3–12.

Nikolaidis, M. G. et al. (2012) 'Exercise as a model to study redox homeostasis in blood: The effect of protocol and sampling point', *Biomarkers*, 17(1), pp. 28–35.

Nikolaidis, M. G., Margaritelis, N. V. and Matsakas, A. (2020) 'Quantitative redox biology of exercise', *International Journal of Sports Medicine*, 41(10), pp. 633–645.

Nossaman, B. D. et al. (2007) 'Analysis of vasodilator responses to peroxynitrite in the hindlimb vascular bed of the cat', *Journal of Cardiovascular Pharmacology*, 50(4), pp. 358–366.

Ochs, M. et al. (2004) 'The number of alveoli in the human lung', *American Journal of Respiratory and Critical Care Medicine*, 169(1), pp. 120–124.

Opondo, M. A., Sarma, S. and Levine, B. D. (2015) 'The cardiovascular physiology of sports and exercise', *Clinics in Sports Medicine*, 34(3), pp. 391–404.

Panday, S., Kar, S. and Kavdia, M. (2021) 'How does ascorbate improve endothelial dysfunction? - A computational analysis', *Free Radical Biology and Medicine*, 165, pp. 111–126.

Parker, L. et al. (2017) 'Exercise and glycemic control: Focus on redox homeostasis and redox-sensitive protein signaling', *Frontiers in Endocrinology*, 8, p. 87.

Petibois, C. and Déléris, G. (2005) 'Evidence that erythrocytes are highly susceptible to exercise oxidative stress: FT-IR spectrometric studies at the molecular level', *Cell Biology International*, 29(8), pp. 709–716.

Phillips, R. et al. (2013) *Physical Biology of the Cell*. Garland Science, New York.

Pias, S. C. (2020) 'How does oxygen diffuse from capillaries to tissue mitochondria? Barriers and pathways', *Journal of Physiology*, 599, pp. 1769–1782.

Piccoli, C. et al. (2005) 'Characterization of mitochondrial and extra-mitochondrial oxygen consuming reactions in human hematopoietic stem cells: Novel evidence of the occurrence of NAD(P)H oxidase activity', *Journal of Biological Chemistry*, 280(28), pp. 26467–26476.

Pittman, R. N. (2016) *Regulation of Tissue Oxygenation*, Second Edition, Colloquium Series on Integrated Systems Physiology: From Molecule to Function.

Poppitt, S. D. et al. (2005) 'Assessment of erythrocyte phospholipid fatty acid composition as a biomarker for dietary MUFA, PUFA or saturated fatty acid intake in a controlled cross-over intervention trial', *Lipids in Health and Disease*, 4, pp. 1–10.

Rakobowchuk, M. et al. (2017) 'Divergent endothelial function but similar platelet microvesicle responses following eccentric and concentric cycling at a similar aerobic power output', *Journal of Applied Physiology*, 122(4), pp. 1031–1039.

Richardson, K. J., Kuck, L. and Simmonds, J. (2020) 'Beyond oxygen transport: active role of erythrocytes in the regulation of blood flow', *American Journal of Physiology - Heart and Circulatory Physiology*, 319(4), pp. H866–H872.

Rifkind, J. M. and Nagababu, E. (2013) 'Hemoglobin redox reactions and red blood cell aging', *Antioxidants and Redox Signaling*, 18, pp. 2274–2283, Mary Ann Liebert, Inc.

Rodriguez-Miguelez, P. et al. (2018) 'Acute tetrahydrobiopterin improves endothelial function in patients with COPD', *Chest*, 154(3), pp. 597–606.

Rogers, C. S. et al. (2009) 'Hypoxia limits antioxidant capacity in red blood cells by altering glycolytic pathway dominance', *The FASEB Journal*, 23(9), pp. 3159–3170.

Romero, E. et al. (2018) 'Same substrate, many reactions: Oxygen activation in flavoenzymes', *Chemical Reviews*, 118, pp. 1742–1769. American Chemical Society.

Saunders, P. U. et al. (2013) 'Relationship between changes in haemoglobin mass and maximal oxygen uptake after hypoxic exposure', *British Journal of Sports Medicine*, 47(SUPPL. 1), pp. i26–i30.

Sawka, M. N. et al. (2000) 'Blood volume: Importance and adaptations to exercise training, environmental stresses, and trauma/sickness', *Medicine and Science in Sports and Exercise*, 32(2), pp. 332–348.

Sentürk, Ü. K. et al. (2001) 'Exercise-induced oxidative stress affects erythrocytes in sedentary rats but not exercise-trained rats', *Journal of Applied Physiology*, 91(5), pp. 1999–2004.

Sies, H. and Jones, D. P. (2020) 'Reactive oxygen species (ROS) as pleiotropic physiological signalling agents', *Nature Reviews Molecular Cell Biology*, 21, pp. 363–383. Nature Research.

Simmonds, M. J., Detterich, J. A. and Connes, P. (2014) 'Nitric oxide, vasodilation and the red blood cell', *Biorheology*, 51(2–3), pp. 121–134.

Singh, S. et al. (2014) 'Cytoglobin modulates myogenic progenitor cell viability and muscle regeneration', *Proceedings of the National Academy of Sciences of the United States of America*, 111(1), p. E129.

Skattebo, O. et al. (2021) 'Effects of 150- and 450-mL acute blood losses on maximal oxygen uptake and exercise capacity', *Medicine & Science in Sports & Exercise*, 53, pp. 1729–1738.

Subudhi, A. W. et al. (2001) 'Antioxidant status and oxidative stress in elite alpine ski racers', *International Journal of Sport Nutrition*, 11(1), pp. 32–41.

Takahashi, M. and Hood, D. A. (1996) 'Protein import into subsarcolemmal and intermyofibrillar skeletal muscle mitochondria: Differential import regulation in distinct subcellular regions', *Journal of Biological Chemistry*, 271(44), pp. 27285–27291.

Theodorou, A. A. et al. (2010) 'Comparison between glucose-6-phosphate dehydrogenase-deficient and normal individuals after eccentric exercise', *Medicine and Science in Sports and Exercise*, 42(6), pp. 1113–1121.

Trinity, J. D., Broxterman, R. M. and Richardson, R. S. (2016) 'Regulation of exercise blood flow: Role of free radicals', *Free Radical Biology and Medicine*, 98, pp. 90–102.

Tsalouhidou, S., Petridou, A. and Mougios, V. (2009) 'Effect of chronic exercise on DNA fragmentation and on lipid profiles in rat skeletal muscle', *Experimental Physiology*, 94(3), pp. 362–370.

Van Wijk, R. and Van Solinge, W. W. (2005) 'The energy-less red blood cell is lost: Erythrocyte enzyme abnormalities of glycolysis', *Blood*, 106(13), pp. 4034–4042.

Vogt, M. et al. (2003) 'Effects of dietary fat on muscle substrates, metabolism, and performance in athletes', *Medicine and Science in Sports and Exercise*, 35(6), pp. 952–960.

Wagner, B. A., Venkataraman, S. and Buettner, G. R. (2011) 'The rate of oxygen utilization by cells', *Free Radical Biology and Medicine*, 51(3), pp. 700–712.

Wang, Y. et al. (2020) 'Downregulated recycling process but not de novo synthesis of glutathione limits antioxidant capacity of erythrocytes in hypoxia', *Oxidative Medicine and Cellular Longevity*, 2020. doi: 10.1155/2020/7834252.

Zerez, C. R. et al. (1988) 'Decreased erythrocyte nicotinamide adenine dinucleotide redox potential and abnormal pyridine nucleotide content in sickle cell disease', *Blood*, 71(2), pp. 512–515.

CHAPTER FIVE

Mitochondrial Redox Regulation in Adaptation to Exercise

Christopher P. Hedges and Troy L. Merry

CONTENTS

Mitochondria and Energy Metabolism / 59
Mitochondrial Production of Reactive Oxygen Species / 59
Skeletal Muscle Mitochondrial ROS Production at Rest / 60
Skeletal Muscle Mitochondrial ROS Production During Exercise / 61
ROS-Mediated Adaptation to Exercise Training / 62
Is There Evidence of Mitochondrial-Derived ROS Act as Exercise Signals? / 63
mtDNA Damage / 63
Peroxiredoxins / 64
Mitochondrial Derived Peptides / 64
Is There Evidence That Mitochondrial-Derived ROS Contribute to Exercise Training-Induced Adaptation? / 65
Concluding Remarks / 65
References / 66

MITOCHONDRIA AND ENERGY METABOLISM

Skeletal muscle contraction requires energy in the form of ATP. Almost all types of exercise will require mitochondrial oxidative phosphorylation (OXPHOS) to supply at least a portion of the ATP used. Production of ATP by mitochondria is dependent on the electron transport system and oxidative phosphorylation machinery. In brief, complexes I and II, mitochondrial glycerophosphate dehydrogenase (mGPDH) and electron transferring flavoprotein dehydrogenase (ETFDH) all reduce the lipophilic mobile electron carrier ubiquinone to ubiquinol. Ubiquinol is then oxidised by complex III, which reduces cytochrome c, which is, in turn, oxidised by complex IV. Complex IV utilises electrons to bind and reduce oxygen and protons (H^+) to form water (H_2O). During this process, each time an electron transitions down energy states the released energy is used by complexes I, III and IV to transport H^+ into the intermembrane space which is against the H^+ concentration gradient (Schultz and Chan, 2001). The net result is the generation of a proton motive force, made up of a concentration gradient of H^+ and an electrochemical membrane potential across the mitochondrial inner membrane. Proton motive force enables production of ATP by ATP synthase (mitochondrial complex V) simultaneous with H^+ return to the mitochondrial matrix (Fillingame, 1997). Thus, ATP production is coupled to oxygen consumption in the mitochondria.

MITOCHONDRIAL PRODUCTION OF REACTIVE OXYGEN SPECIES

Under normal conditions, a very small amount of the oxygen consumed by muscle cells is univalently reduced to superoxide $\left(O_2^{*-}\right)$. Superoxide is very unstable, with a half-life around 10^{-6}s (Giorgio et al., 2007; Pryor, 1986), largely owing

DOI: 10.1201/9781003051619-5

to its reactivity with superoxide dismutase to form hydrogen peroxide (H_2O_2) (Loschen et al., 1974). Superoxide is the "parent radical" for a number of other molecules, which include hydroxyl radicals in addition to H_2O_2, which are collectively known as reactive oxygen species (ROS). H_2O_2 has the longest half-life (thus is most stable) of around 10^{-5}s and thus is a commonly measured ROS (Giorgio et al., 2007; Pryor, 1986). The mechanisms of mitochondrial superoxide production have been previously reviewed (see Murphy, 2009; Cobley, 2020); however, in vivo superoxide production is notoriously difficult to measure accurately. Broadly speaking, mitochondria produce superoxide when the mitochondrial proton motive force is high and oppose the movement of H^+ from the matrix into the intermembrane space, and/or when the NAD/NADH pool remains in a highly reduced state (Korshunov et al., 1997; Lambert and Brand, 2004; Miwa and Brand, 2003). This in turn opposes the normal transitions of electrons through electron transport complexes and between different electron donors/shuttles. As a result, electrons can interact with oxygen at certain sites in electron transport complexes, resulting in single electron reduction of oxygen to superoxide. This can occur when electrons are moving in the normal "forward" direction (i.e. proceeding to complex IV) or can also occur via a phenomenon known as "reverse electron transfer", or RET (Hinkle et al., 1967; Selivanov et al., 2011).

Due to their instability and reactivity, when ROS levels exceed cellular antioxidant capacity, they have the potential to cause widespread cellular damage, potentially affecting organelles, DNA and lipid membranes. Historically, this state was referred to as "oxidative stress" (Sies and Cadenas, 1985), and accordingly, ROS have traditionally garnered a negative reputation. However, it is now recognised that ROS and oxidative stress have beneficial and necessary roles within cells, and that an increase in ROS production does not indicate a pathological situation *per se*. As such, a pathological imbalance where ROS production surpasses antioxidant capacity has more recently been termed "oxidative distress" (Sies et al., 2017). The theory of mitohormesis suggests that a manageable increase in oxidative stress from the mitochondria can promote adaptations to prolong health, a process that has been shown to occur in multiple organisms, and in response to a diverse array of cellular stressors (Schulz et al., 2007; Owusu-Ansah et al., 2013; Scialo et al., 2016; Weimer et al., 2014; Zarse et al., 2012). In addition, both mitochondrial and non-mitochondrial ROS can act acutely in a reversible and context-specific manner to enhance or suppress traditional kinase signalling and multiple transcription factor pathways (Sies and Jones, 2020). Primary targets include the oxidative inhibition of a number of redox-sensitive protein tyrosine phosphatases and oxidation of thiol-specific antioxidants, facilitating their interaction with specific signalling proteins – often acting like a rheostat to fine-tune signalling (Tiganis, 2011; Rhee et al., 2012). Thus, ROS can also act as important signals for cellular adaptation and are essential for the maintenance of cellular homeostasis.

Skeletal Muscle Mitochondrial ROS Production at Rest

There are a number of superoxide production sites in mitochondria, as eloquently summarised by Martin Brand (2016). In brief, these sites are typically where electron transport complexes interact with electron donor molecules (such as NADH, succinate, acyl-CoA) at flavoprotein sites, or at quinone binding sites (Brand, 2016) (Figure 5.1). Because different substrates enter the electron transport system through different complexes, rates and sites of ROS production are dependent on the substrate(s) being oxidised by the mitochondria (Quinlan et al., 2013). For example, oxidation of fatty acids generates ROS through ETFDH (Seifert et al., 2010), and activation of the glycerol-3-phosphate shuttle can generate ROS through mGPDH (Mracek et al., 2014). However, because of the complex interlinked nature of the system, the site of ROS production is not necessarily the site of electron input. For example, ROS production during the oxidation of succinate can be attributed to sites in complexes I and III, despite the fact that succinate is oxidised by complex II (Quinlan et al., 2013). Due to the challenges of measuring specific sites of mitochondrial ROS production in vivo, our understanding of how different sites produce ROS during active and resting states is derived from experiments mimicking substrate concentrations found in muscle during rest and exercise in isolated mitochondria. These show that in mitochondria isolated from rested (non-contracting)

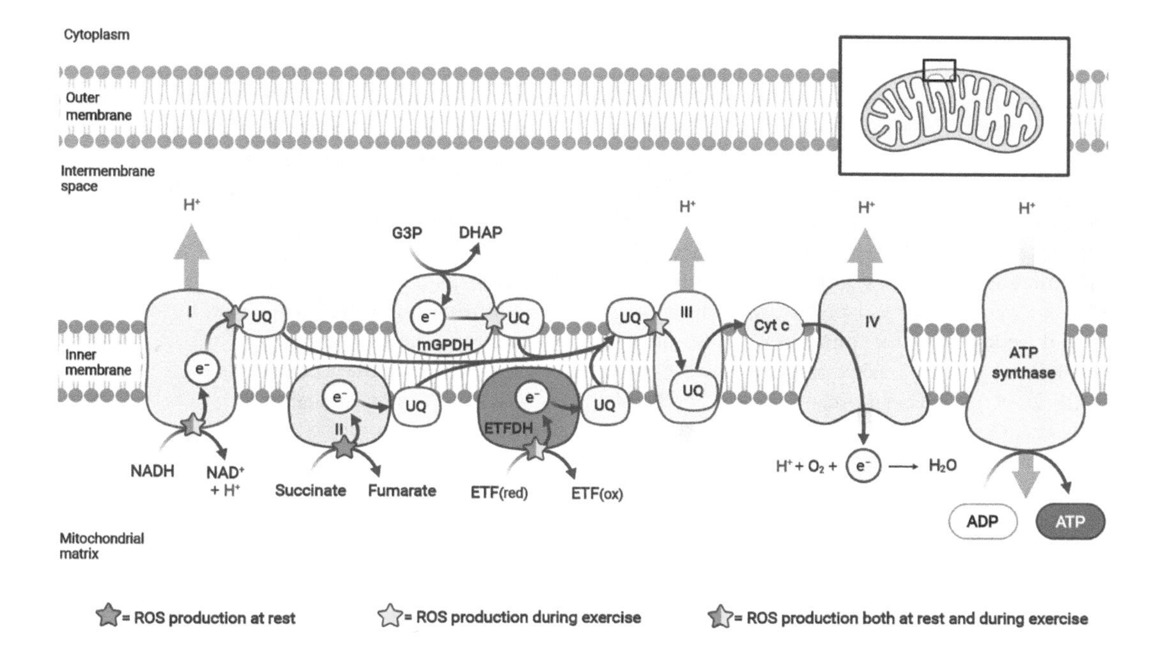

Figure 5.1 Sites of ROS production in skeletal muscle mitochondria during resting and exercising conditions. Derived from (Quinlan et al., 2013; Goncalves et al., 2015). ETFDH, electron transfer flavoprotein dehydrogenase; mGPDH, mitochondrial glycerol-3-phosphate dehydrogenase; UQ, ubiquinone/ubiquinol; Cyt c, cytochrome C; red, reduced; ox, oxidised. Figure created with BioRender.com.

healthy muscle, mitochondrial ROS production occurs from flavoprotein sites in complexes I, II and ETFDH, and from ubiquinol binding sites in complexes I and III (Goncalves et al., 2015; Quinlan et al., 2013) (Figure 5.1).

Skeletal Muscle Mitochondrial ROS Production During Exercise

Exercise is now a well-known stimulus for oxidative stress in muscle, with original observations by Dillard et al. (1978) that exercise results in increased breath pentane levels in humans, and later work by Davies et al. (1982) demonstrating in rats that exercise to exhaustion enhances muscle free radical production. Since these studies, many others have corroborated the finding that exercise of various modes and intensities induce both muscle-specific and systemic oxidative stress, as reviewed in Powers et al. (2011a). Because mitochondrial oxygen consumption increases with exercise, initial opinion favoured the idea that mitochondria were a key source of ROS produced during exercise. However, during exercise, ADP supply to mitochondria increases, supporting oxidative phosphorylation. This allows protons back into the mitochondrial matrix, resulting in lower proton motive force when ADP is increased (Hey-Mogensen et al., 2015). Since a high proton motive force potentiates mitochondrial ROS production, the lowering of proton motive force during exercise results in decreases in absolute mitochondrial ROS production, and the relative amount of ROS produced as a percentage of oxygen consumption – despite increased mitochondrial oxygen consumption (Goncalves et al., 2015). Furthermore, oxidative stress still increases during exercise without high oxygen flux rates, such as isometric exercise (Alessio et al., 2000) and resistance exercise (McBride et al., 1998), suggesting that increased oxygen consumption rates are not a key driver of ROS production. What does change are the sites of mitochondrial ROS production during exercise. When mild aerobic exercise is simulated, over 75% of mitochondrial ROS production is from complexes I and III, with the remainder from ETFDH and mGPDH, and when intense exercise is simulated, mitochondrial ROS production is almost exclusively from complex I (Wong et al., 2017; Goncalves et al., 2015). However, this may only represent a small fraction of overall increase in ROS production that occurs during exercise.

The main producers of ROS during exercise are thought to be enzymes external to mitochondria, such as xanthine oxidase and different NADPH oxidase (NOX) isoforms (Powers et al., 2011b; Powers and Jackson, 2008). NOX enzymes are located in many places in skeletal muscle – including mitochondrial membranes (Sakellariou et al., 2013; Ago et al., 2010; Ferreira and Laitano, 2016), and mice that express a kinase-dead form of NOX2, a cytosolic NOX isoform, show no change in cell redox state or phosphorylation of p38 MAPK in response to exercise (Henriquez-Olguin et al., 2019a). Furthermore, inhibition of NOX2 prevents exercise-induced changes in muscle mitochondrial and antioxidant mRNA expression (Henriquez-Olguin et al., 2016). However, some caution should be taken attributing the effects seen by Henriquez-Olguin et al. (2016) solely to NOX2, as the inhibitor used (apocynin) can act as a general antioxidant rather than specifically as a NOX inhibitor (Ferreira and Laitano, 2016; Heumuller et al., 2008). Xanthine oxidase is another prominent candidate for ROS production during exercise, as it is present in skeletal muscle (Wajner and Harkness, 1989; Hellsten-Westing, 1993), and use of the xanthine oxidase inhibitor allopurinol attenuates oxidative stress and muscle damage in both humans and rodents (Gomez-Cabrera et al., 2003; Vina et al., 2000). It is therefore likely that these non-mitochondrial sources play a key role in cell signalling in response to exercise.

ROS-MEDIATED ADAPTATION TO EXERCISE TRAINING

Seminal studies by Ristow et al. (2009) and Gomez-Cabrera et al. (2005, 2008a) provided early evidence that exercise-induced redox signalling is involved in regulating adaptive responses to exercise training. Through the use of antioxidant supplements, these authors and subsequently others (reviewed in (Merry and Ristow, 2016a) have shown that general and xanthine oxidase targeted antioxidants have the potential to suppress acute protein signalling during exercise, transient transcriptional responses following exercise, and in some cases attenuate beneficial adaptions that occur with long-term exercise training, including improvements in glucose homeostasis, antioxidant defences and mitochondrial biogenesis (Gomez-Cabrera et al., 2008a; Strobel et al., 2011; Silveira et al., 2006; Kang et al., 2009; Ristow et al., 2009; Paulsen et al., 2014; Morrison et al., 2015). While the evidence for antioxidant suppression of acute exercise signalling is strong, the extent that this translates to impaired training adaptions is controversial, as several studies have found no effect of antioxidants on training-induced mitochondrial adaptations (Higashida et al., 2011; Wadley and McConell, 2010; Yfanti et al., 2010). This perhaps indicates redundancy in the system for coordinating long-term training adaptions or questions the effectiveness of prolonged antioxidant supplementation to suppress localised exercise-induced increases in ROS. Furthermore, as reviewed by Cobley et al. (2015), vitamin C and vitamin E, which are the general antioxidants used in many of these studies, do not have any meaningful biological reaction with H_2O_2, meaning that the blunting of adaptive responses seen using these antioxidants may not be fully attributed to the scavenging of exercise-induced ROS.

Activation of classical cytosolic intracellular stress signalling pathways, such as AMPK, p38 MAPK and ERK, can be attenuated by antioxidant supplementation prior to exercise, and result in impaired responses in what is seen as a master regulator of exercise and mitochondrial adaptation, PGC-1α. However, the primary redox-sensitive pathways targets of exercise are not well defined (Merry and Ristow, 2016b), and further insights are expected as redox-proteomics develops as a field. However, antioxidant supplementation and gene knockout models have provided evidence that the transcription factors NF-κB and NFE2L2 (also known as Nrf2) are both exercise and redox sensitive, and potentially contribute to coordinating exercise-induced ROS-driven gene transcription (Gomez-Cabrera et al., 2008a; Gomez-Cabrera et al., 2005; Merry and Ristow, 2016c). While the production site of the exercise-induced ROS that target these transcription factors are difficult to discern from studies using non-site-specific antioxidants, the ability of allopurinol to suppress exercise-induced activation of mitochondrial biogenesis pathways suggests xanthine oxidase is a key redox enzyme regulating exercise adaptations (Wadley et al., 2013; Gomez-Cabrera et al., 2008b). Similarly, recent evidence that ablation of the NADPH oxidase enzyme NOX2 (Henriquez-Olguin et al., 2019b) impairs high-intensity exercise training responses in mice – in terms of both exercise performance and molecular-level adaptation of mitochondrial biogenesis and antioxidant gene expression – suggests cytosolic or extracellular ROS source(s) are required for optimal exercise adaptations.

IS THERE EVIDENCE OF MITOCHONDRIAL-DERIVED ROS ACT AS EXERCISE SIGNALS?

While the major sources of ROS during exercise appear to be mitochondria-independent, this does not preclude a role for discrete localised ROS production from the mitochondria during or immediately following muscle contraction being involved in the adaptive responses initiated locally in muscle, or systemically, in response to repeated bouts of acute exercise (training). Although overall mitochondrial ROS production may diminish during exercise, some sites remain active even during high-intensity exercise (Quinlan et al., 2013); however, the dynamic nature of exercise and muscle contraction makes it is difficult to monitor transient compartmentalised ROS levels in real time and in high resolution. Instead, markers of mitochondrial oxidation state can be used to infer mitochondrial localised ROS production during exercise, with the caveat that ROS do not remain compartmentalised, and changes could be the result of diffusion of non-mitochondrial produced ROS. Genetically encoded redox-sensitive green fluorescent protein sensors (mt-roGFP2-Orp1) (Gutscher et al., 2009), which converts physiological levels of H_2O_2 into a measurable fluorescent signal, or the HyPer family of H_2O_2 reporters (Belousov et al., 2006), may provide tools to monitor subcellular compartmental ROS production during exercise. However, these have so far been unable to detect in vivo exercise-induced changes in mitochondrial ROS levels when electroporated into the tibialis anterior muscle of mice (Henriquez-Olguin et al., 2019a). Other markers of mitochondrial oxidative state have provided evidence of changes in local mitochondria redox signalling with exercise, and mitochondrial-targeted superoxide detection dyes (MitoSOX) indicate there are low levels of mitochondrial ROS production during contraction of isolated muscle fibres (Pearson et al., 2014). It is important to note that this data is generally interpreted under the assumption that the contraction condition does not greatly impact dye uptake which could potentially impact estimations of ROS production. In this section, we will focus on some of the evolving mitochondria-centric redox signalling markers that have been implicated as being responsive to exercise (Figure 5.2).

mtDNA Damage

Oxidative stress can cause damage to mitochondrial DNA (mtDNA), and acute high-intensity exercise increases mtDNA fragmentation (a marker

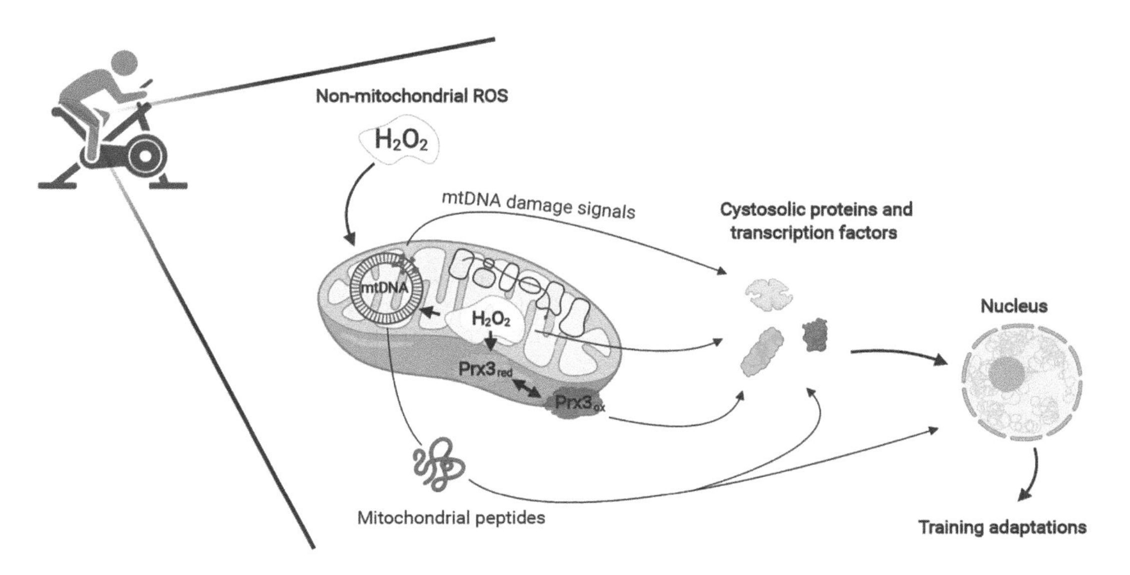

Figure 5.2 Potential pathways through which local mitochondria reactive oxygen species (ROS) may contribute to the coordinating exercise training adaptive responses. mt, mitochondrial; ox, oxidized; red, reduced. Figure created with BioRender.com.

of damage) in both skeletal muscle and blood lymphocytes (Williamson et al., 2020; Williamson and Davison, 2020). Owing to mitochondrial nucleoids (mtDNA-protein complexes) close proximity to sites of oxidative phosphorylation where mitochondrial ROS is generated (Gilkerson et al., 2013), and the observation that mitochondrially targeted antioxidant MitoQ can attenuate exercise-induced mtDNA damage (Williamson et al., 2020), it is likely that mitochondrial-derived ROS are at least partially responsible for exercise-induced mtDNA damage. The accumulation of damaged mtDNA has been associated with impaired mitochondrial bioenergetics and cellular death and dysfunction, contributing to the development of a host of human pathologies. Damaged mtDNA only accumulates when the damage-causing stressor is chronic and/or coupled with insufficient or impaired mtDNA repair mechanisms (Saki and Prakash, 2017). Whether mtDNA damage that occurs during acute exercise accumulates with repeated bouts, and is sustained (as opposed to being rapidly repaired), or is deleterious to cell function, is unclear. Since mtDNA damage is reversible, it is possible that transient damage to the guanine nucleotide (and subsequent 8-oxodG signalling) may serve as a means to promote cytoprotective adaptations associated with regular exercise. This fits with the concept of exercise-induced mitohormesis, a process whereby the exposure of mitochondria to an acute low dose of a stressor that is potentially harmful promotes adaptive changes that minimise the impact of subsequent exposure to the stress (Merry and Ristow, 2016b). Indeed, mtDNA damage can orchestrate nuclear gene expression responses, through the activation of intermediate signalling cascades such as calcium-Nfκβ signalling, alterations in the NAD+/NADH ratio and modulation of classical energy homeostasis regulators AMPK, p53 and HIF1α, via the DNA damage sensor protein ataxia telangiectasia mutated (ATM) (Saki and Prakash, 2017). The vast majority of these signalling intermediates are known to be responsive to acute exercise and to mediate exercise training responses, particularly in terms of mitochondrial biogenesis and antioxidant response (Merry and Ristow, 2016b).

Peroxiredoxins

Mitochondrial ROS produced in response to non-exercise related stressors, such as aging, cancer and metabolic overload, directly target and interact with redox-sensitive signalling proteins and transcription factors that are primarily cytosolic. However, mitochondrial ROS may also directly influencing gene expression by inducing nuclear DNA damage or affecting DNA repair (Srinivas et al., 2019) and epigenetic regulatory systems (Guillaumet-Adkins et al., 2017). While there is strong evidence that cytosolic-initiated signalling pathways are responsive to exercise-induced ROS (Merry and Ristow, 2016b) and can promote exercise-training induced adaptations (Merry and Ristow, 2016a), the source of ROS targeting these pathways is likely non-mitochondrial (Powers and Jackson, 2008). Mitochondrial redox signalling during exercise could involve, or at least be initiated by, more localised events. The oxidation of thiols found in key areas of proteins that control their function or degradation can control distinct and localised signalling events (Forman et al., 2004; Rhee and Kil, 2017). Peroxiredoxins (Prx) are important components of thiol redox signalling reacting with physiological concentrations of H_2O_2 that are below the oxidation threshold for the majority of other redox-sensitive proteins. They are expressed as cellular compartment-specific isoforms and can act as signal transductors or barriers in a reversable and concentration-dependent manner, in part through interactions with thioredoxins (Latimer and Veal, 2016; Rhee and Kil, 2017). As such measuring their oxidation state is becoming popular technique to monitor location-specific redox state. Prx3 is located in the mitochondrial matrix and is oxidised to form dimers in contracted isolated rodent muscle fibres from both young and old mice (Stretton et al., 2020). Interestingly, while the cytosolic Prx's 1 and 2 are also readily oxidised by contraction of muscle fibres from young mice, unlike for mitochondrial Prx3, this response is diminished in older mice potentially implicating Prx's in selective of redox signalling during exercise (Stretton et al., 2020). Understanding whether Prx's are oxidised in muscle during in vivo exercise, and isolating their signalling targets is likely to provide further insight into the role of ROS in regulating exercise training responses.

Mitochondrial Derived Peptides

In addition to encode 13 protein-coding genes that form essential subunits of the oxidative phosphorylation (OXPHOS) complexes, the mitochondrial genome is now recognised to also harbour short open-reading frames that transcribe small regulatory peptides named mitochondrial-derived peptide (MDP's) (Guo et al., 2003; Lee et al., 2015; Cobb et al., 2016). Currently, eight mitochondrial peptides have been described, with the majority being responsive to metabolic perturbations (Merry et al., 2020). Humanin, small humanin-like-peptide 6 (SHLP6) and mitochondrial open-reading frame of the 12S rRNA-c (MOTS-c) have been shown to increase in human plasma or skeletal muscle in response to high-intensity interval exercise (Gidlund et al., 2016; Reynolds et al., 2019; Woodhead et al., 2020). The exercise stimuli that regulates MDP's, or their molecular targets in an exercise context, are yet to be fully elucidated; however, it appears that cell redox status is a driver of at least MOTS-c expression. In cell culture, oxidative stress rapidly induces MOTS-c translocation from mitochondria to the nucleus, where it can bind to nuclear DNA and interact with the known exercise and oxidative stress sensitive transcription factor NFE2L2 (Nrf2) to regulate gene expression (Kim et al., 2018). This potentially provides a pathway whereby redox signalling in mitochondria can directly regulate nuclear gene expression during exercise. Consistent with this, treatment of mice with exogenous MOTS-c enhances exercise capacity and can promote exercise-like adaptations that improve metabolic function in response to ageing and high-calorie diet (Lee et al., 2015; Reynolds et al., 2021).

IS THERE EVIDENCE THAT MITOCHONDRIAL-DERIVED ROS CONTRIBUTE TO EXERCISE TRAINING-INDUCED ADAPTATION?

While there is building evidence of acute exercise-mediated mitochondrial redox signalling, whether this is important in regulating adaptive responses to exercise training has yet to be explored in detail. Tools to specifically examine the causative nature of mitochondrial ROS in exercise are becoming more widely available, and it is expected that this area of research will expand in the coming years. Such tools include mitochondrial-targeted antioxidants Szeto-Schiller-31 (SS31/elamipretide), XJB-5-131, mito-TEMPO, mitochondria-targeted vitamin E and MitoQ (Broome et al., 2018). With the exception of MitoQ, which is available as an over-the-counter supplement in many countries, the use of other mitochondrial-targeted antioxidants has been largely limited to pre-clinical and phase 1/2 clinical trials, perhaps restricting their use in human exercise research. However, the effect of mitochondrial antioxidants in rodents has so far also been limited to a handful of studies. These show SS31 does not greatly effect fatigue of isolated muscle fibres (Katz et al., 2014; Cheng et al., 2015), and the only long-term combined exercise and mitochondrial antioxidant supplementation study in humans showed that MitoQ does not affect improvements in oxidative capacity induced by 3 weeks of endurance training (Shill et al., 2016). Therefore, there is still a lot that is unknown about the effects of targeting mitochondrial ROS in the exercise context. In addition to the use of antioxidant supplements, insights could be gained from the studying of mice with targeted expression or deletion of antioxidants at the mitochondria, such as the mCAT (mitochondrial catalase overexpression) and MnSOD overexpression mice or Prx3 knockout mice. While these mice appear to show acute exercise phenotypes, with mCAT mice having improved exercise capacity (Li et al., 2009) and Prx3 knockout and MnSOD heterozygous knockout mice (Zhang et al., 2016; Kinugawa et al., 2005) have impaired swimming performance with age, how the manipulation of expression of these mitochondrial antioxidants affects exercise training responsiveness has not been examined.

CONCLUDING REMARKS

Mitochondria have long been known as a key organelle in cellular energy provision, but are also an important part of intracellular signalling networks. ROS production is one hypothesised means of mitochondrial signalling for adaptation; however, non-mitochondrial sources are thought to be the key generators of ROS during exercise, and the role of mitochondria in exercise-induced adaptation remains unclear. The advancement of compartmentalised oxidative state monitoring techniques is beginning to provide evidence of exercise-induced mitochondrial redox signalling despite the mitochondria not being the primary source of ROS during exercise. Newly developed mitochondrial-targeted antioxidants and genetic models that manipulate endogenous mitochondria antioxidants expression are now available to be utilised in the investigation of putative mitochondrial redox targets, mtDNA damage, Prx3 or mitochondrial peptides in coordinating adaptive response to exercise training. Such approach will begin to delineate the role of mitochondrial redox signalling in exercise training responses.

REFERENCES

Ago, T., Kuroda, J., Pain, J., Fu, C., Li, H. & Sadoshima, J. 2010. Upregulation of Nox4 by hypertrophic stimuli promotes apoptosis and mitochondrial dysfunction in cardiac myocytes. *Circ Res*, 106, 1253–64.

Alessio, H. M., Hagerman, A. E., Fulkerson, B. K., Ambrose, J., Rice, R. E. & Wiley, R. L. 2000. Generation of reactive oxygen species after exhaustive aerobic and isometric exercise. *Med Sci Sports Exerc*, 32, 1576–81.

Belousov, V. V., Fradkov, A. F., Lukyanov, K. A., Staroverov, D. B., Shakhbazov, K. S., Terskikh, A. V. & Lukyanov, S. 2006. Genetically encoded fluorescent indicator for intracellular hydrogen peroxide. *Nat Methods*, 3, 281–6.

Brand, M. D. 2016. Mitochondrial generation of superoxide and hydrogen peroxide as the source of mitochondrial redox signaling. *Free Radic Biol Med*, 100, 14–31.

Broome, S. C., Woodhead, J. S. T. & Merry, T. L. 2018. Mitochondria-Targeted Antioxidants and Skeletal Muscle Function. *Antioxidants (Basel)*, 7, 107.

Cheng, A. J., Bruton, J. D., Lanner, J. T. & Westerblad, H. 2015. Antioxidant treatments do not improve force recovery after fatiguing stimulation of mouse skeletal muscle fibres. *J Physiol*, 593, 457–72.

Cobb, L. J., Lee, C., Xiao, J., Yen, K., Wong, R. G., Nakamura, H. K., Mehta, H. H., Gao, Q., Ashur, C., Huffman, D. M., Wan, J., Muzumdar, R., Barzilai, N. & Cohen, P. 2016. Naturally occurring mitochondrial-derived peptides are age-dependent regulators of apoptosis, insulin sensitivity, and inflammatory markers. *Aging (Albany NY)*, 8, 796–809.

Cobley, J. N. 2020. Mechanisms of Mitochondrial ROS Production in Assisted Reproduction: The Known, the Unknown, and the Intriguing. *Antioxidants (Basel)*, 9, 933.

Cobley, J. N., Mchardy, H., Morton, J. P., Nikolaidis, M. G. & Close, G. L. 2015. Influence of vitamin C and vitamin E on redox signaling: Implications for exercise adaptations. *Free Radic Biol Med*, 84, 65–76.

Davies, K. J., Quintanilha, A. T., Brooks, G. A. & Packer, L. 1982. Free radicals and tissue damage produced by exercise. *Biochem Biophys Res Commun*, 107, 1198–205.

Dillard, C. J., Litov, R. E., Savin, W. M., Dumelin, E. E. & Tappel, A. L. 1978. Effects of exercise, vitamin E, and ozone on pulmonary function and lipid peroxidation. *J Appl Physiol Respir Environ Exerc Physiol*, 45, 927–32.

Ferreira, L. F. & Laitano, O. 2016. Regulation of NADPH oxidases in skeletal muscle. *Free Radic Biol Med*, 98, 18–28.

Fillingame, R. H. 1997. Coupling H^+ transport and ATP synthesis in F1F(o)-ATP synthases: Glimpses of interacting parts in a dynamic molecular machine. *J Exp Biol*, 200, 217–24.

Forman, H. J., Fukuto, J. M. & Torres, M. 2004. Redox signaling: thiol chemistry defines which reactive oxygen and nitrogen species can act as second messengers. *Am J Physiol Cell Physiol*, 287, C246–56.

Gidlund, E. K., Von Walden, F., Venojarvi, M., Riserus, U., Heinonen, O. J., Norrbom, J. & Sundberg, C. J. 2016. Humanin skeletal muscle protein levels increase after resistance training in men with impaired glucose metabolism. *Physiol Rep*, 4, e13063.

Gilkerson, R., Bravo, L., Garcia, I., Gaytan, N., Herrera, A., Maldonado, A. & Quintanilla, B. 2013. The mitochondrial nucleoid: Integrating mitochondrial DNA into cellular homeostasis. *Cold Spring Harb Perspect Biol*, 5, a011080.

Giorgio, M., Trinei, M., Migliaccio, E. & Pelicci, P. G. 2007. Hydrogen peroxide: A metabolic by-product or a common mediator of ageing signals? *Nat Rev Mol Cell Biol*, 8, 722–8.

Gomez-Cabrera, M. C., Borras, C., Pallardo, F. V., Sastre, J., Ji, L. L. & Vina, J. 2005. Decreasing xanthine oxidase-mediated oxidative stress prevents useful cellular adaptations to exercise in rats. *J Physiol*, 567, 113–20.

Gomez-Cabrera, M. C., Domenech, E., Romagnoli, M., Arduini, A., Borras, C., Pallardo, F. V., Sastre, J. & Vina, J. 2008a. Oral administration of vitamin C decreases muscle mitochondrial biogenesis and hampers training-induced adaptations in endurance performance. *Am J Clin Nutr*, 87, 142–9.

Gomez-Cabrera, M. C., Domenech, E. & Vina, J. 2008b. Moderate exercise is an antioxidant: Upregulation of antioxidant genes by training. *Free Radic Biol Med*, 44, 126–31.

Gomez-Cabrera, M. C., Pallardo, F. V., Sastre, J., Vina, J. & Garcia-Del-Moral, L. 2003. Allopurinol and markers of muscle damage among participants in the Tour de France. *JAMA*, 289, 2503–4.

Goncalves, R. L., Quinlan, C. L., Perevoshchikova, I. V., Hey-Mogensen, M. & Brand, M. D. 2015. Sites of superoxide and hydrogen peroxide production by muscle mitochondria assessed ex vivo under conditions mimicking rest and exercise. *J Biol Chem*, 290, 209–27.

Guillaumet-Adkins, A., Yanez, Y., Peris-Diaz, M. D., Calabria, I., Palanca-Ballester, C. & Sandoval, J. 2017. Epigenetics and Oxidative Stress in Aging. *Oxid Med Cell Longev*, 2017, 9175806.

Guo, B., Zhai, D., Cabezas, E., Welsh, K., Nouraini, S., Satterthwait, A. C. & Reed, J. C. 2003. Humanin peptide suppresses apoptosis by interfering with Bax activation. *Nature*, 423, 456–61.

Gutscher, M., Sobotta, M. C., Wabnitz, G. H., Ballikaya, S., Meyer, A. J., Samstag, Y. & Dick, T. P. 2009. Proximity-based protein thiol oxidation by H_2O_2-scavenging peroxidases. *J Biol Chem*, 284, 31532–40.

Hellsten-Westing, Y. 1993. Immunohistochemical localization of xanthine oxidase in human cardiac and skeletal muscle. *Histochemistry*, 100, 215–22.

Henriquez-Olguin, C., Diaz-Vegas, A., Utreras-Mendoza, Y., Campos, C., Arias-Calderon, M., Llanos, P., Contreras-Ferrat, A., Espinosa, A., Altamirano, F., Jaimovich, E. & Valladares, D. M. 2016. NOX2 inhibition impairs early muscle gene expression induced by a single exercise bout. *Front Physiol*, 7, 282.

Henriquez-Olguin, C., Knudsen, J. R., Raun, S. H., Li, Z., Dalbram, E., Treebak, J. T., Sylow, L., Holmdahl, R., Richter, E. A., Jaimovich, E. & Jensen, T. E. 2019a. Cytosolic ROS production by NADPH oxidase 2 regulates muscle glucose uptake during exercise. *Nat Commun*, 10, 4623.

Henriquez-Olguin, C., Renani, L. B., Arab-Ceschia, L., Raun, S. H., Bhatia, A., Li, Z., Knudsen, J. R., Holmdahl, R. & Jensen, T. E. 2019b. Adaptations to high-intensity interval training in skeletal muscle require NADPH oxidase 2. *Redox Biol*, 24, 101188.

Heumuller, S., Wind, S., Barbosa-Sicard, E., Schmidt, H. H., Busse, R., Schroder, K. & Brandes, R. P. 2008. Apocynin is not an inhibitor of vascular NADPH oxidases but an antioxidant. *Hypertension*, 51, 211–7.

Hey-Mogensen, M., Gram, M., Jensen, M. B., Lund, M. T., Hansen, C. N., Scheibye-Knudsen, M., Bohr, V. A. & Dela, F. 2015. A novel method for determining human ex vivo submaximal skeletal muscle mitochondrial function. *J Physiol*, 593, 3991–4010.

Higashida, K., Kim, S. H., Higuchi, M., Holloszy, J. O. & Han, D. H. 2011. Normal adaptations to exercise despite protection against oxidative stress. *Am J Physiol Endocrinol Metab*, 301, E779.

Hinkle, P. C., Butow, R. A., Racker, E. & Chance, B. 1967. Partial resolution of the enzymes catalyzing oxidative phosphorylation. XV. Reverse electron transfer in the flavin-cytochrome beta region of the respiratory chain of beef heart submitochondrial particles. *J Biol Chem*, 242, 5169–73.

Kang, C., O'Moore, K. M., Dickman, J. R. & Ji, L. L. 2009. Exercise activation of muscle peroxisome proliferator-activated receptor-gamma coactivator-1alpha signaling is redox sensitive. *Free Radic Biol Med*, 47, 1394–400.

Katz, A., Hernandez, A., Caballero, D. M., Briceno, J. F., Amezquita, L. V., Kosterina, N., Bruton, J. D. & Westerblad, H. 2014. Effects of N-acetylcysteine on isolated mouse skeletal muscle: Contractile properties, temperature dependence, and metabolism. *Pflugers Arch*, 466, 577–85.

Kim, K. H., Son, J. M., Benayoun, B. A. & Lee, C. 2018. The mitochondrial-encoded peptide MOTS-c translocates to the nucleus to regulate nuclear gene expression in response to metabolic stress. *Cell Metab*, 28, 516–24.

Kinugawa, S., Wang, Z., Kaminski, P. M., Wolin, M. S., Edwards, J. G., Kaley, G. & Hintze, T. H. 2005. Limited exercise capacity in heterozygous manganese superoxide dismutase gene-knockout mice: Roles of superoxide anion and nitric oxide. *Circulation*, 111, 1480–6.

Korshunov, S. S., Skulachev, V. P. & Starkov, A. A. 1997. High protonic potential actuates a mechanism of production of reactive oxygen species in mitochondria. *FEBS Lett*, 416, 15–8.

Lambert, A. J. & Brand, M. D. 2004. Superoxide production by NADH: Ubiquinone oxidoreductase (complex I) depends on the pH gradient across the mitochondrial inner membrane. *Biochem J*, 382, 511–7.

Latimer, H. R. & Veal, E. A. 2016. Peroxiredoxins in regulation of MAPK signalling pathways; Sensors and barriers to signal transduction. *Mol Cells*, 39, 40–5.

Lee, C., Zeng, J., Drew, B. G., Sallam, T., Martin-Montalvo, A., Wan, J., Kim, S. J., Mehta, H., Hevener, A. L., De Cabo, R. & Cohen, P. 2015. The mitochondrial-derived peptide MOTS-c promotes metabolic homeostasis and reduces obesity and insulin resistance. *Cell Metab*, 21, 443–54.

Li, D., Lai, Y., Yue, Y., Rabinovitch, P. S., Hakim, C. & Duan, D. 2009. Ectopic catalase expression in mitochondria by adeno-associated virus enhances exercise performance in mice. *PLoS One*, 4, e6673.

Loschen, G., Azzi, A., Richter, C. & Flohe, L. 1974. Superoxide radicals as precursors of mitochondrial hydrogen peroxide. *FEBS Lett*, 42, 68–72.

Mcbride, J. M., Kraemer, W. J., Triplett-Mcbride, T. & Sebastianelli, W. 1998. Effect of resistance exercise on free radical production. *Med Sci Sports Exerc*, 30, 67–72.

Merry, T. L., Chan, A., Woodhead, J. S. T., Reynolds, J. C., Kumagai, H., Kim, S. J. & Lee, C. 2020. Mitochondrial-derived peptides in energy metabolism. *Am J Physiol Endocrinol Metab*, 319, E659–66.

Merry, T. L. & Ristow, M. 2016a. Do antioxidant supplements interfere with skeletal muscle adaptation to exercise training? *J Physiol*, 594, 5135–47.

Merry, T. L. & Ristow, M. 2016b. Mitohormesis in exercise training. *Free Radic Biol Med*, 98, 123–30.

Merry, T. L. & Ristow, M. 2016c. Nuclear factor erythroid-derived 2-like 2 (NFE2L2, Nrf2) mediates exercise-induced mitochondrial biogenesis and the anti-oxidant response in mice. *J Physiol*, 594, 5195–207.

Miwa, S. & Brand, M. D. 2003. Mitochondrial matrix reactive oxygen species production is very sensitive to mild uncoupling. *Biochem Soc Trans*, 31, 1300–1.

Morrison, D., Hughes, J., Della Gatta, P. A., Mason, S., Lamon, S., Russell, A. P. & Wadley, G. D. 2015. Vitamin C and E supplementation prevents some of the cellular adaptations to endurance-training in humans. *Free Radic Biol Med*, 89, 852–62.

Mracek, T., Holzerova, E., Drahota, Z., Kovarova, N., Vrbacky, M., Jesina, P. & Houstek, J. 2014. ROS generation and multiple forms of mammalian mitochondrial glycerol-3-phosphate dehydrogenase. *Biochim Biophys Acta*, 1837, 98–111.

Murphy, M. P. 2009. How mitochondria produce reactive oxygen species. *Biochem J*, 417, 1–13.

Owusu-Ansah, E., Song, W. & Perrimon, N. 2013. Muscle mitohormesis promotes longevity via systemic repression of insulin signaling. *Cell*, 155, 699–712.

Paulsen, G., Cumming, K. T., Holden, G., Hallen, J., Ronnestad, B. R., Sveen, O., Skaug, A., Paur, I., Bastani, N. E., Ostgaard, H. N., Buer, C., Midttun, M., Freuchen, F., Wiig, H., Ulseth, E. T., Garthe, I., Blomhoff, R., Benestad, H. B. & Raastad, T. 2014. Vitamin C and E supplementation hampers cellular adaptation to endurance training in humans: A double-blind, randomised, controlled trial. *J Physiol*, 592, 1887–901.

Pearson, T., Kabayo, T., Ng, R., Chamberlain, J., Mcardle, A. & Jackson, M. J. 2014. Skeletal muscle contractions induce acute changes in cytosolic superoxide, but slower responses in mitochondrial superoxide and cellular hydrogen peroxide. *PLoS One*, 9, e96378.

Powers, S. K. & Jackson, M. J. 2008. Exercise-induced oxidative stress: Cellular mechanisms and impact on muscle force production. *Physiol Rev*, 88, 1243–76.

Powers, S. K., Nelson, W. B. & Hudson, M. B. 2011a. Exercise-induced oxidative stress in humans: Cause and consequences. *Free Radic Biol Med*, 51, 942–50.

Powers, S. K., Talbert, E. E. & Adhihetty, P. J. 2011b. Reactive oxygen and nitrogen species as intracellular signals in skeletal muscle. *J Physiol*, 589, 2129–38.

Pryor, W. A. 1986. Oxy-radicals and related species: Their formation, lifetimes, and reactions. *Annu Rev Physiol*, 48, 657–67.

Quinlan, C. L., Perevoshchikova, I. V., Hey-Mogensen, M., Orr, A. L. & Brand, M. D. 2013. Sites of reactive oxygen species generation by mitochondria oxidizing different substrates. *Redox Biol*, 1, 304–12.

Reynolds, J., Lai, R. W., Woodhead, J. S. T., Joly, J. H., Mitchell, C. J., Cameron-Smith, D., Lu, R., Cohen, P., Graham, N. A., Benayoun, B. A., Merry, T. L. & Lee, C. 2019. MOTS-c is an exercise-induced mitochondrial-encoded regulator of age-dependent physical decline and muscle homeostasis. *bioRxiv*, 2019.12.22.886432.

Reynolds, J. C., Lai, R. W., Woodhead, J. S. T., Joly, J. H., Mitchell, C. J., Cameron-Smith, D., Lu, R., Cohen, P., Graham, N. A., Benayoun, B. A., Merry, T. L. & Lee, C. 2021. MOTS-c is an exercise-induced mitochondrial-encoded regulator of age-dependent physical decline and muscle homeostasis. *Nat Commun*, 12, 470.

Rhee, S. G. & Kil, I. S. 2017. Multiple functions and regulation of mammalian peroxiredoxins. *Annu Rev Biochem*, 86, 749–75.

Rhee, S. G., Woo, H. A., Kil, I. S. & Bae, S. H. 2012. Peroxiredoxin functions as a peroxidase and a regulator and sensor of local peroxides. *J Biol Chem*, 287, 4403–10.

Ristow, M., Zarse, K., Oberbach, A., Kloting, N., Birringer, M., Kiehntopf, M., Stumvoll, M., Kahn, C. R. & Bluher, M. 2009. Antioxidants prevent health-promoting effects of physical exercise in humans. *Proc Natl Acad Sci U S A*, 106, 8665–70.

Sakellariou, G. K., Vasilaki, A., Palomero, J., Kayani, A., Zibrik, L., Mcardle, A. & Jackson, M. J. 2013. Studies of mitochondrial and nonmitochondrial sources implicate nicotinamide adenine dinucleotide phosphate oxidase(s) in the increased skeletal muscle superoxide generation that occurs during contractile activity. *Antioxid Redox Signal*, 18, 603–21.

Saki, M. & Prakash, A. 2017. DNA damage related crosstalk between the nucleus and mitochondria. *Free Radic Biol Med*, 107, 216–27.

Schultz, B. E. & Chan, S. I. 2001. Structures and proton-pumping strategies of mitochondrial respiratory enzymes. *Ann Rev Biophys Biomol Struct*, 30, 23–65.

Schulz, T. J., Zarse, K., Voigt, A., Urban, N., Birringer, M. & Ristow, M. 2007. Glucose restriction extends Caenorhabditis elegans life span by inducing mitochondrial respiration and increasing oxidative stress. *Cell Metab*, 6, 280–93.

Scialo, F., Sriram, A., Fernandez-Ayala, D., Gubina, N., Lohmus, M., Nelson, G., Logan, A., Cooper, H. M., Navas, P., Enriquez, J. A., Murphy, M. P. & Sanz, A. 2016. Mitochondrial ROS produced via reverse electron transport extend animal lifespan. *Cell Metab*, 23, 725–34.

Seifert, E. L., Estey, C., Xuan, J. Y. & Harper, M. E. 2010. Electron transport chain-dependent and -independent mechanisms of mitochondrial H_2O_2 emission during long-chain fatty acid oxidation. *J Biol Chem*, 285, 5748–58.

Selivanov, V. A., Votyakova, T. V., Pivtoraiko, V. N., Zeak, J., Sukhomlin, T., Trucco, M., Roca, J. & Cascante, M. 2011. Reactive oxygen species production by forward and reverse electron fluxes in the mitochondrial respiratory chain. *PLoS Comput Biol*, 7, e1001115.

Shill, D. D., Southern, W. M., Willingham, T. B., Lansford, K. A., Mccully, K. K. & Jenkins, N. T. 2016. Mitochondria-specific antioxidant supplementation does not influence endurance exercise training-induced adaptations in circulating angiogenic cells, skeletal muscle oxidative capacity or maximal oxygen uptake. *J Physiol*, 594, 7005–14.

Sies, H., Berndt, C. & Jones, D. P. 2017. Oxidative Stress. *Annu Rev Biochem*, 86, 715–48.

Sies, H. & Cadenas, E. 1985. Oxidative stress: Damage to intact cells and organs. *Philos Trans R Soc Lond B Biol Sci*, 311, 617–31.

Sies, H. & Jones, D. P. 2020. Reactive oxygen species (ROS) as pleiotropic physiological signalling agents. *Nat Rev Mol Cell Biol*, 21, 363–83.

Silveira, L. R., Pilegaard, H., Kusuhara, K., Curi, R. & Hellsten, Y. 2006. The contraction induced increase in gene expression of peroxisome proliferator-activated receptor (PPAR)-gamma coactivator 1alpha (PGC-1alpha), mitochondrial uncoupling protein 3 (UCP3) and hexokinase II (HKII) in primary rat skeletal muscle cells is dependent on reactive oxygen species. *Biochim Biophys Acta*, 1763, 969–76.

Srinivas, U. S., Tan, B. W. Q., Vellayappan, B. A. & Jeyasekharan, A. D. 2019. ROS and the DNA damage response in cancer. *Redox Biol*, 25, 101084.

Stretton, C., Pugh, J. N., Mcdonagh, B., Mcardle, A., Close, G. L. & Jackson, M. J. 2020. 2-Cys peroxiredoxin oxidation in response to hydrogen peroxide and contractile activity in skeletal muscle: A novel insight into exercise-induced redox signalling? *Free Radic Biol Med*, 160, 199–207.

Strobel, N. A., Peake, J. M., Matsumoto, A. Y. A., Marsh, S. A., Coombes, J. S. & Wadley, G. D. 2011. Antioxidant supplementation reduces skeletal muscle mitochondrial biogenesis. *Med Sci Sports Exerc*, 43, 1017–24.

Tiganis, T. 2011. Reactive oxygen species and insulin resistance: The good, the bad and the ugly. *Trends Pharmacol Sci*, 32, 82–9.

Vina, J., Gimeno, A., Sastre, J., Desco, C., Asensi, M., Pallardo, F. V., Cuesta, A., Ferrero, J. A., Terada, L. S. & Repine, J. E. 2000. Mechanism of free radical production in exhaustive exercise in humans and rats; Role of xanthine oxidase and protection by allopurinol. *IUBMB Life*, 49, 539–44.

Wadley, G. D. & Mcconell, G. K. 2010. High-dose antioxidant vitamin C supplementation does not prevent acute exercise-induced increases in markers of skeletal muscle mitochondrial biogenesis in rats. *J Appl Physiol*, 108, 1719–26.

Wadley, G. D., Nicolas, M. A., Hiam, D. S. & Mcconell, G. K. 2013. Xanthine oxidase inhibition attenuates skeletal muscle signaling following acute exercise but does not impair mitochondrial adaptations to endurance training. *Am J Physiol Endocrinol Metab*, 304, E853–62.

Wajner, M. & Harkness, R. A. 1989. Distribution of xanthine dehydrogenase and oxidase activities in human and rabbit tissues. *Biochim Biophys Acta*, 991, 79–84.

Weimer, S., Priebs, J., Kuhlow, D., Groth, M., Priebe, S., Mansfeld, J., Merry, T. L., Dubuis, S., Laube, B., Pfeiffer, A. F., Schulz, T. J., Guthke, R., Platzer, M., Zamboni, N., Zarse, K. & Ristow, M. 2014. D-Glucosamine supplementation extends life span of nematodes and of ageing mice. *Nat Commun*, 5, 3563.

Williamson, J. & Davison, G. 2020. Targeted antioxidants in exercise-induced mitochondrial oxidative stress: Emphasis on DNA damage. *Antioxidants (Basel)*, 9, 1142.

Williamson, J., Hughes, C. M., Cobley, J. N. & Davison, G. W. 2020. The mitochondria-targeted antioxidant MitoQ, attenuates exercise-induced mitochondrial DNA damage. *Redox Biol*, 36, 101673.

Wong, H. S., Dighe, P. A., Mezera, V., Monternier, P. A. & Brand, M. D. 2017. Production of superoxide and hydrogen peroxide from specific mitochondrial sites under different bioenergetic conditions. *J Biol Chem*, 292, 16804–16809.

Woodhead, J. S. T., D'Souza, R. F., Hedges, C. P., Wan, J., Berridge, M. V., Cameron-Smith, D., Cohen, P., Hickey, A. J. R., Mitchell, C. J. & Merry, T. L. 2020. High-intensity interval exercise increases humanin, a mitochondrial encoded peptide, in the plasma and muscle of men. *J Appl Physiol*, 128, 1346–54.

Yfanti, C., Akerstrom, T., Nielsen, S., Nielsen, A. R., Mounier, R., Mortensen, O. H., Lykkesfeldt, J., Rose, A. J., Fischer, C. P. & Pedersen, B. K. 2010. Antioxidant supplementation does not alter endurance training adaptation. *Med Sci Sports Exerc*, 42, 1388–95.

Zarse, K., Schmeisser, S., Groth, M., Priebe, S., Beuster, G., Kuhlow, D., Guthke, R., Platzer, M., Kahn, C. R. & Ristow, M. 2012. Impaired insulin/IGF1 signaling extends life span by promoting mitochondrial L-proline catabolism to induce a transient ROS signal. *Cell Metab*, 15, 451–65.

Zhang, Y. G., Wang, L., Kaifu, T., Li, J., Li, X. & Li, L. 2016. Featured Article: Accelerated decline of physical strength in peroxiredoxin-3 knockout mice. *Exp Biol Med (Maywood)*, 241, 1395–400.

CHAPTER SIX

Basal Redox Status Influences the Adaptive Redox Response to Regular Exercise

Ethan L. Ostrom and Tinna Traustadóttir

CONTENTS

Introduction / 71
Sites of ROS Generation and Chronic Oxidative Distress / 72
Mechanisms of Adaptive Responses to Exercise: Nrf2 / 73
Nrf2 Response to Exercise Training / 74
Role of Basal Redox Status in the Adaptive Response to Exercise / 75
A Paradigm Shift in the Relationship between Antioxidant Enzymes and Redox Signaling / 76
Conclusions and Future Directions / 79
References / 80

INTRODUCTION

Eustress is defined as moderate or normal physiological or environmental stress that is beneficial to the organism. Distress occurs when physiological or environmental stress is of sufficient duration or intensity to overwhelm the system and cause damage or dysfunction to the cell, tissue, organ, or organism. Acute exercise is a powerful physiologic eustress altering metabolic flux to meet the demands of the contracting skeletal muscle. Repeated application of this physiological stimulus, as in regular exercise training, leads to steady-state changes in protein content and beneficial health effects. These favorable adaptions are regulated in part by reactive oxygen species (ROS) production during the exercise bout – oxidative eustress. ROS accumulation initiates redox reactions with protein thiols that act as redox switches, altering protein localization, turnover, interacting partners, or enzyme activity (Nikolaidis et al., 2020; Kramer et al., 2015). Thus, an acute exercise bout can temporarily alter the vicinal milieu within a subcellular compartment to a more oxidized microenvironment leading to constructive changes in redox status, and subsequently resulting in beneficial long-term adaptations through changes in gene expression and protein content (Piantadosi and Suliman, 2006; Henriquez-Olguin et al., 2016, 2019a,b; Cobley et al., 2014, 2015) (Figure 6.1).

ROS act beneficially under physiological concentrations. However, past a certain threshold, ROS can cause hyper-oxidation of protein thiols, formation of DNA adducts, and lipid peroxidation products, potentially leading to accumulation of macromolecular damage and resulting dysfunction (Wadley et al., 2015; Williamson et al., 2020). Oxidative distress can occur in response to acute exercise if the exercise bout is sufficiently intense (Tryfidou et al., 2020). However, the increases in markers of DNA damage in response to high-intensity exercise are often temporary and may

DOI: 10.1201/9781003051619-6

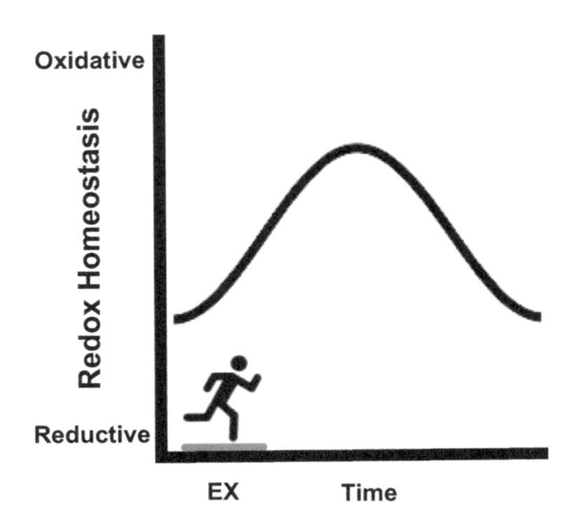

Figure 6.1 Acute exercise (EX) acts as a transient stimulus to produce ROS. This alters the redox microenvironment within the cell, activating redox signaling responses and adaptive changes in cellular capacity and stress resilience.

therefore not be associated with any permanent dysfunction (Tryfidou et al., 2020). Chronic oxidative distress is more commonly seen with aging, sedentary behavior, and overfeeding, driven by elevated basal levels of mitochondrial ROS production (Dai et al., 2014; Campbell et al., 2019; McDonagh et al., 2014; Mansouri et al., 2006; Vasilaki et al., 2006). This increase in basal ROS results in site-specific occupancy of oxidized protein thiols leading to aberrant redox signaling and dysfunctional responses to subsequent redox stressors (Campbell et al., 2019). The key aspect of exercise-induced oxidative signaling is that the signal is constrained by the duration and intensity of the bout – in other words, the oxidative stress is transient. This is in contrast to chronic low-grade oxidative stress seen in aging and overfeeding diseases like type 2 diabetes where the oxidative distress is constant (Mansouri et al., 2006; Vasilaki et al., 2006; McDonagh et al., 2014; Campbell et al., 2019; Anderson et al., 2009; Fisher-Wellman et al., 2014). However, effects of aging may also be due to other mechanisms. Recent proteomic data in a mouse model has demonstrated that aging is associated with reprogramming of redox signaling that is distinct from that of young resulting in the loss of redox networks involved in tissue regulation (Xiao et al., 2020). The differences between acute and chronic ROS production may be relevant for describing the threshold between what constitutes an oxidative eustress and oxidative distress signal. This threshold is intrinsic to the hormesis model of exercise-induced ROS production (Radak et al., 2017), however where this threshold lies is difficult to quantitatively define and will likely vary depending on reactive species (Nikolaidis et al., 2020) and is therefore difficult to assess where the signal stops being an adaptive eustress.

SITES OF ROS GENERATION AND CHRONIC OXIDATIVE DISTRESS

ROS can be generated from many different enzymatic and non-enzymatic sites in the cell, including mitochondria, NADPH oxidases (NOXs), xanthine oxidase (Suh et al., 2004), and lipoxygenases (LOX) (Sakellariou et al., 2014). The compartmentalization of these enzymes constrains the ROS products and can create oxidative or reductive hotspots in certain regions or compartments of the cell (Manford et al., 2020). In vitro, ex vivo, and in vivo work all suggest that under resting conditions in skeletal muscle, the mitochondria are the main source of ROS generation, which contribute to the basal tone of redox signaling (Sakellariou et al., 2014; Goncalves et al., 2015, 2019; Henriquez-Olguin et al., 2020). During exercise (or conditions mimicking exercise in vitro), the sources of ROS generation shift away from mitochondria and toward NOX and XO enzymes (Henriquez-Olguin et al., 2020; Sakellariou et al., 2014). This occurs because energy demand increases, providing ADP stimulated state 3 respiration in mitochondria, and a decline in overall superoxide (O_2^-) leak through decreasing membrane potential and proton motive pressure release (Fisher-Wellman et al., 2014). The concomitant recruitment of NOX subunits to the plasma membrane during exercise activates the complex and increases production of superoxide (Henriquez-Olguin et al., 2016). During resting conditions, chronic overfeeding with high-fat diets increases mitochondrial superoxide leak because of the basal state 4 respiration plus the high driving pressure created by the buildup of NADH from intermediate metabolism (Fisher-Wellman et al., 2014). Together this creates a high backpressure through the mitochondrial membrane potential, and O_2^- / H_2O_2 leak is one release valve mechanism when calorie consumption is high and ATP demand is low (Fisher-Wellman et al., 2014). The chronic levels of mitochondrial oxidative distress from aging, sedentary behavior, and overfeeding stimuli all likely contribute to the increasing dysfunctional redox signaling seen with aging and diseased populations, and consequently in dysfunctional redox responses to exercise (Viña et al., 2013). Together these data suggest the chronic oxidative distress in aging and disease populations are being driven by increased ROS production in the mitochondria and leading to redox signaling dysfunction. Treating with mitochondrial antioxidants reverses the age-related elevation in the oxidized thiol proteome (Campbell et al., 2019). Some evidence suggests that the mitochondrial antioxidant MitoQ decreases exercise-induced mitochondrial DNA damage (Williamson et al., 2020). Together these data indicate that targeting antioxidants to mitochondria may be successful in preventing chronic oxidative distress whether it is from increasing age-related mitochondrial dysfunction or protecting the cell from overexertion/overtraining induced acute oxidative distress. If true, these mitochondrial targeted antioxidants may promote transient oxidative eustress through exercise, where other general antioxidants have failed (Gomez-Cabrera et al., 2005, 2008; Merry and Ristow, 2016; Ristow et al., 2009; Broome et al., 2018). Clearly this is an area where further research is warranted.

MECHANISMS OF ADAPTIVE RESPONSES TO EXERCISE: Nrf2

A critical mechanism involved in the adaptive response to exercise is activating transcription factors that increase mRNA content, driving increases in protein content and exercise capacity (Irrcher et al., 2009; Perry et al., 2010). Many different transcription factors have some level of redox regulation, including NF-kB, AP-1, HIF1, HSF1, SP1, Notch, CREB, and Nrf2 (Marinho et al., 2014). Here we focus on Nrf2 because canonical Nrf2 activation is dependent on oxidative modification of Keap1, Nrf2 directly regulates gene expression for several hundred antioxidant and metabolic enzymes (Hayes and Dinkova-Kostova, 2014), and Nrf2 is a critical component of exercise-induced adaptations (Done and Traustadóttir, 2016).

Nrf2 is an inducible redox stress response transcription factor activated by increases in environmental or endogenous oxidative species (Tebay et al., 2015). Figure 6.2 shows a general schematic of the Nrf2 signaling pathway. Under unstressed conditions, Nrf2 is bound by its negative regulator Keap1, which targets Nrf2 for degradation via the 26S proteasome. Nrf2 would be expected to be lowly expressed under those conditions because the majority will be degraded by the proteasome. During an oxidative stress stimulus, solvent-exposed protein thiols on the backbone of Keap1 become oxidized, impairing Keap1's ability to target Nrf2 for degradation (Fourquet et al., 2010; Eggler et al., 2009; Hu et al., 2011; Zhang and Hannink, 2003). This allows Nrf2 to accumulate in the nucleus, where it heterodimerizes with small Maf proteins, binds to the antioxidant response element (ARE), and increases gene expression responsible for increasing metabolic and redox stress capacity. To our knowledge, direct oxidation of Keap1 in response to acute exercise has not been experimentally verified. However, Nrf2 activation occurs in a duration- and intensity-dependent manner which is associated with the accumulation of reactive species generation (Wang et al., 2015; Li et al., 2015), suggesting that Keap1 is likely oxidatively modified and degraded in response to exercise. Indeed, we have recently found Keap1 to be degraded in an intensity-dependent manner in response to muscle stimulation in mice (Ostrom et al., 2021). In addition, another study demonstrated decreased Keap1 in human skeletal muscle in response to a maximal exercise test, with a concomitant increase in Nrf2 (Gallego-Selles et al., 2020).

Others have shown increases in Nrf2 signaling in response to acute exercise in mouse heart (Gounder et al., 2012; Muthusamy et al., 2012), brain (Tsou et al., 2015; Aguiar et al., 2016), and lung tissue in response to exercise (Kubo et al., 2019). Increases in Nrf2 signaling in response to acute exercise have also been shown in human skeletal muscle (Gallego-Selles et al., 2020) and human PBMCs (Done et al., 2016, 2017; Ostrom and Traustadóttir, 2020). These changes lead to improvements in antioxidant protein content and greater redox stress capacity. Together these data suggest that Nrf2 mediates the redox-dependent adaptations to exercise through generation of ROS during the exercise bout (Done and Traustadóttir, 2016). It should be noted that it has yet to be determined how Nrf2 becomes activated in peripheral tissues other than skeletal muscle. Nrf2 activation in tissues like the brain, lungs, and PBMCs are unlikely to be driven by direct activation of ROS derived from skeletal muscle because: (1) the half-life of reactive species is low,

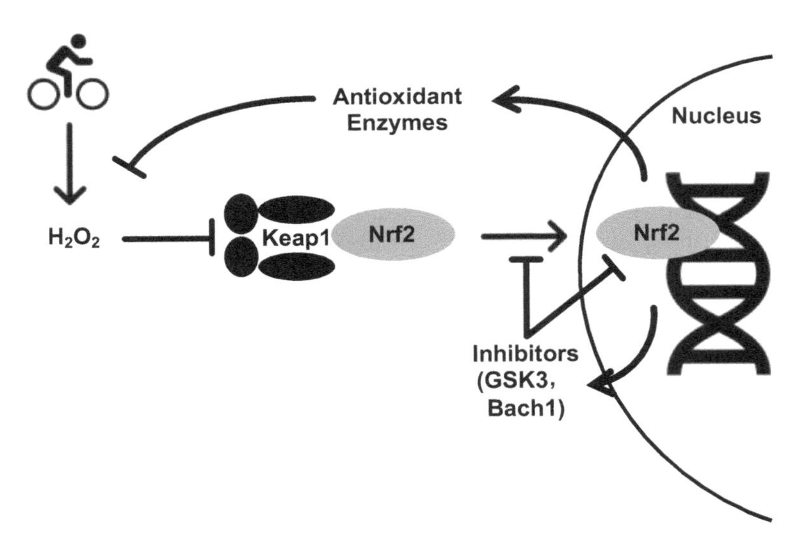

Figure 6.2 ROS-induced Nrf2 activation. An endogenous or environmental signal like exercise increases ROS production which interacts with solvent-exposed cysteine residues on Keap1. Cysteine oxidation activates Nrf2 allowing it to translocate to the nucleus and bind to ARE sequences. This increases gene expression and protein content of antioxidant enzymes, as well as other proteins that negatively regulate Nrf2 like Bach1, which is a competitive inhibitor of ARE binding. Together these antioxidant enzymes and competitive inhibitors quench the signal, completing the negative feedback loop.

(2) the concentration of intracellular and extracellular antioxidant enzymes is very high, and (3) simple diffusion (H_2O_2) across multiple cellular and organellar membranes make the probability of ROS acting as a second messenger over long timescales and distances very low. Unraveling the "Redox Exerkine Signaling" dilemma provides an interesting and challenging future direction for the redox biology and exercise physiology field.

Nrf2 RESPONSE TO EXERCISE TRAINING

We have recently shown that basal levels of Nrf2 nuclear protein after training are unchanged in young healthy individuals, while older individuals who had higher basal levels prior to the intervention, showed a decrease in Nrf2 nuclear content (Ostrom and Traustadóttir, 2020). This decline in Nrf2 protein in response to training has also been shown in mouse skeletal muscle (Crilly et al., 2016). The decline of basal nuclear Nrf2 in response to training in older adults improved their acute response to a single exercise bout, although the improvement was still not as robust as young adults (Ostrom and Traustadóttir, 2020). Together, these data suggest that the important health metric in Nrf2 signaling is the magnitude of the response to a transient oxidative eustress, as opposed to the basal resting levels (Ostrom and Traustadóttir, 2020). Other research using Nrf2 knockout animals has shown that the adaptive response to training requires Nrf2 activation to stimulate mitochondrial biogenesis and increase redox stress capacity (Merry and Ristow, 2016). Furthermore, treating mice with the antioxidants N-acetylcysteine or L-NAME impairs exercise-induced activation of Nrf2 and results in impaired markers of mitochondrial biogenesis (Merry and Ristow, 2016). Together these findings demonstrate that Nrf2 signaling is required for mitochondrial biogenesis in response to training, and that Nrf2 levels mirror the oxidative stress levels in the cell. It would therefore be predicted to observe elevated basal nuclear Nrf2 levels in conditions of elevated basal oxidant production: aging, sedentary behavior, and chronic overfeeding. This increase in basal Nrf2 is initially adaptive; however, a side effect of this increasing chronic oxidative distress is a ceiling effect on Nrf2 inducibility. Any added transient stimulus – such as acute exercise – on top of the chronic oxidative distress results in diminished signaling effects and adaptive responses (Musci et al., 2019; Pomatto and Davies, 2017, 2018). Despite the declining adaptive responses in aging populations, exercise training still elicits responsiveness in some redox-regulated signaling networks (Cobley et al., 2014).

ROLE OF BASAL REDOX STATUS IN THE ADAPTIVE RESPONSE TO EXERCISE

The data above suggest that the antioxidant enzymes and protein thiols that respond to acute exercise are important for exercise training adaptations and are dependent on basal redox status (Margaritelis et al., 2014). Figure 6.3 shows the relationship between basal redox status and acute oxidative eustress induced by exercise in healthy redox homeostasis (left), and in a chronic state of oxidative distress (right). In support of this concept, classifying individuals based on oxidative markers in response to acute exercise significantly predicted an individual's adaptive response to exercise training. Individuals with moderate and high increases in oxidative stress markers in response to acute exercise showed significantly greater adaptations to a 6-week aerobic exercise training protocol compared to individuals with a low acute oxidative response to exercise (Margaritelis et al., 2014, 2018). Additionally, if researchers randomized healthy people to either vitamin C supplementation (antioxidant stimulus) or passive smoking (pro-oxidant stimulus), the F_2-isoprostane and glutathione response to acute exercise was altered. Vitamin C supplementation lowered basal oxidative stress levels, and as a result, those individuals showed appropriate responses to acute exercise. Conversely, if chronic oxidative distress is experimentally elevated using passive cigarette smoking, the redox responses to acute exercise were blunted (Theodorou et al., 2014). In a later study, the same research group performed a targeted antioxidant supplementation of either vitamin C or N-acetylcysteine after assessing basal blood measures. This allowed the researchers to tailor the intervention to reverse a deficiency in either vitamin C levels or glutathione. This targeted approach improved VO_2 peak greater than a non-targeted antioxidant supplementation approach (Margaritelis et al., 2020). Together these data suggest that the threshold between ROS-induced oxidative eustress and oxidative distress is malleable (Radak et al., 2017) and that there is an optimal basal redox status. Furthermore, if that

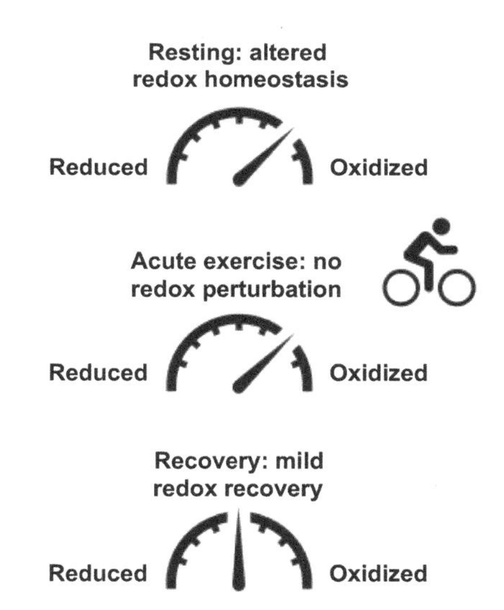

Figure 6.3 The relationship between basal redox state and acute redox responses. Under normal healthy circumstances (left side), redox homeostasis favors the reduced state. During an acute exercise bout, the state shifts to a more oxidized condition, followed by recovery to resting redox homeostasis after completion of the exercise. In unhealthy conditions, there is a shift in redox homeostasis to favor an oxidizing environment creating a chronic oxidative distress. This impairs redox signaling mechanisms in response to acute exercise, preventing the dynamic nature of transient signaling to occur. This is followed by a mild recovery after completion of the exercise bout.

optimal basal redox steady state is not achieved, responses to acute exercise will be altered, potentially impairing exercise adaptations.

An alternative hypothesis has recently been presented, around the concept that the adaptive redox response to exercise is dysfunctional with aging due to a more reduced state of protein thiols in the cytosol, rather than oxidized (Jackson, 2020). The rationale behind this hypothesis is that skeletal muscle denervation that increases with age induces a burst of mitochondrial ROS production that initially stimulates axonal sprouting. This burst in ROS production also increases antioxidant proteins in the cytosol through redox signaling adaptations and over time contributes to an overall shift to a more reduced cytosolic milieu. The more reduced cytosolic environment prevents significant protein thiol oxidation, which results in impaired redox signaling and overall declining adaptations to exercise with increasing age. It is important to note that these two hypotheses are not mutually exclusive. Redox states may differ between cellular compartments or tissues, and therefore, it is possible that there is a shift to a more reduced state in some compartments and a shift to a more oxidized state in others (Duan et al., 2020). Furthermore, the global protein thiol oxidation state is dynamic, and therefore, thiol oxidation states are likely dependent on the timeframe of the tissue sample collection, which may explain some of the differences seen in the field (McDonagh et al., 2014; Campbell et al., 2019; Michailidis et al., 2007; Jackson, 2020). Regardless of the discrepancies, collectively the data suggest that the redox homeostatic setpoint changes with a combination of age, sedentary behavior, and overfeeding, which can alter redox-specific adaptations to exercise.

A PARADIGM SHIFT IN THE RELATIONSHIP BETWEEN ANTIOXIDANT ENZYMES AND REDOX SIGNALING

The generally accepted idea of redox signaling and adaptive responses to exercise is that acute increase in ROS will stimulate Nrf2 activation, which consequently leads to upregulation of antioxidant gene transcription. With regular temporal Nrf2 activation, the adaptive response will increase the antioxidant enzyme capacity, resulting in the oxidative stimulus being quenched quicker (Figure 6.2), given the same ROS production.

In this model, one would predict that the redox stress response capacity would increase after exercise training such that the same oxidative stimulus would be quenched quicker and more efficiently upon subsequent stressors of the same magnitude. In other words, at the same given oxidative stimulus intensity, the signal transduction response would be diminished. However, we have recently shown that Nrf2 signaling response to acute exercise increases in magnitude after exercise training (Ostrom and Traustadóttir, 2020), suggesting the oxidant-induced negative feedback mechanism does not provide a complete picture of exercise-induced redox signaling (Figure 6.4). While it is certainly possible that exercise training increases the ability to produce ROS at the same relative exercise intensity compared to pretraining, it is also likely that our implicit assumptions about the relationship between the antioxidant defense system and redox signaling mechanisms are flawed.

Hydrogen peroxide (H_2O_2) has been viewed as the main signaling molecule in redox signal transduction, where it was assumed that H_2O_2 reacts directly with protein thiols to initiate a signaling cascade. However, there are several problems with this assumption. Peroxiredoxins (Prxs) are some of the most abundantly and ubiquitously expressed antioxidant proteins in mammalian cells (Stocker et al., 2018), and they react with H_2O_2 on orders of magnitude faster than even the most reactive solvent-exposed protein thiols. Therefore, it would appear that Prxs are able to quench the oxidative stress signal from endogenous and exogenous signals before it can initiate a signal transduction cascade through other protein thiols (Stocker et al., 2018). Thus, Prxs likely outcompete most if not all protein thiols in the cell that utilize redox signaling mechanisms. The floodgate hypothesis proposes that H_2O_2 signaling persists by first causing hyperoxidation-induced inhibition of Prx enzymes, which then allows H_2O_2 to accumulate and diffuse throughout the cell to react with other protein thiols to initiate a redox signaling cascade. This model, if true, would circumvent the problems with kinetic efficiencies between H_2O_2 and Prxs versus H_2O_2 and protein thiols (Wood et al., 2003). While this may occur as there is some experimental evidence to support it (Hanzen et al., 2016), the floodgate hypothesis still does not explain redox signaling specificity. How does H_2O_2 react with some protein thiols and not others? More recently, it has been shown that Prxs may act as redox signaling intermediates rather than just peroxide quenchers, overcoming the kinetic and reaction specificity problems of the floodgate model.

Prxs have been shown to form mixed disulfide intermediates with well-known stress response signaling cascades. One study used an eloquent series of experiments to demonstrate that Prx2 can take oxidizing equivalents from hydrogen peroxide and transfer those oxidizing equivalents to the transcription factor STAT3, causing STAT3 dimerization and tetramerization, which impairs STAT3 transcriptional activity (Sobotta et al., 2015). These data indicate that Prx2 is acting as a redox middleman by taking oxidizing equivalents from H_2O_2 and transferring them to other redox-sensitive proteins (Sobotta et al., 2015). Prx1 has also been shown to act as a H_2O_2 signaling intermediate by forming mixed disulfide links with the protein ASK1 (Jarvis et al., 2012). ASK1 is required for phosphorylation and activation of p38 MAPK. ASK1 oxidation through Prx1 mediated oxidation promoted p38 phosphorylation and activation, while knockdown of Prx1 impaired p38 phosphorylation and activation (Jarvis et al., 2012). MAPK p38 signaling is an important redox-sensitive mediator of the adaptive responses to exercise (Henriquez-Olguin et al., 2016; Parker et al., 2017). Prx-mediated oxidation of ASK1 is likely a viable redox-mediated adaptive signaling process in exercise adaptations, although this has yet to be empirically determined. In addition to these findings, previous studies have demonstrated increased levels of oxidized Prx dimers in isolated PBMCs following acute exercise and increased abundance of Prx2 following an exercise intervention (Moghaddam et al., 2011). The mechanism described by the "redox relay" alleviates the inconsistencies with the floodgate model by relying on the superior kinetic efficiency of Prxs, as well as providing a mechanism for specificity through steric hindrances in protein-protein interactions between Prxs and their target redox-sensitive proteins.

Together these data provide evidence that Prxs are involved in gene transcriptional regulation and protein chaperone behavior in addition to their traditional antioxidant capacities. Based on these findings, it would be reasonable to predict that the increase in antioxidant enzymes in response to exercise training would not only prevent unwanted hyperoxidation and oxidative distress but also increase redox signaling flux through specific signaling pathways in response to oxidative eustress stimuli. In this way, these antioxidant enzymes are simultaneously acting as true antioxidants (quenching the signal) and as redox relay intermediates. We suspect that the increases in Nrf2 signaling after exercise training in our recent work may be mediated by these redox relay mechanisms. The negative feedback loop model (Figure 6.2) would not explain an increase in Nrf2 signaling after training unless the increase in antioxidant enzymes also worked to direct the oxidative signal through these signaling pathways; however, this still needs to be confirmed. For a more complete review of Prx-mediated redox signaling in exercise, we refer the reader to this excellent review by Wadley et al. (2016).

Another redox signaling mechanism is protein glutathionylation. This is where glutathione is added to a cysteine residue creating a mixed disulfide (–SSG) modification which can also alter protein function. This reaction is catalyzed by the family of glutathione synthetase transferases (GSTs) while glutaredoxins are responsible for removing glutathione from protein thiols. Glutathionylation occupancy increases in mouse skeletal muscle subjected to fatiguing contractions (Kramer et al., 2018). This increase is likely a protective mechanism to desensitize the thiol proteome to hyperoxidation induced by overexertion and prevent further oxidation through inhibiting metabolic pathways that generate more ROS (Mailloux, 2020; Gill et al., 2018). Additionally, glutathione modification of Keap1 cysteines has been shown to induce Nrf2 activation (Carvalho et al., 2016). This suggests that ROS-induced increases in protein glutathionylation is a positive feed-forward loop that enhances Nrf2 signaling in addition to minimizing ROS production through inhibiting enzymes that generate ROS and quench the signal through GPx-mediated reduction of H_2O_2. Taken together, it seems likely that the traditional view that antioxidant enzymes are pure antioxidants (i.e., they quench ROS signals) needs to be refined. A more nuanced model about the relationship between the antioxidant defense system and Nrf2 activation in response to exercise is shown in Figure 6.5, along with the chronic oxidative distress stemming from sedentary behavior and overfeeding induced mitochondrial dysfunction seen in aging populations. However, more

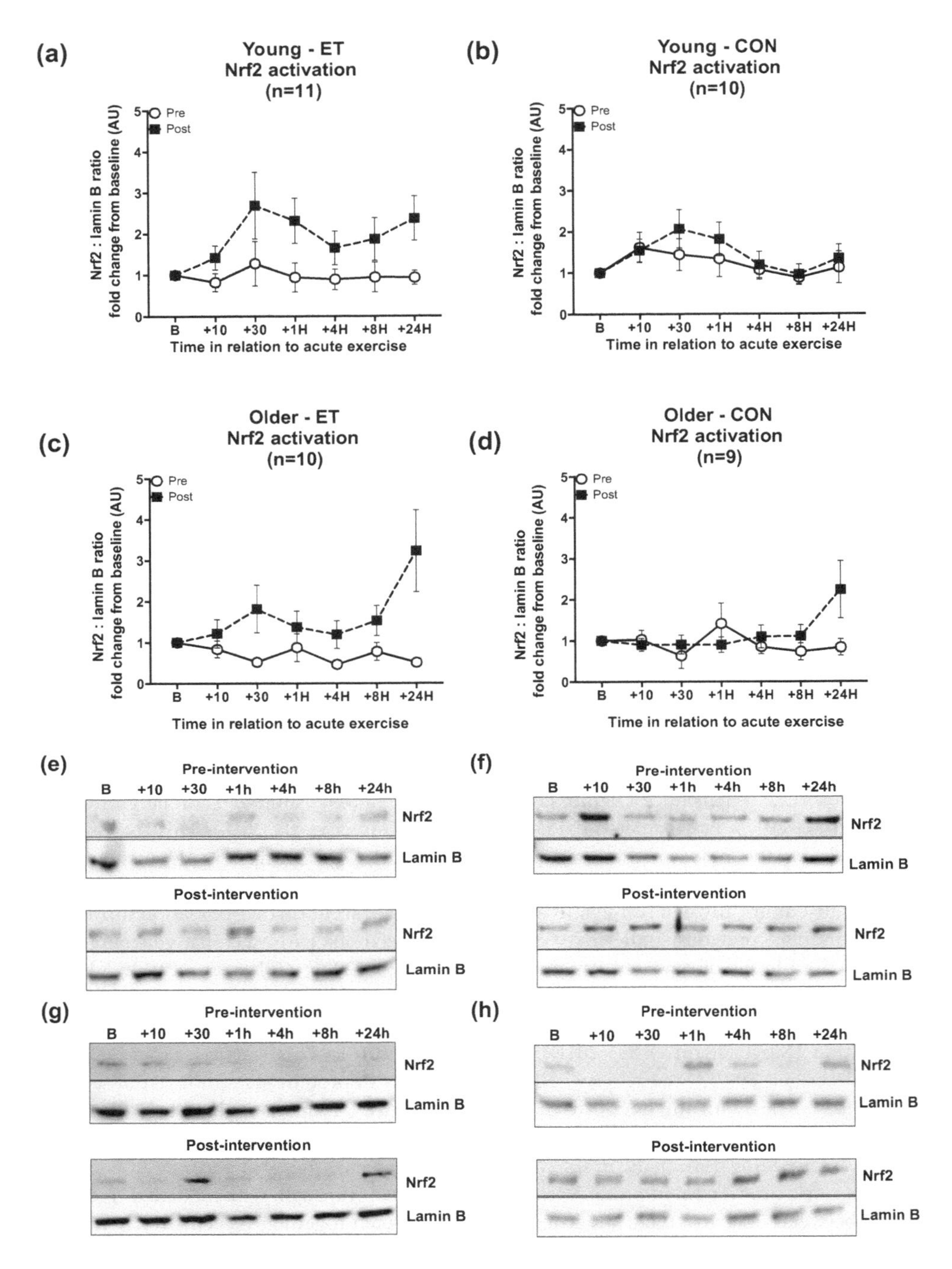

Figure 6.4 Nrf2 signaling responses to acute exercise before and after exercise training or inactive control intervention by age. Open circles indicate Nrf2 activation pre-intervention, and black squares indicate Nrf2 activation post-intervention. Panels (a) and (c) are ET subjects while panels (b) and (d) are CON subjects. (a) and (b) Eight weeks of aerobic exercise training increased the Nrf2 signaling response to an acute exercise bout in the young-trained cohort improved, while the Nrf2 signaling in the control group did not change. (c) and (d) Aerobic training in the older group showed improvements, while the older control group showed no changes. The increase in Nrf2 signaling after training was similar in pattern for young and older adults, although young improved their responses to a greater degree than older ET subjects (a) vs (d) post-intervention, main effect of age, $p = 0.014$. Representative Western blots for (e) Young-ET, (f) Young-CON, (g) Older-ET, and (h) Older-CON. ET – Endurance Trained, CON – Control intervention. Data shown as Mean±SEM.

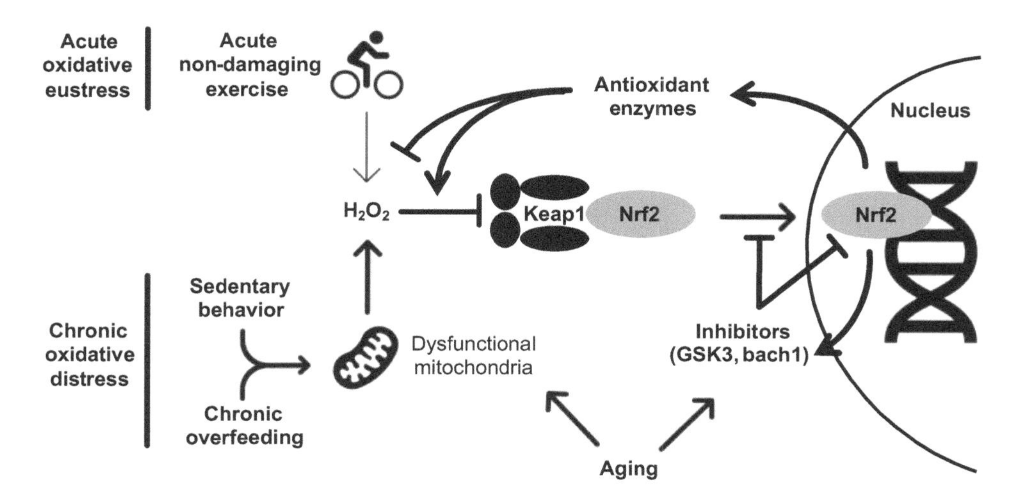

Figure 6.5 Effects of acute and chronic oxidative stimuli on adaptive redox signaling responses mediated by Nrf2. Chronic oxidative distress from sedentary behavior, overfeeding, or aging leads to increase in chronic low levels of ROS production through mitochondria. This activates Nrf2, increasing antioxidant enzyme capacity, but also increasing Nrf2 negative regulators Bach1 and GSK3. Acute exercise generates ROS, but in a system that is already saturated with increased protein thiol oxidation and high levels of negative regulators, the ROS signal from exercise becomes less effective (depicted by thin arrow). Regular exercise training can restore some of the signaling homeostasis but not to the same degree as young individuals. The traditional view that antioxidant enzymes like PRDXs and GSTs act to quench the ROS signal (negative feedback loop) is not supported entirely by new literature in the field. Rather these enzymes may be acting as redox signaling mediators (positive feed-forward loop) transferring oxidizing equivalents from ROS to protein thiols and using glutathionylation as a signaling mediator or protective mechanism in response to the overproduction of ROS. This relationship is depicted by antioxidant enzymes enhancing the effect of hydrogen peroxide on Keap1 ( ) in addition to quenching the signal (---|).

research is needed to know if this redox relay signaling mechanism occurs in response to acute exercise-induced oxidative eustress.

CONCLUSIONS AND FUTURE DIRECTIONS

Oxidative eustress initiates adaptive processes in response to exercise that stem from multiple enzymatic sources in the cell. These processes initiate cell signaling cascades through protein thiol oxidation/reduction reactions that improve cellular metabolism and increase redox stress response capacity. These increases in mitochondrial biogenesis and the antioxidant defense system are governed partly by the inducible redox stress response transcription factor, Nrf2. Increases in Nrf2 activation and downstream target genes occur in response to acute exercise through a combination of ROS sources. The chronic increases in basal or resting levels of mitochondrial ROS from sedentary behavior, overfeeding, and age-related mitochondrial dysfunction increase basal levels of Nrf2, which likely impairs the Nrf2 signaling system from responding further to an added transient oxidative eustress like exercise. This is exacerbated by the feed-forward redox signaling mechanisms elicited by the initial adaptive response to increasing basal ROS production, simultaneously protecting the cell from its own dysfunctional redox circuitry and making it harder for transient oxidative eustress signals like exercise to be effective.

Overall, these concepts – particularly the redox relay signaling mechanisms of the antioxidant defense system – should be tested further in young healthy, aging, and diseased populations, as well as in appropriate model systems. This is an exciting area of research, with many questions still unanswered. The progression of several new assays and measurement techniques to delineate these challenging questions, including immunological techniques to assess protein thiol redox status (Cobley et al., 2019a,b; Cobley and Husi, 2020), is exciting. These methods provide promising orthogonal approaches to redox proteomic techniques and will likely advance the exercise physiology field dramatically when applied with appropriate study designs.

REFERENCES

Aguiar, A. S., Jr., Duzzioni, M., Remor, A. P., Tristao, F. S., Matheus, F. C., Raisman-Vozari, R., Latini, A. & Prediger, R. D. 2016. Moderate-intensity physical exercise protects against experimental 6-hydroxydopamine-induced hemiparkinsonism through Nrf2-antioxidant response element pathway. *Neurochem Res*, 41, 64–72.

Anderson, E. J., Lustig, M. E., Boyle, K. E., Woodlief, T. L., Kane, D. A., Lin, C. T., Price, J. W., 3rd, Kang, L., Rabinovitch, P. S., Szeto, H. H., Houmard, J. A., Cortright, R. N., Wasserman, D. H. & Neufer, P. D. 2009. Mitochondrial H_2O_2 emission and cellular redox state link excess fat intake to insulin resistance in both rodents and humans. *J Clin Invest*, 119, 573–81.

Broome, S. C., Woodhead, J. S. T. & Merry, T. L. 2018. Mitochondria-targeted antioxidants and skeletal muscle function. *Antioxidants (Basel, Switzerland)*, 7, 107.

Campbell, M. D., Duan, J., Samuelson, A. T., Gaffrey, M. J., Merrihew, G. E., Egertson, J. D., Wang, L., Bammler, T. K., Moore, R. J., White, C. C., Kavanagh, T. J., Voss, J. G., Szeto, H. H., Rabinovitch, P. S., Maccoss, M. J., Qian, W. J. & Marcinek, D. J. 2019. Improving mitochondrial function with SS-31 reverses age-related redox stress and improves exercise tolerance in aged mice. *Free Radical Biol Med*, 134, 268–81.

Carvalho, A. N., Marques, C., Guedes, R. C., Castro-Caldas, M., Rodrigues, E., Van Horssen, J. & Gama, M. J. 2016. S-glutathionylation of Keap1: A new role for glutathione S-transferase pi in neuronal protection. *FEBS Lett*, 590, 1455–66.

Cobley, J. N. & Husi, H. 2020. Immunological techniques to assess protein thiol redox state: Opportunities, challenges and solutions. *Antioxidants (Basel, Switzerland)*, 9, 315.

Cobley, J. N., Moult, P. R., Burniston, J. G., Morton, J. P. & Close, G. L. 2015. Exercise improves mitochondrial and redox-regulated stress responses in the elderly: Better late than never! *Biogerontology*, 16, 249–64.

Cobley, J. N., Noble, A., Jimenez-Fernandez, E., Valdivia Moya, M. T., Guille, M. & Husi, H. 2019a. Catalyst-free click pegylation reveals substantial mitochondrial ATP synthase subunit alpha oxidation before and after fertilisation. *Redox Biol*, 26, 101258.

Cobley, J. N., Sakellariou, G. K., Husi, H. & Mcdonagh, B. 2019b. Proteomic strategies to unravel age-related redox signalling defects in skeletal muscle. *Free Radical Biol Med*, 132, 24–32.

Cobley, J. N., Sakellariou, G. K., Owens, D. J., Murray, S., Waldron, S., Gregson, W., Fraser, W. D., Burniston, J. G., Iwanejko, L. A., Mcardle, A., Morton, J. P., Jackson, M. J. & Close, G. L. 2014. Lifelong training preserves some redox-regulated adaptive responses after an acute exercise stimulus in aged human skeletal muscle. *Free Radical Biol Med*, 70, 23–32.

Crilly, M. J., Tryon, L. D., Erlich, A. T. & Hood, D. A. 2016. The role of Nrf2 in skeletal muscle contractile and mitochondrial function. *J Appl Physiol*, 121, 730–40.

Dai, D. F., Chiao, Y. A., Marcinek, D. J., Szeto, H. H. & Rabinovitch, P. S. 2014. Mitochondrial oxidative stress in aging and healthspan. *Longevity Healthspan*, 3, 6.

Done, A. J. & Traustadóttir, T. 2016. Nrf2 mediates redox adaptations to exercise. *Redox Biol*, 10, 191–99.

Done, A. J., Gage, M. J., Nieto, N. C. & Traustadóttir, T. 2016. Exercise-induced Nrf2-signaling is impaired in aging. *Free Radical Biol Med*, 96, 130–8.

Done, A. J., Newell, M. J. & Traustadóttir, T. 2017. Effect of exercise intensity on Nrf2 signalling in young men. *Free Radical Res*, 51, 646–55.

Duan, J., Zhang, T., Gaffrey, M. J., Weitz, K. K., Moore, R. J., Li, X., Xian, M., Thrall, B. D. & Qian, W.-J. 2020. Stochiometric quantification of the thiol redox proteome of macrophages reveals subcellular compartmentalization and susceptibility to oxidative perturbations. *Redox Biol*, 36(5), 101649.

Eggler, A. L., Small, E., Hannink, M. & Mesecar, A. D. 2009. Cul3-mediated Nrf2 ubiquitination and antioxidant response element (ARE) activation are dependent on the partial molar volume at position 151 of Keap1. *Biochem J*, 422, 171–80.

Fisher-Wellman, K. H., Weber, T. M., Cathey, B. L., Brophy, P. M., Gilliam, L. A., Kane, C. L., Maples, J. M., Gavin, T. P., Houmard, J. A. & Neufer, P. D. 2014. Mitochondrial respiratory capacity and content are normal in young insulin-resistant obese humans. *Diabetes*, 63, 132–41.

Fourquet, S., Guerois, R., Biard, D. & Toledano, M. B. 2010. Activation of Nrf2 by nitrosative agents and H_2O_2 involves Keap1 disulfide formation. *J Biol Chem*, 285, 8463–71.

Gallego-Selles, A., Martin-Rincon, M., Martinez-Canton, M., Perez-Valera, M., Martin-Rodriguez, S., Gelabert-Rebato, M., Santana, A., Morales-Alamo, D., Dorado, C. & Calbet, J. A. L. 2020. Regulation of Nrf2/Keap1 signalling in human skeletal muscle during exercise to exhaustion in normoxia, severe acute hypoxia and post-exercise ischaemia: Influence of metabolite accumulation and oxygenation. *Redox Biol*, 36, 101627.

Gill, R. M., O'brien, M., Young, A., Gardiner, D. & Mailloux, R. J. 2018. Protein S-glutathionylation lowers superoxide/hydrogen peroxide release from skeletal muscle mitochondria through modification of complex I and inhibition of pyruvate uptake. *PLoS One*, 13, e0192801.

Gomez-Cabrera, M. C., Borras, C., Pallardo, F. V., Sastre, J., Ji, L. L. & Vina, J. 2005. Decreasing xanthine oxidase-mediated oxidative stress prevents useful cellular adaptations to exercise in rats. *J Physiol*, 567, 113–20.

Gomez-Cabrera, M. C., Domenech, E., Romagnoli, M., Arduini, A., Borras, C., Pallardo, F. V., Sastre, J. & Vina, J. 2008. Oral administration of vitamin C decreases muscle mitochondrial biogenesis and hampers training-induced adaptations in endurance performance. *Am J Clin Nutr*, 87, 142–9.

Goncalves, R. L., Quinlan, C. L., Perevoshchikova, I. V., Hey-Mogensen, M. & Brand, M. D. 2015. Sites of superoxide and hydrogen peroxide production by muscle mitochondria assessed ex vivo under conditions mimicking rest and exercise. *J Biol Chem*, 290, 209–27.

Goncalves, R. L. S., Watson, M. A., Wong, H.-S., Orr, A. L. & Brand, M. D. 2019. The use of site-specific suppressors to measure the relative contributions of different mitochondrial sites to skeletal muscle superoxide and hydrogen peroxide production. *Redox Biol*, 28, 101341.

Gounder, S. S., Kannan, S., Devadoss, D., Miller, C. J., Whitehead, K. J., Odelberg, S. J., Firpo, M. A., Paine, R., 3rd, Hoidal, J. R., Abel, E. D. & Rajasekaran, N. S. 2012. Impaired transcriptional activity of Nrf2 in age-related myocardial oxidative stress is reversible by moderate exercise training. *PLoS One*, 7, e45697.

Hanzen, S., Vielfort, K., Yang, J., Roger, F., Andersson, V., Zamarbide-Fores, S., Andersson, R., Malm, L., Palais, G., Biteau, B., Liu, B., Toledano, M. B., Molin, M. & Nystrom, T. 2016. Lifespan control by redox-dependent recruitment of chaperones to misfolded proteins. *Cell*, 166, 140–51.

Hayes, J. D. & Dinkova-Kostova, A. T. 2014. The Nrf2 regulatory network provides an interface between redox and intermediary metabolism. *Trends Biochem Sci*, 39, 199–218.

Henriquez-Olguin, C., Diaz-Vegas, A., Utreras-Mendoza, Y., Campos, C., Arias-Calderon, M., Llanos, P., Contreras-Ferrat, A., Espinosa, A., Altamirano, F., Jaimovich, E. & Valladares, D. M. 2016. NOX2 inhibition impairs early muscle gene expression induced by a single exercise bout. *Front Physiol*, 7, 282.

Henriquez-Olguin, C., Knudsen, J. R., Raun, S. H., Li, Z., Dalbram, E., Treebak, J. T., Sylow, L., Holmdahl, R., Richter, E. A., Jaimovich, E. & Jensen, T. E. 2019a. Cytosolic ros production by NADPH oxidase 2 regulates muscle glucose uptake during exercise. *Nat Commun*, 10, 4623.

Henriquez-Olguin, C., Renani, L. B., Arab-Ceschia, L., Raun, S. H., Bhatia, A., Li, Z., Knudsen, J. R., Holmdahl, R. & Jensen, T. E. 2019b. Adaptations to high-intensity interval training in skeletal muscle require NADPH oxidase 2. *Redox Biol*, 24, 101188.

Henriquez-Olguin, C., Meneses-Valdes, R. & Jensen, T. E. 2020. Compartmentalized muscle redox signals controlling exercise metabolism: Current state, future challenges. *Redox Biol*, 35, 101473.

Hu, C., Eggler, A. L., Mesecar, A. D. & Van Breemen, R. B. 2011. Modification of keap1 cysteine residues by sulforaphane. *Chem Res Toxicol*, 24, 515–21.

Irrcher, I., Ljubicic, V. & Hood, D. A. 2009. Interactions between ROS and AMP kinase activity in the regulation of PGC-1α transcription in skeletal muscle cells. *Am J Physiol-Cell Physiol*, 296, C116–23.

Jackson, M. J. 2020. On the mechanisms underlying attenuated redox responses to exercise in older individuals: A hypothesis. *Free Radical Biol Med*, 161, 326–38.

Jarvis, R. M., Hughes, S. M. & Ledgerwood, E. C. 2012. Peroxiredoxin 1 functions as a signal peroxidase to receive, transduce, and transmit peroxide signals in mammalian cells. *Free Radical Biol Med*, 53, 1522–30.

Kramer, P. A., Duan, J., Qian, W.-J. & Marcinek, D. J. 2015. The measurement of reversible redox dependent post-translational modifications and their regulation of mitochondrial and skeletal muscle function. *Front Physiol*, 6, 347.

Kramer, P. A., Duan, J., Gaffrey, M. J., Shukla, A. K., Wang, L., Bammler, T. K., Qian, W. J. & Marcinek, D. J. 2018. Fatiguing contractions increase protein S-glutathionylation occupancy in mouse skeletal muscle. *Redox Biol*, 17, 367–76.

Kubo, H., Asai, K., Kojima, K., Sugitani, A., Kyomoto, Y., Okamoto, A., Yamada, K., Ijiri, N., Watanabe, T., Hirata, K. & Kawaguchi, T. 2019. Exercise ameliorates emphysema of cigarette smoke-induced copd in mice through the exercise-irisin-Nrf2 axis. *Int J Chron Obstruct Pulmon Dis*, 14, 2507–16.

Li, T., He, S., Liu, S., Kong, Z., Wang, J. & Zhang, Y. 2015. Effects of different exercise durations on Keap1-Nrf2-are pathway activation in mouse skeletal muscle. *Free Radical Res*, 49, 1269–74.

Mailloux, R. J. 2020. Protein S-glutathionylation reactions as a global inhibitor of cell metabolism for the desensitization of hydrogen peroxide signals. *Redox Biol*, 32, 101472.

Manford, A. G., Rodriguez-Perez, F., Shih, K. Y., Shi, Z., Berdan, C. A., Choe, M., Titov, D. V., Nomura, D. K. & Rape, M. 2020. A cellular mechanism to detect and alleviate reductive stress. *Cell*, 183, 46–61 e21.

Mansouri, A., Muller, F. L., Liu, Y., Ng, R., Faulkner, J., Hamilton, M., Richardson, A., Huang, T.-T., Epstein, C. J. & Van Remmen, H. 2006. Alterations in mitochondrial function, hydrogen peroxide release and oxidative damage in mouse hind-limb skeletal muscle during aging. *Mech Ageing Dev*, 127, 298–306.

Margaritelis, N. V., Kyparos, A., Paschalis, V., Theodorou, A. A., Panayiotou, G., Zafeiridis, A., Dipla, K., Nikolaidis, M. G. & Vrabas, I. S. 2014. Reductive stress after exercise: The issue of redox individuality. *Redox Biol*, 2, 520–8.

Margaritelis, N. V., Theodorou, A. A., Paschalis, V., Veskoukis, A. S., Dipla, K., Zafeiridis, A., Panayiotou, G., Vrabas, I. S., Kyparos, A. & Nikolaidis, M. G. 2018. Adaptations to endurance training depend on exercise-induced oxidative stress: exploiting redox interindividual variability. *Acta Physiol (Oxford)*, 222.

Margaritelis, N. V., Paschalis, V., Theodorou, A. A., Kyparos, A. & Nikolaidis, M. G. 2020. Antioxidant supplementation, redox deficiencies and exercise performance: A falsification design. *Free Radical Biol Med.*, 158, 44–52.

Marinho, H. S., Real, C., Cyrne, L., Soares, H. & Antunes, F. 2014. Hydrogen peroxide sensing, signaling and regulation of transcription factors. *Redox Biol*, 2, 535–62.

Mcdonagh, B., Sakellariou, G. K., Smith, N. T., Brownridge, P. & Jackson, M. J. 2014. Differential cysteine labeling and global label-free proteomics reveals an altered metabolic state in skeletal muscle aging. *J Proteome Res*, 13, 5008–21.

Merry, T. L. & Ristow, M. 2016. Nuclear factor erythroid-derived 2-like 2 (Nfe2L2, Nrf2) mediates exercise-induced mitochondrial biogenesis and the anti-oxidant response in mice. *J Physiol*, 594, 5195–207.

Michailidis, Y., Jamurtas, A. Z., Nikolaidis, M. G., Fatouros, I. G., Koutedakis, Y., Papassotiriou, I. & Kouretas, D. 2007. Sampling time is crucial for measurement of aerobic exercise-induced oxidative stress. *Med Sci Sports Exerc*, 39, 1107–13.

Moghaddam, D. A., Heber, A., Capin, D., Kreutz, T., Opitz, D., Lenzen, E., Bloch, W., Brixius, K. & Brinkmann, C. 2011. Training increases peroxiredoxin 2 contents in the erythrocytes of overweight/obese men suffering from type 2 diabetes. *Wien Med Wochenschr*, 161, 511–8.

Musci, R. V., Hamilton, K. L. & Linden, M. A. 2019. Exercise-induced mitohormesis for the maintenance of skeletal muscle and healthspan extension. *Sports (Basel)*, 7(7), 170.

Muthusamy, V. R., Kannan, S., Sadhaasivam, K., Gounder, S. S., Davidson, C. J., Boeheme, C., Hoidal, J. R., Wang, L. & Rajasekaran, N. S. 2012. Acute exercise stress activates Nrf2/ARE signaling and promotes antioxidant mechanisms in the myocardium. *Free Radical Biol Med*, 52, 366–76.

Nikolaidis, M. G., Margaritelis, N. V. & Matsakas, A. 2020. Quantitative redox biology of exercise. *Int J Sports Med.*, 41(10), 633–645.

Ostrom, E. L. & Traustadóttir, T. 2020. Aerobic exercise training partially reverses the impairment of Nrf2 activation in older humans. *Free Radical Biol Med*, 160, 418–32.

Ostrom, E. L., Valencia, A. P., Marcinek, D. J. & Traustadóttir, T. 2021. High intensity muscle stimulation activates a systemic Nrf2-mediated redox stress response. *Free Radical Biol Med*, 172, 82–9.

Parker, L., Shaw, C. S., Stepto, N. K. & Levinger, I. 2017. Exercise and glycemic control: Focus on redox homeostasis and redox-sensitive protein signaling. *Front Endocrinol*, 8, 87.

Perry, C. G., Lally, J., Holloway, G. P., Heigenhauser, G. J., Bonen, A. & Spriet, L. L. 2010. Repeated transient MRNA bursts precede increases in transcriptional and mitochondrial proteins during training in human skeletal muscle. *J Physiol*, 588, 4795–810.

Piantadosi, C. A. & Suliman, H. B. 2006. Mitochondrial transcription factor A induction by redox activation of nuclear respiratory factor 1. *J Biol Chem*, 281, 324–33.

Pomatto, L. C. D. & Davies, K. J. A. 2017. The role of declining adaptive homeostasis in ageing. *J Physiol*, 595, 7275–309.

Pomatto, L. C. D. & Davies, K. J. A. 2018. Adaptive homeostasis and the free radical theory of ageing. *Free Radical Biol Med*, 124, 420–40.

Radak, Z., Ishihara, K., Tekus, E., Varga, C., Posa, A., Balogh, L., Boldogh, I. & Koltai, E. 2017. Exercise, oxidants, and antioxidants change the shape of the bell-shaped hormesis curve. *Redox Biol*, 12, 285–90.

Ristow, M., Zarse, K., Oberbach, A., Kloting, N., Birringer, M., Kiehntopf, M., Stumvoll, M., Kahn, C. R. & Bluher, M. 2009. Antioxidants prevent health-promoting effects of physical exercise in humans. *Proc Natl Acad Sci U S A*, 106, 8665–70.

Sakellariou, G. K., Jackson, M. J. & Vasilaki, A. 2014. Redefining the major contributors to superoxide production in contracting skeletal muscle. The role of Nad(P)H oxidases. *Free Radical Res*, 48, 12–29.

Sobotta, M. C., Liou, W., Stöcker, S., Talwar, D., Oehler, M., Ruppert, T., Scharf, A. N. & Dick, T. P. 2015. Peroxiredoxin-2 and STAT3 form a redox relay for H_2O_2 signaling. *Nat Chem Biol*, 11, 64–70.

Stocker, S., Van Laer, K., Mijuskovic, A. & Dick, T. P. 2018. The conundrum of hydrogen peroxide signaling and the emerging role of peroxiredoxins as redox relay hubs. *Antioxid Redox Signal*, 28, 558–73.

Suh, J. H., Shenvi, S. V., Dixon, B. M., Liu, H., Jaiswal, A. K., Liu, R. M. & Hagen, T. M. 2004. Decline in transcriptional activity of Nrf2 causes age-related loss of glutathione synthesis, which is reversible with lipoic acid. *Proc Natl Acad Sci U S A*, 101, 3381–6.

Tebay, L. E., Robertson, H., Durant, S. T., Vitale, S. R., Penning, T. M., Dinkova-Kostova, A. T. & Hayes, J. D. 2015. Mechanisms of activation of the transcription factor Nrf2 by redox stressors, nutrient cues, and energy status and the pathways through which it attenuates degenerative disease. *Free Radical Biol Med*, 88, 108–46.

Theodorou, A. A., Paschalis, V., Kyparos, A., Panayiotou, G. & Nikolaidis, M. G. 2014. Passive smoking reduces and vitamin C increases exercise-induced oxidative stress: Does this make passive smoking an anti-oxidant and vitamin C a pro-oxidant stimulus? *Biochem Biophys Res Commun*, 454, 131–6.

Tryfidou, D. V., Mcclean, C., Nikolaidis, M. G. & Davison, G. W. 2020. DNA Damage Following Acute Aerobic Exercise: A Systematic Review and Meta-analysis. *Sports Med*, 50, 103–127.

Tsou, Y. H., Shih, C. T., Ching, C. H., Huang, J. Y., Jen, C. J., Yu, L., Kuo, Y. M., Wu, F. S. & Chuang, J. I. 2015. Treadmill exercise activates Nrf2 antioxidant system to protect the nigrostriatal dopaminergic neurons from MPP+ toxicity. *Exp Neurol*, 263, 50–62.

Vasilaki, A., Mansouri, A., Van Remmen, H., Van Der Meulen, J. H., Larkin, L., Richardson, A. G., Mcardle, A., Faulkner, J. A. & Jackson, M. J. 2006. Free radical generation by skeletal muscle of adult and old mice: Effect of contractile activity. *Aging Cell*, 5, 109–17.

Viña, J., Borras, C., Abdelaziz, K. M., Garcia-Valles, R. & Gomez-Cabrera, M. C. 2013. The free radical theory of aging revisited: The cell signaling disruption theory of aging. *Antioxid Redox Signaling*, 19, 779–87.

Wadley, A. J., Aldred, S. & Coles, S. J. 2016. An unexplored role for peroxiredoxin in exercise-induced redox signalling? *Redox Biol*, 8, 51–8.

Wadley, A. J., Chen, Y. W., Bennett, S. J., Lip, G. Y., Turner, J. E., Fisher, J. P. & Aldred, S. 2015. Monitoring changes in thioredoxin and over-oxidised peroxiredoxin in response to exercise in humans. *Free Radical Res*, 49, 290–8.

Wang, P., Li, C. G., Qi, Z., Cui, D. & Ding, S. 2015. Acute exercise induced mitochondrial H_2O_2 production in mouse skeletal muscle: Association with p66Shc and Foxo3a signaling and antioxidant enzymes. *Oxid Med Cell Longevity*, 2015, 10.

Williamson, J., Hughes, C. M., Cobley, J. N. & Davison, G. W. 2020. The mitochondria-targeted antioxidant MitoQ, attenuates exercise-induced mitochondrial DNA damage. *Redox Biol*, 36, 101673.

Wood, Z. A., Poole, L. B. & Karplus, P. A. 2003. Peroxiredoxin evolution and the regulation of hydrogen peroxide signaling. *Science*, 300, 650–3.

Xiao, H., Jedrychowski, M. P., Schweppe, D. K., Huttlin, E. L., Yu, Q., Heppner, D. E., Li, J., Long, J., Mills, E. L., Szpyt, J., He, Z., Du, G., Garrity, R., Reddy, A., Vaites, L. P., Paulo, J. A., Zhang, T., Gray, N. S., Gygi, S. P. & Chouchani, E. T. 2020. A Quantitative Tissue-Specific Landscape of Protein Redox Regulation during Aging. *Cell*, 180, 968–983 e24.

Zhang, D. D. & Hannink, M. 2003. Distinct cysteine residues in Keap1 are required for Keap1-dependent ubiquitination of Nrf2 and for stabilization of Nrf2 by chemopreventive agents and oxidative stress. *Mol Cell Biol*, 23, 8137–51.

CHAPTER SEVEN

Time to 'Couple' Redox Biology with Exercise Immunology

Alex J. Wadley and Steven J. Coles

CONTENTS

Introduction / 85
Redox Reactions and Immunity / 86
Global Oxidation in Immune Cells after Single Bouts of Exercise / 86
Evaluating Immune Cell Thiol Redox State after Exercise / 88
Single Cell Approaches / 89
Future Perspectives / 89
  Appreciation for Oxidative Eustress / 89
  Immunometabolism / 90
  Extracellular Environment / 90
Conclusion / 91
References / 91

INTRODUCTION

Exercise immunology is a continually expanding discipline, with relevance to both sport performance (Walsh, 2018) and the management of chronic inflammatory disease (Gleeson et al., 2011). Regular bouts of moderate-to-vigorous intensity exercise confer protection against common upper respiratory tract infections (e.g., rhinovirus and influenza) (Matthews et al., 2002; Nieman et al., 2011), as well as offsetting the development of multiple immune-related chronic diseases (Peake et al., 2017; Campbell and Turner, 2018; Simpson et al., 2020). The benefits of regular exercise relate to changes in both body composition (i.e., reduced central fat and/or increased skeletal muscle mass) and a steady summation of changes to the immune system after each individual session (Gleeson et al., 2011). The immune system is exquisitely sensitive to physical activity. Physiological changes in cardiac output, blood flow and shear stress, as well as biochemical changes in catecholamine (e.g., adrenaline) (Krüger et al., 2008; Graff et al., 2018) and cytokine concentrations (Bay et al., 2020), are proposed mechanisms mediating the preferential mobilisation of highly functional immune cells into peripheral blood during exercise. This non-uniform response appears to be highly coordinated and is thought to prime the immune system to survey the body for tissue damage, infection and malignant transformation after each bout (Rooney et al., 2018). The mechanisms underlying these events are continually being established. Notably, recent evidence has highlighted that shifts in the redox environment within immune cells influence immune cell trafficking after exercise (Petersen et al., 2012; Michailidis et al., 2013; Sakelliou et al., 2016; Wadley et al., 2018a,b; Spanidis et al., 2018).

DOI: 10.1201/9781003051619-7

The fields of exercise immunology and redox biology have been regularly intertwined over the last 30 years, primarily in the context of single bouts of exhaustive exercise or periods of high-volume training (collectively termed 'arduous exercise'). A recurring and contentious topic in the field of exercise immunology is that arduous exercise might suppress immune function (Simpson et al., 2020). Comparably, a notion in the field of redox biology posits that excessive reactive oxygen and nitrogen species (RONS) production after arduous exercise elicits *oxidative stress*, which has been associated with skeletal muscle fatigue, pain and delayed recovery (Powers, Talbert and Adhihetty, 2011). The aforementioned relationship is therefore highly intuitive. A body of evidence supports a role of redox reactions in governing multiple aspects of innate and adaptive immunity (Sies and Jones, 2020); however, the marked physiological changes that exercise causes to the immune system, coupled with the multiple challenges in quantitatively investigating cellular redox state after exercise (Cobley et al., 2017; Nikolaidis, Margaritelis and Matsakas, 2020), make redox exercise immunology and challenging field of investigation.

This chapter will summarise the body of literature on the topic of redox exercise immunology and suggest how modern technological advances in biomedical science offer scope to delineate the molecular pathways underpinning redox-driven changes in immunity after exercise. The concept of *oxidative eustress* will be emphasised following recent evidence from our laboratory highlighting that exhaustive exercise can elicit *reductive stress* in cytotoxic T cells. Given the unique metabolic demands of individual subsets of immune cells, including cytotoxic T cells, the field of *immunometabolism* will be briefly discussed as a topic for future investigation. By integrating these topics, major advances are permissible in the field of exercise physiology to benefit individuals across the spectrum of human health.

REDOX REACTIONS AND IMMUNITY

Redox reactions play a central role in modulating aspects of both innate and adaptive immunity (Nathan and Cunningham-Bussel, 2013; Pei and Wallace, 2018). RONS like hydrogen peroxide recruit phagocytes (e.g., neutrophils) to the site of infection, which then produce large amounts of superoxide (via NADPH oxidases), nitric oxide (via nitric oxide synthases) and hypochlorous acid (via myeloperoxidase) that can destroy foreign pathogens (Winterbourn, Kettle and Hampton, 2016). These RONS also support the release of extracellular traps, which further facilitate this process (Brinkmann et al., 2004; Kenny et al., 2017). Redox reactions have also been implicated in humoral immunity, with disulphide bond formation in the endoplasmic reticulum of plasma cells facilitating the correct folding of immunoglobulins prior to release (Anelli, Sannino and Sitia, 2015) and the process of wound healing, by remodelling the extracellular matrix to seal a wound (Cano Sanchez et al., 2018). Interestingly, there is evidence that immune processes can be impaired by the blocking (Lévigne et al., 2016) and overproduction of RONS. This highlights the critical balance between the production and scavenging of RONS in governing immune function. In this respect, there is evidence to support a role for glutathione and redox-sensitive proteins (e.g., redox enzymes and transcriptions factors) in governing the proliferation, survival and function of various immune cells (e.g., T cells, B cells and macrophages) (Sies and Jones, 2020).

There is data to support a beneficial role of single and regular bouts of exercise in modulating immunological processes, such as wound healing (Emery et al., 2005) and antibody production (Eskola et al., 1978), which are redox-regulated pathways as highlighted above. The next sections will evaluate the research investigating the effects of exercise on the redox status of immune cells, with a particular focus on the effects of single bouts of exercise.

GLOBAL OXIDATION IN IMMUNE CELLS AFTER SINGLE BOUTS OF EXERCISE

Single bouts of exercise have been shown to increase indices of *oxidative stress* in leukocytes isolated from whole blood after exercise cessation. For example, it has been reported that protein carbonylation and lipid peroxidation (i.e., broad global surrogates of oxidative damage) were higher in peripheral blood mononuclear cells (PBMCs) and neutrophils, respectively, after prolonged cycling (Sureda et al., 2005; Tauler et al., 2006). Furthermore, DNA strand breaks have been reported to be higher in PBMCs after brief exhaustive exercise (Peres et al., 2020). Separate studies investigating various antioxidant enzymes (i.e., superoxide dismutase, glutathione peroxidase and peroxiredoxin) have indicated that their activity and protein expression also increase following different modes and intensities of exercise (Tauler et al., 2006; Ferrer et al., 2009; Turner et al., 2011, 2013; Wadley et al., 2015a,b). These studies collectively indicate that proteins, lipids and DNA within bloodborne innate and adaptive immune cells are sensitive to oxidation after exercise, with a concurrent antioxidant enzyme response. However, when drawing comparisons between blood samples taken before, immediately and in the hours following exercise (Figure 7.1a and b), it is important to appreciate the vast heterogeneity of immune cells within blood, since the effects of oxidation may vary depending on the immune cell type.

Bouts of exercise evoke profound changes in the number and composition of immune cells within the bloodstream. During exercise, natural killer cells and lymphocytes with high functional capacity (i.e., high cytotoxicity and tissue migration potential) are preferentially drawn into the circulation (Figure 7.1a and b) through dislodgment from the vascular endothelium and recruitment from marginal pools (Gustafson et al., 2017). Depending on the type of exercise (i.e., intensity, duration), these immune cells can then leave the circulation ≈2–3 hours following cessation by migrating into various tissues (e.g., gut, lungs), thus lowering the number, and again, altering the composition of immune cells in the blood (Krüger et al., 2008). Although the aforementioned studies indicate that the immune cell compartment of blood is more oxidised after exercise, they don't

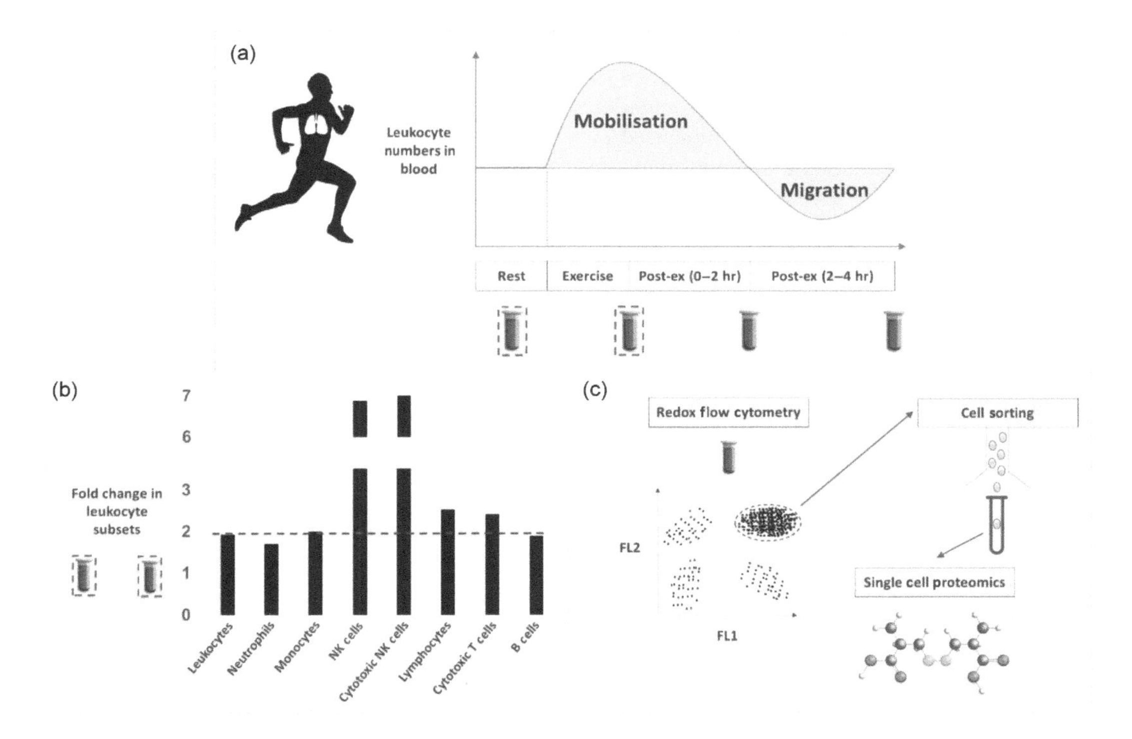

Figure 7.1 Reductionist approach to conduct single-cell analysis of immune cell redox state after exercise in humans. (a) Exercise evokes powerful physiological changes on the number of immune cells in peripheral blood. Notably, leukocyte number increases during and after moderate-to-vigorous exercise (mobilisation) and then falls below resting levels in the hours after exercise cessation (migration). (b) Graphical figure indicating how exercise-induced leucocytosis is not a uniform response. Fold change in blood cell concentrations immediately after a high-intensity bout of cycling (20 minutes at 80% VO_2 max) relative to resting values has been extracted from three published articles (variance not included due to graphical representation) (Wadley et al., 2015a,b). NK: Natural Killer (c) By employing suitable cell sorting methods, single-cell proteomics can be used to determine exercise-driven redox modifications and their relevance to immune function.

account for the well-documented compositional changes in immune cells resulting from exercise, or identify how perturbations in the redox state of immune cells can functionally impact different aspects of cellular immunity (Figure 7.1).

EVALUATING IMMUNE CELL THIOL REDOX STATE AFTER EXERCISE

It is now well established that 2-electron oxidants, such as hydrogen peroxide (H_2O_2) and peroxynitrite ($ONOO^-$), mediate a broad range of cellular processes (Ferreira and Reid, 2008; Reid, 2008; Lamb and Westerblad, 2011). These processes occur through reversible oxidation of redox-sensitive amino acids, such as tyrosine, methionine and cysteine. The latter has received a lot of attention over the last 10 years, with reversible cysteine oxidation implicated with a broad range of cellular processes, notably metabolism (Kolossov et al., 2015) and signal transduction (Sobotta et al., 2014). Cysteine contains a terminal sulphhydryl (–SH) or 'thiol' group that is highly electronegative in nature. Cysteine oxidation is dependent on the pKa of the thiol, which itself is influenced by the amino acid sidechain microenvironment in which that 'particular' thiol resides. This means that for some proteins, solvent accessible cysteine thiols are 'redox active', conferring an ability to partake in cellular redox reactions, which may include the reduction of oxidising agents such as H_2O_2 and $ONOO^-$ (Flohé et al., 2011; Daiber et al., 2013; Poole, 2015). Notably, reversible oxidation of cysteine thiols has been implicated with signal transduction in human skeletal muscle during exercise. Redox enzymes such as peroxiredoxin (PRDX) are particularly important in this context and are thought to act as transducers of oxidant signals ('redox relays') due to their high turnover rate of cellular H_2O_2 (Winterbourn, 2013, 2015; Wadley, Aldred and Coles, 2016). Recent data has also demonstrated that PRDX thiols can be oxidised in PBMCs after different types of exercise (Turner et al., 2013; Wadley et al., 2015a,b).

The role of cysteine thiols in regulating immune responses to exercise is a recent area of investigation, with emerging data indicating that exercise might cause *reductive stress* in immune cells after exercise (Wadley et al., 2018a,b; Spanidis et al., 2018). A series of dietary intervention studies have indicated that supplementation with the thiol donor, N-acetylcysteine (NAC) can blunt immune cell mobilisation patterns in response to both muscle-damaging (Michailidis et al., 2013; Sakelliou et al., 2016) and exhaustive aerobic exercise (Petersen et al., 2012). Although the exact tissue uptake kinetics are unclear, this data suggests that alterations in the thiol redox state of immune cells or other tissues (i.e., increased reductive capacity) may regulate immunity after exercise. Spanidis et al. (2018) stratified participants into 'oxidative' and 'reductive' groups based on their individual changes from rest in erythrocyte reduced glutathione (GSH) concentration following a muscle-damaging protocol (i.e., increase in GSH was termed 'reductive') (Spanidis et al., 2018). PBMCs isolated at the same timepoints from the 'oxidative' participants were more sensitive to in vitro oxidation and had lower catalase activity compared to 'reduced' participants. This suggests that the reductive capacity of PBMCs could be important for governing their function, although the exact nature is unclear. Furthermore, both the above studies did not account for compositional shifts in immune cell populations after exercise.

Recently, our group validated a novel method to investigate changes in the thiol redox state of different T cell populations using digital flow cytometry (Wadley et al., 2018a,b). By incorporating the thiol specific probe, fluorescein-5 maleimide (F5M) into routine flow cytometry panels (excitation: λ488 nm; emission: λ525 nm) (Wadley et al., 2018a,b, 2019b), we were able to label and identify intracellular proteins with solvent accessible reduced cysteines in distinct human immune cell populations before and after exercise. Within each cell population, a loss of F5M signal indicated thiol oxidation, whereas a gain in F5M signal indicated thiol reduction. By exploring changes in T cell populations in response to a cycling ramp test to exhaustion (N = 20), we reported that T-cytotoxic ($CD8^+$) lymphocyte thiols became transiently *reductive* after exercise to a greater extent than T-helper ($CD4^+$) lymphocytes, both cell types indicating a state of *reductive stress*. We also identified a marked increase in the concentration of $CD8^+$ T cells with highly reduced thiols in post-exercise blood (termed $CD8^+$ *Reduced*$^+$). Flow cytometric exploration of the $CD8^+$ *Reduced*$^+$ cell population showed less expression of the lymphoid homing receptor, C–C Chemokine Receptor-7 (CCR7) compared to $CD8^+$ *Reduced*$^-$, an immunophenotype suggestive of tissue migratory properties (Sallusto et al., 1999). With our expanding knowledge of how specific immune populations with rapid effector functions are preferentially mobilised into the circulation during exercise, these data suggest that a more *reduced* cellular redox state could, in part, mediate immune cell mobilisation and possibly extravasation. The appearance of *CD8$^+$ Reduced$^+$* in blood was very transient, falling back to resting concentrations within 15 minutes, which is difficult to interpret without further investigation. The response could signify a subtle, but transient deviation from *oxidative eustress* that elicits CD8$^+$ T cell functions or the rapid migration of *CD8$^+$ Reduced$^+$* from the circulation after exercise. It must be noted that CD8$^+$ T cells are also a heterogeneous population of lymphocytes, composed of naïve, memory and effector T cells. These cells all have differing basal redox states (Turner et al., 2011) and metabolic phenotypes (discussed later) (Fox, Hammerman and Thompson, 2005; van der Windt and Pearce, 2012). Nonetheless, this method provides a platform for future studies to interrogate the redox environment within cells of the immune system after exercise to provide important mechanistic insight. This could involve staining for different immune cell populations (e.g., natural killer cells), other cysteine modifications (e.g., sulfenic acid) or other redox-sensitive amino acids (e.g., methionine and tyrosine).

SINGLE CELL APPROACHES

Multiparameter flow cytometry is a well-established method for evaluating immunophenotypic and functional changes in PBMCs. Workflow protocols that incorporate standard immunophenotyping with intracellular cytokine analysis (e.g., Interleukin (IL)-2, IL-4, IL-10 and Interferon Gamma) have successfully identified transient changes in T cell function in response to different exercise intensities (LaVoy et al., 2017). A similar approach has been used to evaluate natural killer and CD8$^+$ T cell functional activity in response to various exercise types. Here, lysosome-associated membrane protein-1 (CD107a) on natural killer and CD8$^+$ T cells, as a marker cellular degranulation, displayed differential expression levels depending on exercise frequency and intensity (Broadbent and Coutts, 2017). Our approach uses multiparameter flow cytometry to evaluate the intracellular redox environment in immunophenotypically distinct T cell populations (redox flow cytometry, see Figure 7.1c) (Wadley et al., 2019b). However, the experimental limit has been reached using conventional flow cytometry. As such cell immunophenotyping, functional and intracellular redox environment analysis have yet to be united in a single experimental platform.

The development of CyTOF™ mass cytometry may be a solution to current experimental limitations. This technology has permitted high-resolution assessment of natural killer/T cell immunophenotype and functional analysis at the single-cell level (Kay, Strauss-Albee and Blish, 2016; Brodie and Tosevski, 2018). Like traditional flow cytometry, CyTOF™ mass cytometry utilises specific antibodies (labelled with metals in this instance) to detect the expression of various cellular antigens (Heck, Bishop and Ellis, 2019). Because the technology is based on time-of-flight mass spectrometry as a detection method, the limitations/complications associated with standard flow cytometry (e.g., spectral overlap) are ameliorated. Therefore, upwards of 50 different immunophenotypic and(or) functional molecules for a single cell may be detected simultaneously. So far protocols have been established to concurrently evaluate cytokines, transcription factors and immunophenotypic markers (Lin, Gupta and Maecker, 2015; Simoni et al., 2018). In theory, the analysis of protein/peptide markers of the intracellular redox environment (e.g., peroxiredoxin, thioredoxin, glutaredoxin, glutathione) as well as redox-sensitive transcription factors implicated with *oxidative* and *reductive stress* (Bellezza et al., 2018) could easily be incorporated into CyTOF™ mass cytometry workflows.

FUTURE PERSPECTIVES

Appreciation for Oxidative Eustress

The evidence in this chapter highlights the need for further investigations in the field of redox exercise immunology. Dysregulation of the immune system is central to many chronic human diseases, with the role of *oxidative distress* now widely characterised (Nathan and Cunningham-Bussel, 2013; Sies and Jones, 2020). The role of *reductive stress* is less clear, but some recent evidence implicates *reductive stress* with enhanced T cell autoimmunity in rheumatoid arthritis (Weyand, Shen and Goronzy, 2018). Subtle or acute *reductive* shifts in immune cells are even less well characterised, with our findings following a single bout of exercise the first to our knowledge. New techniques, as well as an appreciation for the scientific disciplines highlighted below, could shed light on the functional significance of these results. It is important to highlight that acute shifts in both *oxidative* and *reductive* redox state may be important for changes in immunity after exercise.

Immunometabolism

Immunometabolism is a rapidly expanding field in immunological research, with rewiring of cellular metabolism now linked with modulating multiple immune processes (Dimeloe et al., 2017). For example, T cells become highly reliant on glycolysis upon activation to support their effector functions (Fox, Hammerman and Thompson, 2005; van der Windt and Pearce, 2012). Interestingly, emerging data indicate that shifts in cellular redox state also support this metabolic reprogramming (Muri and Kopf, 2020). Indeed, T cell activation, which involves antigen presentation to the T cell receptor (TCR), is enhanced and sustained by mitochondrial and NADPH oxidase-derived ROS (Jackson et al., 2004; Kamiński et al., 2012). Following TCR signalling, CD28 ligation is the costimulatory signal needed to evoke the pronounced glycolytic shift. This is sustained through rapid glucose transport (GLUT1) and *reduction* of accumulating pyruvate to lactate, which maintains the cellular NAD^+: NADH ratio. Despite a lower ATP yield per molecule of glucose, aerobic glycolysis drives more rapid metabolism of glucose than mitochondrial oxidation. The metabolic shift also supports the provision of biosynthetic precursors (nucleotides, amino acids and fatty acids) via the pentose phosphate pathway (PPP) that sustain the formation of effector molecules for T cell functions (Vander Heiden, Cantley and Thompson, 2009; Macintyre and Rathmell, 2013). The PPP enhances nicotinamide adenine dinucleotide phosphate (NADPH) supply (van der Windt and Pearce, 2012), the crucial cofactor needed to provide reducing equivalent for multiple cellular antioxidant enzymes (e.g., peroxiredoxin, thioredoxin, glutaredoxin) and the abundant tripeptide, glutathione. This activation-induced shift in T cell metabolism, therefore, provides reductive capacity for these cells to modulate cellular RONS and thus maintain redox homeostasis and functional capacity (Ma et al., 2018). Interestingly, this evidence indirectly supports our findings of more reductive $CD8^+$ T cells in the circulation after exercise (*$CD8^+$ Reduced$^+$*) (Wadley et al., 2018a,b). Given that we know that exercise evokes a marked and preferential increase in immune cells that are highly primed for their effector functions, we can intuitively suggest that these cells would have a more glycolytic phenotype. Future work should intertwine the fields of immunology, metabolism and redox biochemistry to further understand the mobilisation of specific immune cells after exercise.

Extracellular Environment

This chapter has exclusively focused on changes in redox state of immune cells on a cell-by-cell basis, but it is important to consider how the extracellular redox environment might be important in regulating immunity after exercise. Some cytosolic redox enzymes are released via non-classical pathways, associated with extracellular vesicles (EVs), such as exosomes and nanoparticles (Léveillard and Aït-Ali, 2017). EV's are important mediators of cellular communication during exercise (Whitham et al., 2018), and it is conceivable that 'redox cross talk' occurs between cells/tissues after exercise. A series of redox enzymes (i.e., PRDX-1, PRDX-2, PRDX-5, PRDX-6, thioredoxin (TRX)-1, superoxide dismutase (SOD)1 and SOD2) are known to be secreted in EVs via a non-classical route in response to stress. Interestingly, there is data to suggest that redox enzymes secreted via the classical secretory pathway (e.g., SOD3 and PRDX-4), but not the non-classical secretory pathway (e.g., TRX-1, PRDX-2 and TRX-Reductase) are released into plasma in response to exercise (Wadley et al., 2019a). This suggests that non-classically secreted redox enzymes may be contained within EV's. In the context of immunity, PRDX-1 and PRDX-2 have been shown to be released in exosomes following exposure to inflammatory stimuli in vitro (Mullen et al., 2015), with PRDX-1 also linked with cytokine production (Riddell et al., 2010). The cargo of EV's needs to be explored after bouts of exercise to discern whether intercellular 'redox cross talk' occurs between cells of the immune system after exercise. It is important to note that other plasma proteins (Guseh et al., 2020) and indeed metabolites (e.g., lactate) (Ratter et al., 2018; Zwaag et al., 2020) that are released during exercise may also play important roles in regulating redox-driven immunity after exercise.

CONCLUSION

The redox environment of immune cells is sensitive to changes in both *oxidative* and *reductive stress*; however, accurate evaluation of these changes and their functional significance in vivo is a challenge. In the field of exercise physiology, this is further complicated by changes in the composition of leukocytes measured before and after exercise. The use of flow cytometric methods and single-cell technologies such as CyTOF™ mass cytometry offer huge potential to unravel this complex picture, placing redox exercise immunology at the dawn of a new era.

REFERENCES

Anelli, T., Sannino, S. and Sitia, R. (2015) 'Proteostasis and "redoxtasis" in the secretory pathway: Tales of tails from ERp44 and immunoglobulins.' *Free Radical Biology and Medicine*, 83, pp. 323–30. doi: 10.1016/j.freeradbiomed.2015.02.020.

Bay, M. L. et al. (2020) 'Human immune cell mobilization during exercise: Effect of IL-6 receptor blockade.' *Experimental Physiology*, 105(12), pp. 2086–98. doi: 10.1113/EP088864.

Bellezza, I. et al. (2018) 'Nrf2-Keap1 signaling in oxidative and reductive stress.' *Biochimica et Biophysica Acta Molecular Cell Research*, 1865(5), pp. 721–33. doi: 10.1016/j.bbamcr.2018.02.010.

Brinkmann, V. et al. (2004) 'Neutrophil extracellular traps kill bacteria.' *Science (New York, N.Y.)*, 303(5663), pp. 1532–5. doi: 10.1126/science.1092385.

Broadbent, S. and Coutts, R. (2017) 'Intermittent and graded exercise effects on NK cell degranulation markers LAMP-1/LAMP-2 and $CD8^+CD38^+$ in chronic fatigue syndrome/myalgic encephalomyelitis.' *Physiological Reports*, 5(5). doi: 10.14814/phy2.13091.

Brodie, T. M. and Tosevski, V. (2018) 'Broad immune monitoring and profiling of T cell subsets with mass cytometry.' *Methods in Molecular Biology (Clifton, N.J.)*, 1745, pp. 67–82. doi: 10.1007/978-1-4939-7680-5_4.

Campbell, J. P. and Turner, J. E. (2018) 'Debunking the myth of exercise-induced immune suppression: Redefining the impact of exercise on immunological health across the lifespan.' *Frontiers in Immunology*, 9(APR), pp. 1–21. doi: 10.3389/fimmu.2018.00648.

Cano Sanchez, M. et al. (2018) 'Targeting oxidative stress and mitochondrial dysfunction in the treatment of impaired wound healing: A systematic review.' *Antioxidants (Basel, Switzerland)*, 7(8). doi: 10.3390/antiox7080098.

Cobley, J. N. et al. (2017) 'Exercise redox biochemistry: Conceptual, methodological and technical recommendations.' *Redox Biology*. Elsevier B.V., 12(March), pp. 540–8. doi: 10.1016/j.redox.2017.03.022.

Daiber, A. et al. (2013) 'Protein tyrosine nitration and thiol oxidation by peroxynitrite-strategies to prevent these oxidative modifications.' *International Journal of Molecular Sciences*, 14(4), pp. 7542–70. doi: 10.3390/ijms14047542.

Dimeloe, S. et al. (2017) 'T-cell metabolism governing activation, proliferation and differentiation; a modular view.' *Immunology*, 150(1), pp. 35–44. doi: 10.1111/imm.12655.

Emery, C. F. et al. (2005) 'Exercise accelerates wound healing among healthy older adults: A preliminary investigation.' *The Journals of Gerontology: Series A, Biological Sciences and Medical Sciences*, 60(11), pp. 1432–6. doi: 10.1093/gerona/60.11.1432.

Eskola, J. et al. (1978) 'Effect of sport stress on lymphocyte transformation and antibody formation.' *Clinical and Experimental Immunology*, 32(2), pp. 339–45.

Ferreira, L. F. and Reid, M. B. (2008) 'Muscle-derived ROS and thiol regulation in muscle fatigue.' *Journal of Applied Physiology (Bethesda, Md. : 1985)*, 104(3), pp. 853–60. doi: 10.1152/japplphysiol.00953.2007.

Ferrer, M. D. et al. (2009) 'Antioxidant regulatory mechanisms in neutrophils and lymphocytes after intense exercise.' *Journal of Sports Sciences*, 27(1), pp. 49–58. doi: 10.1080/02640410802409683.

Flohé, L. et al. (2011) 'A comparison of thiol peroxidase mechanisms.' *Antioxidants & Redox Signaling*, 15(3), pp. 763–80. doi: 10.1089/ars.2010.3397.

Fox, C. J., Hammerman, P. S. and Thompson, C. B. (2005) 'Fuel feeds function: Energy metabolism and the T-cell response.' *Nature Reviews Immunology*, 5(11), pp. 844–52. doi: 10.1038/nri1710.

Gleeson, M. et al. (2011) 'The anti-inflammatory effects of exercise: Mechanisms and implications for the prevention and treatment of disease.' *Nature Reviews Immunology*, 11, pp. 607–15.

Graff, R. M. et al. (2018) 'β 2-Adrenergic receptor signaling mediates the preferential mobilization of differentiated subsets of $CD8^+$ T-cells, NK-cells and non-classical monocytes in response to acute exercise in humans.' *Brain, Behavior, and Immunity*, 74(July), pp. 143–53. doi: 10.1016/j.bbi.2018.08.017.

Guseh, J. S. et al. (2020) 'An expanded repertoire of intensity-dependent exercise-responsive plasma proteins tied to loci of human disease risk.' *Scientific Reports*, 10(1), p. 10831. doi: 10.1038/s41598-020-67669-0.

Gustafson, M. P. et al. (2017) 'A systems biology approach to investigating the influence of exercise and fitness on the composition of leukocytes in peripheral blood.' *Journal of Immunother Cancer*, 5(30), pp. 1–14.

Heck, S., Bishop, C. J. and Ellis, R. J. (2019) 'Immunophenotyping of human peripheral blood mononuclear cells by mass cytometry.' *Methods in Molecular Biology (Clifton, N.J.)*, 1979, pp. 285–303. doi: 10.1007/978-1-4939-9240-9_18.

Jackson, S. H. et al. (2004) 'T cells express a phagocyte-type NADPH oxidase that is activated after T cell receptor stimulation.' *Nature Immunology*, 5(8), pp. 818–27. doi: 10.1038/ni1096.

Kamiński, M. M. et al. (2012) 'T cell activation is driven by an ADP-dependent glucokinase linking enhanced glycolysis with mitochondrial reactive oxygen species generation.' *Cell Reports*, 2(5), pp. 1300–15. doi: 10.1016/j.celrep.2012.10.009.

Kay, A. W., Strauss-Albee, D. M. and Blish, C. A. (2016) 'Application of mass cytometry (CyTOF) for functional and phenotypic analysis of natural killer cells.' *Methods in Molecular Biology (Clifton, N.J.)*, 1441, pp. 13–26. doi: 10.1007/978-1-4939-3684-7_2.

Kenny, E. F. et al. (2017) 'Diverse stimuli engage different neutrophil extracellular trap pathways.' *eLife*, 6. doi: 10.7554/eLife.24437.

Kolossov, V. L. et al. (2015) 'Thiol-based antioxidants elicit mitochondrial oxidation via respiratory complex III.' *American Journal of Physiology: Cell Physiology*. doi: 10.1152/ajpcell.00006.2015.

Krüger, K. et al. (2008) 'Exercise-induced redistribution of T lymphocytes is regulated by adrenergic mechanisms.' *Brain, Behavior, and Immunity*, 22(3), pp. 324–38. doi: 10.1016/j.bbi.2007.08.008.

Lamb, G. D. and Westerblad, H. (2011) 'Acute effects of reactive oxygen and nitrogen species on the contractile function of skeletal muscle.' *The Journal of Physiology*, 589, pp. 2119–27. doi: 10.1113/jphysiol.2010.199059.

LaVoy, E. C. et al. (2017) 'T-cell redeployment and intracellular cytokine expression following exercise: Effects of exercise intensity and cytomegalovirus infection.' *Physiological Reports*, 5(1). doi: 10.14814/phy2.13070.

Lévigne, D. et al. (2016) 'NADPH oxidase 4 deficiency leads to impaired wound repair and reduced dityrosine-crosslinking, but does not affect myofibroblast formation.' *Free Radical Biology and Medicine*, 96, pp. 374–84. doi: 10.1016/j.freeradbiomed.2016.04.194.

Lin, D., Gupta, S. and Maecker, H. T. (2015) 'Intracellular cytokine staining on PBMCs using CyTOF™ mass cytometry.' *Bio-Protocol*, 5(1). doi: 10.21769/BioProtoc.1370.

Ma, R. et al. (2018) 'A Pck1-directed glycogen metabolic program regulates formation and maintenance of memory CD8$^+$ T cells.' *Nature Cell Biology*, 20(1), pp. 21–7. doi: 10.1038/s41556-017-0002-2.

Macintyre, A. N. and Rathmell, J. C. (2013) 'Activated lymphocytes as a metabolic model for carcinogenesis.' *Cancer and Metabolism*, 1(1), p. 5. doi: 10.1186/2049-3002-1-5.

Matthews, C. E. et al. (2002) 'Moderate to vigorous physical activity and risk of upper-respiratory tract infection.' *Medicine and Science in Sports and Exercise*, 34(8), pp. 1242–8. doi: 10.1097/00005768-200208000-00003.

Michailidis, Y. et al. (2013) 'Thiol-based antioxidant supplementation alters human skeletal muscle signaling and attenuates its inflammatory response... signaling and attenuates its inflammatory response and recovery after.' *The American Journal of Clinical Nutrition*, 98, pp. 233–45. doi: 10.3945/ajcn.112.049163.

Mullen, L. et al. (2015) 'Cysteine oxidation targets peroxiredoxins 1 and 2 for exosomal release through a novel mechanism of redox-dependent secretion.' 10, pp. 98–108. doi: 10.2119/molmed.2015.00033.

Muri, J. and Kopf, M. (2020) 'Redox regulation of immunometabolism.' *Nature Reviews Immunology*. doi: 10.1038/s41577-020-00478-8.

Nathan, C. and Cunningham-Bussel, A. (2013) 'Beyond oxidative stress: An immunologist's guide to reactive oxygen species.' *Nature Reviews Immunology*, 13(5), pp. 349–61. doi: 10.1038/nri3423.

Nieman, D. C. et al. (2011) 'Upper respiratory tract infection is reduced in physically fit and active adults.' *British Journal of Sports Medicine*, 45(12), pp. 987–92. doi: 10.1136/bjsm.2010.077875.

Nikolaidis, M. G., Margaritelis, N. V and Matsakas, A. (2020) 'Quantitative redox biology of exercise.' *International Journal of Sports Medicine*, 41(10), pp. 633–45. doi: 10.1055/a-1157-9043.

Peake, J. M. et al. (2017) 'Recovery of the immune system after exercise.' *Journal of Applied Physiology (Bethesda, Md. : 1985)*, 122(5), pp. 1077–87. doi: 10.1152/japplphysiol.00622.2016.

Pei, L. and Wallace, D. C. (2018) 'Mitochondrial etiology of neuropsychiatric disorders.' *Biological Psychiatry*, 83(9), pp. 722–30. doi: 10.1016/j.biopsych.2017.11.018.

Peres, A. et al. (2020) 'DNA damage in mononuclear cells following maximal exercise in sedentary and physically active lean and obese men.' *European Journal of Sport Science*, pp. 1–10. doi: 10.1080/17461391.2020.1801850.

Petersen, A. C. et al. (2012) 'Infusion with the antioxidant N-acetylcysteine attenuates early adaptive responses to exercise in human skeletal muscle.' *Acta Ohysiol*, 204, pp. 382–92. doi: 10.1111/j.1748-1716.2011.02344.x.

Poole, L. B. (2015) 'The basics of thiols and cysteines in redox biology and chemistry.' *Free Radical Biology and Medicine*, 80, pp. 148–57. doi: 10.1016/j.freeradbiomed.2014.11.013.

Powers, S. K., Talbert, E. E. and Adhihetty, P. J. (2011) 'Reactive oxygen and nitrogen species as intracellular signals in skeletal muscle.' *The Journal of Physiology*, 589(Pt 9), pp. 2129–38. doi: 10.1113/jphysiol.2010.201327.

Ratter, J. M. et al. (2018) 'In vitro and in vivo effects of lactate on metabolism and cytokine production of human primary PBMCs and monocytes.' *Frontiers in Immunology*, 9, p. 2564. doi: 10.3389/fimmu.2018.02564.

Reid, M. (2008) 'Free radicals and muscle fatigue: Of ROS, canaries, and the IOC.' *Free Radical Biology and Medicine*, 44(2), pp. 169–79.

Riddell, J. R. et al. (2010) 'Peroxiredoxin 1 stimulates secretion of proinflammatory cytokines by binding to TLR4.' *Journal of Immunology (Baltimore, Md. : 1950)*, 184(2), pp. 1022–30. doi: 10.4049/jimmunol.0901945.

Rooney, B. V. et al. (2018) 'Lymphocytes and monocytes egress peripheral blood within minutes after cessation of steady state exercise: A detailed temporal analysis of leukocyte extravasation.' *Physiology and Behavior*, 194, pp. 260–67. doi: 10.1016/j.physbeh.2018.06.008.

Sakelliou, A. et al. (2016) 'Evidence of a redox-dependent regulation of immune responses to exercise-induced inflammation.' *Oxidative Medicine and Cellular Longevity*, 2016, pp. 1–19.

Sallusto, F. et al. (1999) 'Two subsets of memory T lymphocytes with distinct homing potentials and effector functions.' *Nature*, 401, pp. 708–12.

Sies, H. and Jones, D. P. (2020) 'Reactive oxygen species (ROS) as pleiotropic physiological signalling agents.' *Nature Reviews Molecular Cell Biology*, 21(7), pp. 363–83. doi: 10.1038/s41580-020-0230-3.

Simoni, Y. et al. (2018) 'Mass cytometry: A powerful tool for dissecting the immune landscape.' *Current Opinion in Immunology*, 51, pp. 187–96. doi: 10.1016/j.coi.2018.03.023.

Simpson, R. J. et al. (2020) 'Can exercise affect immune function to increase susceptibility to infection?' *Exercise Immunology Review*, 26, pp. 8–22.

Sobotta, M. C. et al. (2014) 'Peroxiredoxin-2 and STAT3 form a redox relay for H_2O_2 signaling.' *Nature Chemical Biology*, 11(1), pp. 64–70. doi: 10.1038/nchembio.1695.

Spanidis, Y. et al. (2018) 'Exercise-induced reductive stress is a protective mechanism against oxidative stress in peripheral blood mononuclear cells.' *Oxidative Medicine and Cellular Longevity*, 2018, p. 3053704. doi: 10.1155/2018/3053704.

Sureda, A. et al. (2005) 'Relation between oxidative stress markers and antioxidant endogenous defences during exhaustive exercise.' *Free Radical Research*, 39(12), pp. 1317–24. doi: 10.1080/10715760500177500.

Tauler, P. et al. (2006) 'Increased lymphocyte antioxidant defences in response to exhaustive exercise do not prevent oxidative damage.' *The Journal of Nutritional Biochemistry*, 17(10), pp. 665–71. doi: 10.1016/j.jnutbio.2005.10.013.

Léveillard, T. and Aït-Ali, N. (2017) 'Cell signaling with extracellular thioredoxin and thioredoxin-like proteins: Insight into their mechanisms of action.' *Oxidative Medicine and Cellular Longevity*, 2017, p. 11. doi: 10.1155/2017/8475125.

Turner, J. E. et al. (2011) 'Assessment of oxidative stress in lymphocytes with exercise.' *Journal of Applied Physiology (Bethesda, Md. : 1985)*, 111(1), pp. 206–11.

Turner, J. E. et al. (2013) 'The antioxidant enzyme peroxiredoxin-2 is depleted in lymphocytes seven days after ultra-endurance exercise.' *Free Radical Research*, 47(10), pp. 821–8.

Vander Heiden, M. G., Cantley, L. C. and Thompson, C. B. (2009) 'Understanding the Warburg effect: The metabolic requirements of cell proliferation.' *Science (New York, N.Y.)*, 324(5930), pp. 1029–33. doi: 10.1126/science.1160809.

Wadley, A. J. et al. (2015a) 'Monitoring changes in thioredoxin and over-oxidised peroxiredoxin in response to exercise in humans.' *Free Radical Research*, 49(3), pp. 290–8. doi: 10.3109/10715762.2014.1000890.

Wadley, A. J. et al. (2015b) 'The impact of intensified training with a high or moderate carbohydrate feeding strategy on resting and exercise-induced oxidative stress.' *European Journal of Applied Physiology*. 115(8), pp. 1757–67. doi: 10.1007/s00421-015-3162-4.

Wadley, A. J., Aldred, S. and Coles, S. J. (2016) 'An unexplored role for peroxiredoxin in exercise-induced redox signalling?' *Redox Biology*, 8. doi: 10.1016/j.redox.2015.10.003.

Wadley, A. J. et al. (2018a) 'Detecting intracellular thiol redox state in leukaemia and heterogeneous immune cell populations: An optimised protocol for digital flow cytometers.' *MethodsX*, 5, pp. 1473–83. doi: 10.1016/j.mex.2018.10.013.

Wadley, A. J. et al. (2018b) 'Preliminary evidence of reductive stress in human cytotoxic T-cells following exercise.' *Journal of Applied Physiology*. 125, pp. 586–95.

Wadley, A. J. et al. (2019a) 'Characterisation of extracellular redox enzyme concentrations in response to exercise in humans.' *Journal of Applied Physiology*, 127(3), pp. 858–66.

Wadley, A. J. et al. (2019b) Using flow cytometry to detect and measure intracellular thiol redox status in viable T cells from heterogeneous populations, *Methods in Molecular Biology*. doi: 10.1007/978-1-4939–9463-2_5.

Walsh, N. P. (2018) 'Recommendations to maintain immune health in athletes.' *European Journal of Sport Science*, 18(6), pp. 820–31. doi: 10.1080/17461391.2018.1449895.

Weyand, C. M., Shen, Y. and Goronzy, J. J. (2018) 'Redox-sensitive signaling in inflammatory T cells and in autoimmune disease.' *Free Radical Biology and Medicine*, 125, pp. 36–43. doi: 10.1016/j.freeradbiomed.2018.03.004.

Whitham, M. et al. (2018) 'Extracellular vesicles provide a means for tissue crosstalk during exercise.' *Cell Metabolism*, Elsevier, 27(1), pp. 237–51. e4. doi: 10.1016/j.cmet.2017.12.001.

Winterbourn, C. C. (2013) 'The biological chemistry of hydrogen peroxide.' *Methods in Enzymology*, Elsevier Inc, 528, pp. 3–25. doi: 10.1016/B978-0-12-405881-1.00001-X.

Winterbourn, C. C. (2015) 'Are free radicals involved in thiol-based redox signaling?' *Free Radical Biology and Medicine*, Elsevier, 80, pp. 164–70. doi: 10.1016/j.freeradbiomed.2014.08.017.

Winterbourn, C. C., Kettle, A. J. and Hampton, M. B. (2016) 'Reactive oxygen species and neutrophil function.' *Annual Review of Biochemistry*, 85, pp. 765–92. doi: 10.1146/annurev-biochem-060815-014442.

van der Windt, G. J. W. and Pearce, E. L. (2012) 'Metabolic switching and fuel choice during T-cell differentiation and memory development.' *Immunological Reviews*, 249(1), pp. 27–42. doi: 10.1111/j.1600-065X.2012.01150.x.

Zwaag, J. et al. (2020) 'Involvement of lactate and pyruvate in the anti-inflammatory effects exerted by voluntary activation of the sympathetic nervous system.' *Metabolites*, 10(4). doi: 10.3390/metabo10040148.

CHAPTER EIGHT

Exercise and RNA Oxidation

Emil List Larsen, Kristian Karstoft and Henrik Enghusen Poulsen

CONTENTS

Introduction / 95
Epitranscriptomic Changes / 95
RNA Oxidation in an In Vivo Setting / 97
Acute Exercise and RNA Oxidation / 98
Regular Exercise and RNA Oxidation / 98
Conclusion / 98
Acknowledgment / 99
References / 99

INTRODUCTION

Skeletal muscle is a unique tissue that can increase oxygen consumption rate from the resting state by a factor of 10 or more (Hill, Long and Lupton, 1924). Oxygen is reduced by the mitochondrial respiratory chain to generate water and energy, but during this process, reactive oxygen species (ROS), like superoxide and hydrogen peroxide, are also formed (Chance, Sies and Boveris, 1979). Although the mitochondrial respiratory rate is markedly increased during exercise, current evidence suggest that the production of mitochondria ROS is not increased during exercise, but instead post-exercise (Goncalves et al., 2015; Laker et al., 2017; Henriquez-Olguin, Meneses-Valdes and Jensen, 2020). This may be explained by a shift in mitochondrial respiratory state – from state 4 to state 3 during exercise – with less ROS generated in state 3 (Powers et al., 2020). In addition, exercise may change the kinetic properties of the mitochondrial sites known to produce ROS (Goncalves et al., 2015). Nonetheless, ROS are formed during exercise in the cytosol of skeletal muscle cells, but other sources than mitochondria seem to mediate the production, e.g., phospholipase A2 and NADPH-oxidases (Henriquez-Olguin, Meneses-Valdes and Jensen, 2020; Powers et al., 2020).

Even though ROS are associated with disease development (Halliwell, 1991), sedentary individuals are at higher risk of disease than regularly exercising individuals (Arem et al., 2015). This poses a paradox that is incompletely explained but consistently found in epidemiology (Powers et al., 2020). One theory is that exercise induces respiratory and resistant mechanisms with an overshoot that protect against disease development during non-exercising periods (Powers et al., 2020). This is consistent with the findings that during exercise there is an increased risk of cardiovascular death (Ferreira et al., 2010); counteracted by a larger risk reduction during non-exercising periods (Mandsager et al., 2018).

EPITRANSCRIPTOMIC CHANGES

During the last decades, much interest has been on genetic predisposition; however, the post-gene regulatory mechanisms have also attracted attention. Especially, the biological functions of RNA have emerged during the discoveries of functional

DOI: 10.1201/9781003051619-8

non-coding types of RNA, and new intricate regulatory roles are continually being discovered. Alongside, the consequences of chemical modifications of RNA are revealed and considered epitranscriptomic changes; important for protein synthesis, signaling, and cellular homeostasis (Li, Xiong and Yi, 2016; Jonkhout et al., 2017). Several different oxidative modifications of DNA have been discovered and given the close resemblance with the chemical structure of RNA, it is assumed that similar oxidation products of RNA exist (Poulsen et al., 2012). Nonetheless, only a few in vivo RNA oxidation products have been identified (Weimann et al., 2012, 2019). The primary focus has been on the oxidation product 8-oxo-7,8-dihydroguanosine (8-oxoGuo) that corresponds to the DNA oxidation product; 8-oxo-7,8-dihydro-2′-deoxyguanosine (8-oxodG). These oxidation products are chemically well described, and the low redox potential of guanosine compared with the other nucleotides makes guanosine more prone to oxidation. In addition, validated analysis methods for detection of 8-oxoGuo and 8-oxodG exist (Poulsen et al., 2014). Several oxidative species (e.g., hydroxyl radical or carbonate radical) may be responsible for the oxidative modifications of nucleic acids. It is likely to assume that the different cellular locations of RNA and DNA presume different sources of ROS. ROS generated in the cytosol (e.g., mitochondria ROS) may oxidize RNA to a greater extent than nuclear DNA. In addition to the cellular location, the chemical structure of RNA permit oxidation more than DNA. RNA is to a large extent found single-stranded and typically lacks protective proteins (Poulsen et al., 2014). DNA-repair mechanisms are essential and well-described. In contrast, RNA lacks a template strand for repair. Thus, if RNA repair mechanism exists, then the mechanism has to be different from DNA repair (i.e., without using a template strand) (Yan and Zaher, 2019), and it seems more likely that RNA is degraded instead of repaired (Poulsen et al., 2012). Due to the difference in structure and location between RNA and DNA, it is not a surprise that the amount of oxidized RNA is greater than oxidized DNA (Weimann, Belling and Poulsen, 2002).

Different types of messenger RNA (mRNA) and micro-RNA (miRNA) as well as different sites show different susceptibility to oxidation (Shan, Tashiro and Lin, 2003; Poulsen et al., 2012; Wang et al., 2015). Thus, regulatory effects of oxidative RNA modifications may exist. However, oxidative RNA modifications are found positively associated with disease development and progression in epidemiological studies (Nunomura et al., 2006; Broedbaek et al., 2011, 2013; Liu et al., 2016; Kjær et al., 2017), which is supported by mechanistic in vitro work suggesting potential pathophysiological consequences of RNA oxidation (see Table 8.1). In vitro studies have shown that oxidation of mRNA alters the protein synthesis. Modified or truncated proteins are synthesized, or the translational process may be stalled (Tanaka, Chock and Stadtman, 2007) with the possibility of protein aggregation (Shan, Tashiro and Lin, 2003). The altered protein synthesis may explain, why oxidatively RNA modifications are positively associated with development of neurodegenerative diseases (e.g., Alzheimer's disease) (Nunomura et al., 2001). Furthermore, oxidation of miRNAs disrupts intracellular signaling, which may be important in different pathogeneses (Wang et al., 2015). As such, the current evidence supports only pathophysiological roles of RNA oxidation. Future studies will hopefully investigate whether physiological effects of RNA oxidation exist.

TABLE 8.1
Biological consequences of oxidative RNA modifications.

RNA type	Consequence	Citation
miRNA (miR-184)	Initiation of cellular apoptosis	Wang et al. (2015)
mRNA	Translational stunning, modified proteins, and truncated proteins	Tanaka, Chock and Stadtman (2007)
mRNA	Protein dysfunction and protein aggregation	Shan, Tashiro and Lin (2003)
mRNA	Reduced protein synthesis	Shan, Chang and Lin (2007)
rRNA	Disrupted translation process	Willi et al. (2018)

miRNA, micro-RNA; mRNA, messenger RNA; rRNA, ribosomal RNA.

Investigations focused on specific RNA sites oxidized and the consequences hereof are important, but not possible using the current methodology for measuring 8-oxoGuo in plasma and urine.

RNA OXIDATION IN AN IN VIVO SETTING

Measuring oxidative stress definitively and not artifacts has always been one of the greatest challenges in the field of redox biology. Likewise, measuring oxidative RNA modifications poses challenges. 8-oxoGuo can be measured in urine, plasma/serum, or tissues; however, the interpretation and potential limitations differ greatly. Urinary excretion of 8-oxoGuo is frequently used, because it is easy to collect and quantify. The positive associations between urinary excretion of 8-oxoGuo and mortality in patients with type 2 diabetes have defined the urinary excretion rate as a prognostic and clinically relevant biomarker (Broedbaek et al., 2011, 2013; Kjær et al., 2017). However, the measurement of 8-oxoGuo in urine has limitations. The measure does not discriminate between organ of origin; thus, tissue-specific changes in oxidatively generated RNA modifications may go undetected (Poulsen et al., 2019). The urine concentration of 8-oxoGuo must be corrected for urine dilution, e.g., using urine volume (24 hours), creatinine, or density (Larsen, Weimann and Poulsen, 2019). Especially, creatinine adjustment may be problematic, when investigating exercise interventions, since urinary excretion of creatinine may increase up to 50% (Calles-Escandon et al., 1984). Plasma concentrations of oxidized nucleic acids are frequently used; however, the small water-soluble molecules are assumed not to be reabsorbed in the kidneys. If so, oxidized nucleic acids will freely filtrate through the glomerulus, and plasma concentrations will dominantly be determined by the kidney function (like plasma creatinine) instead of the intracellular level of oxidized nucleic acids. A model to overcome this challenge has been published and may make plasma samples a superior material to measure oxidized nucleic acids (Poulsen et al., 2019). Measurement of oxidized nucleic acids in tissues allows the interpretation of tissue-specific changes unlike plasma and urine samples. The concentration of 8-oxoGuo in tissues reflects an accumulation of oxidized RNA based on the cellular formation and removal. The quantification is challenging, and there is a great risk of spurious oxidation (Poulsen et al., 2012; Weimann et al., 2012). In addition, only a few tissues are available for measurement in a human in vivo setting.

The vast majority of intervention studies exploring effects on oxidatively generated nucleic acid modifications have focused on DNA oxidation and only very little evidence regarding RNA oxidation exists (Larsen, Weimann and Poulsen, 2019). ELISA methods for the determination of 8-oxodG have attracted much awareness due to low specificity. Nonetheless, it is widely used, because it does not require expert chemical experience or expensive equipment. However, most ELISA kits are unable to distinguish between the oxidative RNA product (i.e., 8-oxoGuo) and the oxidative DNA product (i.e., 8-oxodG) (Wu et al., 2004; Song et al., 2009); thus, many studies exploring the effects of exercise on DNA oxidation using ELISA methods may be greatly influenced by RNA oxidation (Larsen, Weimann and Poulsen, 2019). As such, intervention studies exploring effects of exercise on 8-oxodG measured by ELISA methods may very well reflect a combined measurement of 8-oxodG and 8-oxoGuo (i.e., reflecting both DNA and RNA oxidation). During exercise, both NADPH-oxidase and phospolipase A2 increase ROS formation in the cytosol. In addition, increased oxidative phosphorylation in mitochondria particular after exercise generates ROS (Goncalves et al., 2015; Laker et al., 2017; Henriquez-Olguin, Meneses-Valdes and Jensen, 2020; Powers et al., 2020). As such, the cytosolic concentration of ROS is increased and presumably increases the oxidation of RNA to a greater extent than DNA oxidation due to compartmentalization of the ROS. The ROS production during and following exercise seems to be transient and may mediate beneficial adaptions of the organism to exercise (Margaritelis et al., 2018). An upregulation of antioxidative enzymes occurs in response to regular exercise (Radak et al., 2008), and it may be speculated that this decreases oxidatively generated RNA modifications. This hypothesis is supported by a large cross-sectional questionnaire study that included individuals with type 2 diabetes (n = 1,992). The study revealed that individuals who exercised regularly, generally present with lower oxidative RNA modifications, measured as urinary excretion of 8-oxoGuo, than individuals who did not exercise regularly. In contrast, oxidative DNA modifications were similar between the groups (Kofoed Kjaer et al., 2019).

ACUTE EXERCISE AND RNA OXIDATION

A single study has investigated the effects of an acute bout of exercise on RNA oxidation. The study was observational and examined changes in 24-hour urinary excretion of 8-oxoGuo using ultra-performance liquid chromatography tandem mass-spectrometry in healthy males following a marathon. In addition, the study also determined changes in DNA oxidation, measured as urinary excretion of 8-oxodG. Both RNA and DNA oxidation remained unchanged immediately after the marathon. DNA oxidation was very stable, whereas RNA oxidation trended toward an increased formation rate (Larsen et al., 2020). In contrast to this finding, previous studies that measured urinary excretion of 8-oxodG following a marathon found increased excretion rates of 8-oxodG. However, the studies used ELISA for the determination of 8-oxodG (Radak et al., 2000; Tsai et al., 2001; Mrakic-Sposta et al., 2015). As mentioned above, due to the low specificity and lack of discrimination between 8-oxodG and 8-oxoGuo, this may reflect an increase in 8-oxoGuo instead of 8-oxodG (Larsen et al., 2020), or cross-reactivity with unknown substances. The studies included individuals with a cardiorespiratory fitness level above average, thus the ROS production may be smaller and/or the antioxidative defense may be greater than in untrained individuals. The study that explored effects on RNA oxidation following a marathon also revealed reduced RNA and DNA oxidation ~4 days after the marathon that normalized ~7 days post-exercise (Larsen et al., 2020). The study was uncontrolled; nonetheless, the finding indicates adaptative effects of exercise on RNA oxidation, but also DNA oxidation, which is in contrast to the findings in the cross-sectional study (Kofoed Kjaer et al., 2019).

REGULAR EXERCISE AND RNA OXIDATION

Only a few studies have investigated the impact of regular physical activity on RNA oxidation. The Vienna Active Aging Study randomized elderly from a nursing home (83 ± 6 years, n = 117) to 6 months of resistance training ± protein and vitamin supplementation ± cognitive training. The exercise protocol was matched to the American College of Sports Medicine guidelines for older subjects; therefore, the exercise protocol was of low intensity. No significant changes in RNA oxidation, measured as urinary excretion of 8-oxoGuo, were found during the study period in any intervention group. Nor were any changes in DNA oxidation, measured as urinary excretion of 8-oxodG, evident (Franzke et al., 2018). Hofer et al. randomized 29 healthy adults (BMI: 23.5–29.9 kg/m^2; age 50–60 years) to 1-year energy deficit of 20% through exercise or caloric restriction. None of the intervention groups significantly changed RNA oxidation, measured as urinary excretion of 8-oxoGuo. However, both groups decreased 8-oxoGuo/Guo in white blood cells indicating tissue-specific improvement of RNA oxidation following regular exercise (Hofer et al., 2008). This finding pinpoints the previous mentioned limitation when measuring 8-oxoGuo in urine and plasma samples; no discrimination is made between tissue origin of RNA oxidation. During exercise, it is expected that specific tissues increase ROS formation more than others; similarly, following regular exercise the antioxidant system may be changed differently in different tissues. Plasma and urine 8-oxoGuo concentrations are accumulated from the entire organism and, thus, may obscure tissue-specific changes. Organ-specific measurement of 8-oxoGuo as well as other markers of oxidative stress may change the redox field greatly, and analyses of extracellular microvesicles may be a promising avenue by which such measurements may be reached.

CONCLUSION

The limited literature exploring effects of exercise on RNA oxidation suggest that muscle tissue increase oxygen consumption markedly without increasing RNA oxidation notably. This supports that mitochondrial ROS may not increase during exercise. Decreased excretion rates of 8-oxoGuo were found ~4 days after a marathon (Larsen et al., 2020). In agreement, a cross-sectional study showed lower excretion rates of 8-oxoGuo in individuals with type 2 diabetes that answered they exercised regularly (Kofoed Kjaer et al., 2019). The randomized controlled studies did not demonstrate any changes in RNA oxidation, measured as urinary excretion of 8-oxoGuo, following regular exercise (Hofer et al., 2008; Franzke et al., 2018). However, the ratio of 8-oxoGuo/Guo was decreased in white blood cells, suggesting tissue-specific improvements in RNA oxidation (Hofer et al., 2008). This suggests that exercise may

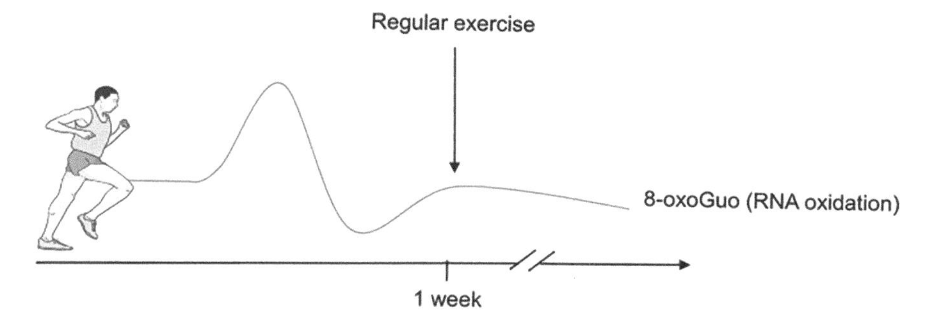

Figure 8.1 The current evidence suggests that an acute exercise bout generates a trend toward increased RNA oxidation immediately after cessation of the exercise bout. Hereafter, a period with reduced RNA oxidation follows, which is possibly mediated through an upregulation of antioxidant factors, decreased formation of reactive oxygen species, or perhaps increased RNA repair. Theoretically, this induces lower steady-state RNA oxidation in individuals that perform regular exercise, something which however is yet to be confirmed in a randomized controlled setting. Abbreviation: 8-oxoGuo, 8-oxo-7,8-dihydroguanosine.

mediate beneficial effects in specific tissues by improving antioxidant defense and/or by reducing pro-oxidant formation and, thus, reduce the formation of oxidatively generated RNA modifications (Figure 8.1). If regularly performed exercise decreases oxidative RNA modifications, then it may potentially contribute to the protective effects of physical activity against development and progression of neurodegenerative diseases (e.g., Alzheimer's disease) and cardiovascular diseases in patients with type 2 diabetes (Thompson et al., 2003; Erickson, Weinstein and Lopez, 2012).

ACKNOWLEDGMENT

We thank Servier Medical Art (http://smart.servier.com/) for the image that has been modified and used in Figure 8.1 under the CC-BY license.

REFERENCES

Arem, H., Moore, S. C., Patel, A., Hartge, P., Berrington de Gonzalez, A., Visvanathan, K., Campbell, P. T., Freedman, M., Weiderpass, E., Adami, H. O., Linet, M. S., Lee, I.-M. and Matthews, C. E. (2015) 'Leisure time physical activity and mortality: a detailed pooled analysis of the dose-response relationship.' *JAMA Internal Medicine*, United States, 175(6), pp. 959–967. doi: 10.1001/jamainternmed.2015.0533.

Broedbaek, K., Siersma, V., Henriksen, T., Weimann, A., Petersen, M., Andersen, J. T., Jimenez-Solem, E., Stovgaard, E. S., Hansen, L. J., Henriksen, J. E., Bonnema, S. J., de Fine Olivarius, N. and Poulsen, H. E. (2011) 'Urinary markers of nucleic acid oxidation and long-term mortality of newly diagnosed type 2 diabetic patients.' *Diabetes Care*, 34, pp. 2594–2596. doi: 10.2337/dc11-1620.

Broedbaek, K., Siersma, V., Henriksen, T., Weimann, A., Petersen, M., Andersen, J. T., Jimenez-Solem, E., Hansen, L. J., Henriksen, J. E., Bonnema, S. J., de Fine Olivarius, N. and Poulsen, H. E. (2013) 'Association between urinary markers of nucleic acid oxidation and mortality in type 2 diabetes: A population-based cohort study.' *Diabetes Care*, 36, pp. 669–676. doi: 10.2337/dc12-0998.

Calles-Escandon, J., Cunningham, J. J., Snyder, P., Jacob, R., Huszar, G., Loke, J. and Felig, P. (1984) 'Influence of exercise on urea, creatinine, and 3-methylhistidine excretion in normal human subjects.' *American Journal of Physiology-Endocrinology and Metabolism*, American Physiological Society, 246(4), pp. E334–E338. doi: 10.1152/ajpendo.1984.246.4.E334.

Chance, B., Sies, H. and Boveris, A. (1979) 'Hydroperoxide metabolism in mammalian organs.' *Physiological Reviews*, 59(3), pp. 527–605.

Erickson, K. I., Weinstein, A. M. and Lopez, O. L. (2012) 'Physical activity, brain plasticity, and Alzheimer's disease.' *Archives of Medical Research*, 43(8), pp. 615–621. doi: 10.1016/j.arcmed.2012.09.008.

Ferreira, M., Santos-Silva, P. R., de Abreu, L. C., Valenti, V. E., Crispim, V., Imaizumi, C., Filho, C. F., Murad, N., Meneghini, A., Riera, A. R. P., de Carvalho, T. D., Vanderlei, L. C. M., Valenti, E. E., Cisternas, J. R., Moura Filho, O. F. and Ferreira, C. (2010) 'Sudden cardiac death athletes: A systematic review.' *Sports Medicine, Arthroscopy, Rehabilitation, Therapy & Technology : SMARTT*, 2, p. 19. doi: 10.1186/1758-2555-2-19.

Franzke, B., Schober-Halper, B., Hofmann, M., Oesen, S., Tosevska, A., Henriksen, T., Poulsen, H. E., Strasser, E.-M., Wessner, B. and Wagner, K.-H. (2018) 'Age and the effect of exercise, nutrition and cognitive training on oxidative stress: The Vienna Active Aging Study (VAAS), a randomized controlled trial.' *Free Radical Biology and Medicine*, 121, pp. 69–77. doi: 10.1016/j.freeradbiomed.2018.04.565.

Goncalves, R. L. S., Quinlan, C. L., Perevoshchikova, I. V., Hey-Mogensen, M. and Brand, M. D. (2015) 'Sites of superoxide and hydrogen peroxide production by muscle mitochondria assessed ex vivo under conditions mimicking rest and exercise.' *The Journal of Biological Chemistry*, 290(1), pp. 209–227. doi: 10.1074/jbc.M114.619072.

Halliwell, B. (1991) 'Reactive oxygen species in living systems: Source, biochemistry, and role in human disease.' *The American Journal of Medicine*, United States, 91(3C), pp. 14S–22S. doi: 10.1016/0002-9343(91)90279-7.

Henriquez-Olguin, C., Meneses-Valdes, R. and Jensen, T. E. (2020) 'Compartmentalized muscle redox signals controlling exercise metabolism: Current state, future challenges.' *Redox Biology*, 35, p. 101473. doi: 10.1016/j.redox.2020.101473.

Hill, A. V., Long, C. N. H. and Lupton, H. (1924) 'Muscular exercise, lactic acid and the supply and utilisation of oxygen: Parts VII–VIII.' *Proceedings of the Royal Society of London: Series B, Containing Papers of a Biological Character*, The Royal Society London, 97(682), pp. 155–176.

Hofer, T., Fontana, L., Anton, S. D., Weiss, E. P., Villareal, D., Malayappan, B. and Leeuwenburgh, C. (2008) 'Long-term effects of caloric restriction or exercise on DNA and RNA oxidation levels in white blood cells and urine in humans.' *Rejuvenation Research*, United States, 11(4), pp. 793–799. doi: 10.1089/rej.2008.0712.

Jonkhout, N., Tran, J., Smith, M. A., Schonrock, N., Mattick, J. S. and Novoa, E. M. (2017) 'The RNA modification landscape in human disease.' *RNA (New York, N.Y.)*, 23(12), pp. 1754–1769. doi: 10.1261/rna.063503.117.

Kjær, L. K., Cejvanovic, V., Henriksen, T., Petersen, K. M., Christensen, C. K., Torp-Pedersen, C., Gerds, T. A., Brandslund, I., Mandrup-Poulsen, T. and Poulsen, H. E. (2017) 'Cardiovascular and all-cause mortality risk associated with urinary excretion of 8-oxoGuo, a biomarker for RNA oxidation, in patients with type 2 diabetes: A prospective cohort study.' *Diabetes Care*, 40(12), pp. 1771–1778. doi: https://doi.org/10.2337/dc17-1150.

Kofoed Kjaer, L., Cejvanovic, V., Henriksen, T., Hansen, T., Pedersen, O., Kjeldahl Christensen, C., Torp-Pedersen, C., Alexander Gerds, T., Brandslund, I., Mandrup-Poulsen, T. and Enghusen Poulsen, H. (2019) 'Urinary nucleic acid oxidation product levels show differential associations with pharmacological treatment in patients with type 2 diabetes.' *Free Radical Research*, England, 53(6), pp. 694–703. doi: 10.1080/10715762.2019.1622011.

Laker, R. C., Drake, J. C., Wilson, R. J., Lira, V. A., Lewellen, B. M., Ryall, K. A., Fisher, C. C., Zhang, M., Saucerman, J. J., Goodyear, L. J., Kundu, M. and Yan, Z. (2017) 'Ampk phosphorylation of Ulk1 is required for targeting of mitochondria to lysosomes in exercise-induced mitophagy.' *Nature Communications*, 8(1), p. 548. doi: 10.1038/s41467-017-00520-9.

Larsen, E. L., Weimann, A. and Poulsen, H. E. (2019) 'Interventions targeted at oxidatively generated modifications of nucleic acids focused on urine and plasma markers.' *Free Radical Biology and Medicine*, pp. 256–283. doi: 10.1016/j.freeradbiomed.2019.09.030.

Larsen, E. L., Poulsen, H. E., Michaelsen, C., Kjær, L. K., Lyngbæk, M., Andersen, E. S., Petersen-Bønding, C., Lemoine, C., Gillum, M., Jørgensen, N. R., Ploug, T., Vilsbøll, T., Knop, F. K. and Karstoft, K. (2020) 'Differential time responses in inflammatory and oxidative stress markers after a marathon: An observational study.' *Journal of Sports Sciences*, England, 38(18), pp. 2080–2091. doi: 10.1080/02640414.2020.1770918.

Li, X., Xiong, X. and Yi, C. (2016) 'Epitranscriptome sequencing technologies: Decoding RNA modifications.' *Nature Methods*, United States, 14(1), pp. 23–31. doi: 10.1038/nmeth.4110.

Liu, X., Gan, W., Zou, Y., Yang, B., Su, Z., Deng, J., Wang, L. and Cai, J. (2016) 'Elevated levels of urinary markers of oxidative DNA and RNA damage in type 2 diabetes with complications.' *Oxidative Medicine and Cellular Longevity*, pp. 1–7. doi: 10.1155/2016/4323198.

Mandsager, K., Harb, S., Cremer, P., Phelan, D., Nissen, S. E. and Jaber, W. (2018) 'Association of cardiorespiratory fitness with long-term mortality among adults undergoing exercise treadmill testing.' *JAMA Network Open*, 1(6), pp. 1–12. doi: 10.1001/jamanetworkopen.2018.3605.

Margaritelis, N. V, Theodorou, A. A., Paschalis, V., Veskoukis, A. S., Dipla, K., Zafeiridis, A., Panayiotou, G., Vrabas, I. S., Kyparos, A. and Nikolaidis, M. G. (2018) 'Adaptations to endurance training depend on exercise-induced oxidative stress: Exploiting redox interindividual variability.' *Acta Physiologica (Oxford, England)*, 222(2). doi: 10.1111/apha.12898.

Mrakic-Sposta, S., Gussoni, M., Moretti, S., Pratali, L., Giardini, G., Tacchini, P., Dellanoce, C., Tonacci, A., Mastorci, F., Borghini, A., Montorsi, M. and Vezzoli, A. (2015) 'Effects of mountain ultra-marathon running on ROS production and oxidative damage by micro-invasive analytic techniques.' *PloS One*, United States, 10(11), p. e0141780. doi: 10.1371/journal.pone.0141780.

Nunomura, A., Perry, G., Aliev, G., Hirai, K., Takeda, A., Balraj, E. K., Jones, P. K., Ghanbari, H., Wataya, T., Shimohama, S., Chiba, S., Atwood, C. S., Petersen, R. B. and Smith, M. A. (2001) 'Oxidative damage is the earliest event in Alzheimer disease.' *Journal of Neuropathology and Experimental Neurology*, 60(8), pp. 759–767. doi: 10.1093/jnen/60.8.759.

Nunomura, A., Honda, K., Takeda, A., Hirai, K., Zhu, X., Smith, M. A. and Perry, G. (2006) 'Oxidative damage to RNA in neurodegenerative diseases.' *Journal of Biomedicine and Biotechnology*, pp. 1–6. doi: 10.1155/JBB/2006/82323.

Poulsen, H. E., Specht, E., Broedbaek, K., Henriksen, T., Ellervik, C., Mandrup-Poulsen, T., Tonnesen, M., Nielsen, P. E., Andersen, H. U. and Weimann, A. (2012) 'RNA modifications by oxidation: A novel disease mechanism?' *Free Radical Biology and Medicine*, 52(8), pp. 1353–1361. doi: 10.1016/j.freeradbiomed.2012.01.009.

Poulsen, H. E., Nadal, L. L., Broedbaek, K., Nielsen, P. E. and Weimann, A. (2014) 'Detection and interpretation of 8-oxodG and 8-oxoGua in urine, plasma and cerebrospinal fluid.' *Biochimica et Biophysica Acta*, Elsevier B.V., 1840(2), pp. 801–808. doi: 10.1016/j.bbagen.2013.06.009.

Poulsen, H. E., Weimann, A., Henriksen, T., Kjær, L. K., Larsen, E. L., Carlsson, E. R., Christensen, C. K., Brandslund, I. and Fenger, M. (2019) 'Oxidatively generated modifications to nucleic acids in vivo: Measurement in urine and plasma.' *Free Radical Biology and Medicine*, pp. 336–341. doi: 10.1016/j.freeradbiomed.2019.10.001.

Powers, S. K., Deminice, R., Ozdemir, M., Yoshihara, T., Bomkamp, M. P. and Hyatt, H. (2020) 'Exercise-induced oxidative stress: Friend or foe?' *Journal of Sport and Health Science*, 9(5), pp. 415–425. doi: 10.1016/j.jshs.2020.04.001.

Radak, Z., Pucsuk, J., Boros, S., Josfai, L. and Taylor, A. W. (2000) 'Changes in urine 8-hydroxydeoxyguanosine levels of super-marathon runners during a four-day race period.' *Life Sciences*, Netherlands, 66(18), pp. 1763–1767. doi: 10.1016/s0024–3205(00)00499-9.

Radak, Z., Chung, H. Y., Koltai, E., Taylor, A. W. and Goto, S. (2008) 'Exercise, oxidative stress and hormesis.' *Ageing Research Reviews*, 7(1), pp. 34–42. doi: 10.1016/j.arr.2007.04.004.

Shan, X., Tashiro, H. and Lin, C. G. (2003) 'The identification and characterization of oxidized RNAs in Alzheimer's disease.' *The Journal of Neuroscience*, 23(12), pp. 4913 LP–4921. doi: 10.1523/JNEUROSCI.23-12-04913.2003.

Shan, X., Chang, Y. and Lin, C. G. (2007) 'Messenger RNA oxidation is an early event preceding cell death and causes reduced protein expression.' *FASEB Journal : Official Publication of the Federation of American Societies for Experimental Biology*, United States, 21(11), pp. 2753–2764. doi: 10.1096/fj.07-8200com.

Song, M.-F., Li, Y.-S., Ootsuyama, Y., Kasai, H., Kawai, K., Ohta, M., Eguchi, Y., Yamato, H., Matsumoto, Y., Yoshida, R. and Ogawa, Y. (2009) 'Urea, the most abundant component in urine, cross-reacts with a commercial 8-OH-dG ELISA kit and contributes to overestimation of urinary 8-OH-dG.' *Free Radical Biology & Medicine*, United States, 47(1), pp. 41–46. doi: 10.1016/j.freeradbiomed.2009.02.017.

Tanaka, M., Chock, P. B. and Stadtman, E. R. (2007) 'Oxidized messenger RNA induces translation errors.' *Proceedings of the National Academy of Sciences*, 104(1), pp. 66–71. doi: 10.1073/pnas.0609737104.

Thompson, P. D., Buchner, D., Pina, I. L., Balady, G. J., Williams, M. A., Marcus, B. H., Berra, K., Blair, S. N., Costa, F., Franklin, B., Fletcher, G. F., Gordon, N. F., Pate, R. R., Rodriguez, B. L., Yancey, A. K. and Wenger, N. K. (2003) 'Exercise and physical activity in the prevention and treatment of atherosclerotic cardiovascular disease: a statement from the council on clinical cardiology (subcommittee on exercise, rehabilitation, and prevention) and the council on nutrition, physica.' *Circulation*, United States, 107(24), pp. 3109–3116. doi: 10.1161/01.CIR.0000075572.40158.77.

Tsai, K., Hsu, T. G., Hsu, K. M., Cheng, H., Liu, T. Y., Hsu, C. F. and Kong, C. W. (2001) 'Oxidative DNA damage in human peripheral leukocytes induced by massive aerobic exercise.' *Free Radical Biology & Medicine*, United States, 31(11), pp. 1465–1472. doi: 10.1016/s0891-5849(01)00729-8.

Wang, J.-X., Gao, J., Ding, S.-L., Wang, K., Jiao, J.-Q., Wang, Y., Sun, T., Zhou, L.-Y., Long, B., Zhang, X.-J., Li, Q., Liu, J.-P., Feng, C., Liu, J., Gong, Y., Zhou, Z. and Li, P.-F. (2015) 'Oxidative modification of miR-184 enables it to target Bcl-xL and Bcl-w.' *Molecular Cell*, Elsevier Inc., 59(1), pp. 50–61. doi: 10.1016/j.molcel.2015.05.003.

Weimann, A., Belling, D. and Poulsen, H. E. (2002) 'Quantification of 8-oxo-guanine and guanine as the nucleobase, nucleoside and deoxynucleoside forms in human urine by high-performance liquid chromatography-electrospray tandem mass spectrometry.' *Nucleic Acids Research*, 30(2), p. E7. doi: 10.1093/nar/30.2.e7.

Weimann, A., Broedbaek, K., Henriksen, T., Stovgaard, E. S. and Poulsen, H. E. (2012) 'Assays for urinary biomarkers of oxidatively damaged nucleic acids.' *Free Radical Research*, 46(4), pp. 531–540. doi: 10.3109/10715762.2011.647693.

Weimann, A., McLeod, G., Henriksen, T., Cejvanovic, V. and Poulsen, H. E. (2019) 'Identification and quantification of isoguanosine in humans and mice.' *Scandinavian Journal of Clinical and Laboratory Investigation*, Taylor & Francis, pp. 1–8. doi: 10.1080/00365513.2019.1585566.

Willi, J., Küpfer, P., Evéquoz, D., Fernandez, G., Katz, A., Leumann, C. and Polacek, N. (2018) 'Oxidative stress damages rRNA inside the ribosome and differentially affects the catalytic center.' *Nucleic Acids Research*, 46(4), pp. 1945–1957. doi: 10.1093/nar/gkx1308.

Wu, L. L., Chiou, C. C., Chang, P. Y. and Wu, J. T. (2004) 'Urinary 8-OHdG: A marker of oxidative stress to DNA and a risk factor for cancer, atherosclerosis and diabetics.' *Clinica Chimica Acta*, 339(1–2), pp. 1–9. doi: 10.1016/j.cccn.2003.09.010.

Yan, L. L. and Zaher, H. S. (2019) 'How do cells cope with RNA damage and its consequences?' *The Journal of Biological Chemistry*, 294(41), pp. 15158–15171. doi: 10.1074/jbc.REV119.006513.

CHAPTER NINE

Exercise and DNA Damage

CONSIDERATIONS FOR THE NUCLEAR AND MITOCHONDRIAL GENOME

Josh Williamson and Gareth W. Davison

CONTENTS

Introduction / 103
Sources of Exercise-Induced RONS / 106
Exercise and DNA Damage / 107
nDNA Damage / 107
mtDNA Damage / 108
Contributing Variables / 109
Conclusion and Future Perspectives / 109
References / 110

INTRODUCTION

Regular exercise has a plethora of health benefits associated with human longevity, primarily through the prevention and management of noncommunicable diseases such as cancer, diabetes, and cardiovascular disease (Booth et al., 2017). These beneficial phenotypic adaptations are a result of acute and chronic responses, which are largely thought to be governed by redox-sensitive triggers (e.g., mitochondrial adaptation: sarcoplasmic calcium, ATP:ADP, NAD+:NADH, RONS). However, the metabolic challenge of sporadic and/or strenuous exercise augments the production of reactive oxygen and nitrogen species (RONS), eliciting a state of oxidative stress and subsequent damage to deoxyribonucleic acid (DNA) and potentially resulting in mutagenic, clastogenic and carcinogenic effects (Williamson and Davison, 2020). Current evidence demonstrates that high-intensity exercise is a stimulus for nuclear (nDNA) and mitochondrial (mtDNA) DNA damage (Tryfidou et al., 2020). Yet, the potential beneficial and/or detrimental responses of exercise-induced DNA damage are still to be fully elucidated. It is conceivable that an imbalance between DNA damage and repair may result in pathological outcomes as outlined in Box 9.1.

The majority of data on exercise-induced DNA damage has focused on the nuclear genome (Davison, 2016). This is partly explained by the limited availability of analytical techniques to assess DNA damage within the mitochondrial genome (Gonzalez-Hunt et al., 2018). For one, attempting to prepare, isolate, and extract mtDNA has potential to induce artefactual oxidation, likely through (1) exposure to elevated oxygen levels, (2) liberated metals from tissues or contained within extraction reagents, and (3) technique-specific variables such as long incubation periods, high temperatures, and mitochondrial isolation. The number of genomic copies of mtDNA (10–100 per cell) exceeds the

DOI: 10.1201/9781003051619-9

BOX 9.1: The Potential Downstream Beneficial and Pathological Signals of DNA Damage (Williamson and Davison, 2020)

Beneficial Effects: Regular, moderate-intensity exercise produces RONS inciting damage to DNA (and other macromolecules). Although complex, it has been proposed that the increase in RONS (and potentially the oxidative damage itself) can activate redox signals leading to an enhanced capacity of the organism to overcome future stress. Over time, these exercise-mediated signals trigger beneficial adaptations to RONS handling by influencing antioxidant enzyme capacity and DNA repair.

Detrimental Effects: High-intensity and sporadic exercise increase DNA damage; however, the complete molecular effects on health are complex and not fully understood. Nevertheless, the premise involves the generation of cell DNA oxidation products such as 8-oxo-2′-deoxyguanosine (8-oxo-dG), which is normally removed and repaired by a DNA glycosylase (such as 8-oxoguanine DNA glycosylase – OGG1). However, unpaired 8-oxo-dG can lead to a stalling in DNA replication forks and G > T transversion mutations.

capabilities of most methodologies to accurately quantify mtDNA damage, and the relatively small amount of mtDNA to nDNA ascertains that mtDNA is practically undetectable in total DNA extracts (Furda et al., 2014; Gonzalez-Hunt et al., 2018). As a result of these technical limitations, research examining DNA damage following exercise has focused on the nuclear genome. Although the measurements for the quantification of nDNA damage can be subject to the same artefacts synonymous with mtDNA, the comet assay attempts to overcome many of these problems and is generally considered the gold standard in measuring DNA strand breaks in eukaryotic cells. Additionally, due to the versatility, sensitivity and simplicity of the comet assay, it can be adapted (i.e., alkaline single-cell gel electrophoresis, neutral single-cell gel electrophoresis, lesion-specific enzymes, comet-fluorescence in situ hybridisation, etc.) to detect single- and double-strand damage, pyrimidine dimers, oxidised bases, and alkylation damage. Although there may be limited resolution, and a degree of variation between laboratories (Collins, 2002; Valavanidis et al., 2009), the comet assay remains the primary method for assessing DNA damage in exercise redox research. An overview of the common techniques to assess nuclear and mitochondrial DNA damage is found in Table 9.1 (Gonzalez-Hunt et al., 2018).

While the narrative surrounding the sources of RONS attributable to DNA damage following exercise are not well understood, it is known that physiologically relevant levels of superoxide, nitric oxide, hydrogen peroxide, or organic peroxides do not react at appreciable rates (or at all) with any DNA base or deoxyribose sugar. Although the identity of the reactive species generated from Fenton reactions remains equivocal, historically, the hydroxyl free radical ($^{\bullet}OH$) has been linked to exercise-induced DNA damage (Reaction 1 & 2: Cobley et al., 2015; Davison, 2016).

$$\text{Reaction 1: } H_2O_2 + Fe^{2+} \rightarrow Fe^{3+} + {}^{-}OH + {}^{\bullet}OH\left[k \sim 76\,M^{-1}\,s^{-1}\right]$$

$$\text{Reaction 2: } H_2O_2 + Cu^{+} \rightarrow Cu^{2+} + {}^{-}OH + {}^{\bullet}OH\left[k \sim 4.7\times10^{3}\,M^{-1}\,s^{-1}\right]$$

As a result, the hydroxyl free radical reacts appreciably with DNA bases at diffusion-controlled rates ($k \sim 3$–10×10^9 $M^{-1}s^{-1}$) with the 4 nucleobases and the 2-deoxyribose moiety; the only exception to this is the methyl group of thymine and the 2-amino group of guanine as these reactions are driven by competitive hydrogen abstraction ($\sim 2\times10^9$ $M^{-1}s^{-1}$) (Cadet and Davies, 2017; Chatgilialoglu et al., 2011). The most common reaction associated with DNA and the hydroxyl radical is via addition reactions to the 5,6-pyrimidine and 7,8-purine double bonds of nucleobases (von Sonntag, 1987; Cadet et al., 2017). Guanine is the most easily oxidised base due to its oxidation potential (1.29 V), which following an addition reaction by the hydroxyl free radical generates the promutagenic products, 8-oxo-7,8-dihydro-20-deoxyguanosine (8-oxodG) and

TABLE 9.1
Common approaches to quantifying nuclear and mitochondrial DNA damage.

Lesion	Assay	Description	Limitations
		Nuclear DNA	
Base/AP sites	Lesion-specific ELISA	The use of lesion-specific enzymes such as FPG, ENDOIII and hOGG1 aid in the detection oxidised purines and pyrimidines An enzyme immunoassay that uses a colour reaction of the enzyme coupled to the antibody. A commonly used technique due to its versatility, selectivity, specificity of reactions, the possibility of multiple repetitions, relatively short analysis time and relatively low cost	The sensitivity of certain enzymes has been questioned; for example, FPG also recognises alkylation damage. Additionally as per the ESCODD project, results between laboratories can vary A limited number of commercially available specific antibodies and high cost of purchase, reduced specificity of antibodies labelled with enzymes and the need to obtain appropriate dilutions of reagents in case of indirect techniques. More specifically for the case of DNA damage detection, the possible cross-reactivity of applied antibodies with DNA bases
Strand breaks	Single cell gel electrophoresis Fluorescence in situ hybridisation (FISH) Immunofluorescence	A simple and rapid method for the quantification of DNA damage and breaks. 'Comets' are made by the fragmented DNA being pulled by the electric current; allowing for the measurement of strand breaks FISH is a variant of the SCGE assay with the additional sensitivity of examining a particular gene of interest Antibodies are used to target proteins along the DNA damage-repair response cascade such as γ-H2AX, XRCC1 or Ku proteins	The assay loses accuracy for small deletions. Limited by the amount of samples that can be analysed simultaneously; although, modern automated systems are making this process more efficient. Due to the 10 kB resolution of the SCGE assay, for FISH detection of gene-specific loci, targeting probes need to be at least 10 kB or longer Possibility of false positives. Further, some markers (e.g., γ-H2AX) may occur due to apoptosis or as intermediates to repair and/or replication. Although automated systems are being used, many researchers count antibody foci manually
Varied	PCR-based assays (e.g. LORD-Q, LA-qPCR)	Long-run real-time (rt) PCR for DNA damage quantification allows sensitive and robust detection of the number of lesions (including double-stranded breaks, abasic sites, thymine dimers and 5-hydroxymethyl dC) within a defined sequence	Genome coverage is often limited to ~3–4 kb segments/amplicons. Specificity is limited to any lesions capable of stalling a polymerase. Optimisation requires

(Continued)

TABLE 9.1 (***Continued***)
Common approaches to quantifying nuclear and mitochondrial DNA damage.

Lesion	Assay	Description	Limitations
		Mitochondrial DNA	
Abasic sites	Aldehyde reactive probe immunoflourescence	This reacts with a ring-opened sugar moiety resulting from the loss of a base in the DNA (abasic sites)	Low throughput
Varied	PCR-based assays	Quantification of lesion frequency capable of stalling a polymerase during PCR. Long amplicons can be used resulting in almost full coverage of the mitochondrial genome	Specific lesions are not detectable

2,6-diamino-5-formamido4-hydroxypyrimidine (FAPy-G) (Kawanishi et al., 2001). The relative amount (and indeed mechanisms) of end products of hydroxyl-derived reactions (i.e., addition, ring-opening, reduction, protonation) to the purine and pyrimidine bases can be affected by the presence of oxygen and/or transition metal ions.

More recently, a working hypothesis suggests that due to the physiological concentration of bicarbonate (25 and 14.4 mM in serum and intracellular media, respectively), hydroxyl free radicals are not formed through Fenton reactions in mammalian cells (Illés et al., 2020). Instead, it has been proposed that the carbonate radical anion is the primary product from the Fenton reaction (Reaction 3: Illés et al., 2019; Fleming and Burrows, 2020):

$$\text{Reaction 3}: \left(Fe^{II}(CO_3)(OOH)(H_2O)_2 \rightarrow Fe^{III}(OH)_3(H_2O) + CO_3^{\bullet-}\right)$$

Given the reduction potentials of both the hydroxyl and carbonate radicals (pH 7.0, $E° = 2.4$ and 1.6 V respectively), it is not surprising that 2′-deoxyguanosine is a common outcome for these potent one-electron oxidants (Medinas et al., 2007). Hydroxyl attack on DNA generates an array of products, resulting from hydrogen abstraction from 2′ deoxyribose (H1′ abstraction), 5,8′-cyclopurines (H5′ abstraction), or any other organic compounds or biomolecules in its vicinity (Cadet et al., 2017). Instead, the carbonate radical primarily functions as a one-electron oxidant resulting in a guanine radical cation and subsequent products of C8 or C5 oxidation (Fleming and Burrows, 2020). With that being said, the biological relevance and implications of the carbonate radical in Fenton-mediated chemistry and subsequent oxidative DNA damage are yet to be ascertained, and readers are directed to recent work underlying the chemistry of carbonate attack on DNA (Fleming and Burrows, 2020).

SOURCES OF EXERCISE-INDUCED RONS

While the contribution surrounding the mechanisms associated with RONS generation remain an active area of investigation, NADPH oxidase, the mitochondrial electron transport chain and xanthine oxidase, are often cited as predominant sources during exercise. Prototypic NADPH oxidase (NOX) enzymes are thought to dominate exercise-induced $O_2^{\bullet-}$ production, in part, because several factors (notably ATP demand) should decrease mitochondrial $O_2^{\bullet-}$ production. More specifically, it appears NOX_2 with its access to the cytosolic domain and the localisation of NOX_4 to the mitochondria are the potential suspects of exercise-mediated generation of superoxide and basal hydrogen peroxide (NOX_4 only). Furthermore, contraction-induced superoxide production is largely derived from non-mitochondrial sources, specifically NOX_2 due to agonist activation (e.g., angiotensin II, cytokines and mechanical stress) (Sakellariou et al., 2014). Traditionally, it was understood that exercise-induced superoxide originated from the mitochondrial respiratory chain (Boveris and Chance,

1973; St-Pierre et al., 2002); however recent estimations suggest greater concentrations of mitochondrial superoxide are produced during State 4 respiration (i.e., basal) in comparison to State 3 (i.e., exercise). The reduced flavin mononucleotide or the $N-1a$ and $N-1b$ iron–sulphur clusters from complex I ($NADH + FMN \rightarrow FMNH^- + NAD^+$; $FMNH + O_2 \rightarrow FMN + O_2^{\bullet-}$) and the ubiquinol oxidation site of complex III ($SQ^{\bullet-} + O_2 \rightarrow Q + O_2^{\bullet-}$; Q pool- and membrane potential-dependent) are understood to be major sources of $O_2^{\bullet-} / H_2O_2$ in the respiratory chain. Additional sources of mitochondrial $O_2^{\bullet-} / H_2O_2$ include complex II ($FADH^- + O_2 \rightarrow FAD + O_2^{\bullet-}$) and reverse electron transfer; however, conjecture still exists as to the specific mechanisms involved. It is also worth noting that nitric oxide synthase catalyses the generation of nitric oxide, an important signalling molecule that can readily react with superoxide to form the strong oxidising agent peroxynitrite ($O_2^{\bullet-} + NO \rightarrow ONOO^-[k \sim 7 \times 10^9\ M^{-1} s^{-1}]$), thus increasing the potential for hydroxyl (Reaction 4) and/or carbonate (Reaction 5) free radical formation via peroxynitrous acid or nitrosoperoxocarbonate decomposition respectively. For an appreciable insight to the discourse regarding the potential sources of $O_2^{\bullet-} / H_2O_2$ during exercise, readers are directed to the review by Powers et al. (2020).

$$\text{Reaction 4}: ONOO^{\bullet} + H^+ \leftrightarrow ONOOH$$

$$\rightarrow \left[ONO^{\bullet}\ {}^{\bullet}OH\right] \rightarrow {}^{\bullet}NO_2 + {}^{\bullet}OH$$

$$\text{Reaction 5}: ONOO^{\bullet} + CO_2 / HCO_3^- \rightarrow ONOOCO_2^-$$

$$\rightarrow {}^{\bullet}NO_2 + CO_3^{\bullet-}$$

EXERCISE AND DNA DAMAGE

The link between DNA and exercise is of interest due to the established relationship between DNA damage and a range of pathological diseases (Pham-Huy et al., 2008). The ability of exercise to produce hydroxyl free radicals (and possibly inhibit DNA repair) potentiates the generation of multiple deleterious products capable of further propagating genomic damage (Figure 9.1). Given the common use of single DNA damage biomarkers in exercise redox research, this suggests there is a high probability of underestimating the total extent of oxidative damage as a function of exercise. Not to mention, the current analytical techniques available can only quantify a fraction of the oxidised modified biomarkers associated with DNA damage (Nikitaki et al., 2015).

nDNA DAMAGE

Oxidised DNA bases are common biomarkers used to quantify nDNA damage as a consequent of exercise; specifically, oxidised guanine in the form of 8-oxoGua and 8-oxodGua is the most commonly measured end products. Okamura et al. (1997) reported an increase in urinary 8-hydroxy-deoxyguanosine over an 8-day training camp consisting of 30±3 km/day. These results have been confirmed in sub-elite runners following 42-km of running; interestingly, urinary 8-hydroxy-2′-deoxyguanosine concentrations still had not returned to baseline at the 1-week follow-up time point (Tsai et al., 2001). It is also worth noting that prolonged, strenuous ultra-endurance exercise has confirmed the increase in urinary 8-hydroxy-deoxyguanosine following a 6-day period of 11 hours of exercise per day (Poulsen et al., 1996), a 2-day ultramarathon (Miyata et al., 2008), and 4-day supra marathon (Radák et al., 2000). Although these data would suggest exercise of varying intensities and modalities can increase DNA damage, which is then circulated and ultimately excreted in the urine, there are a number of discrepancies between methodologies which can contribute to indirect and/or unreliable results (Poulsen et al., 1999; Knasmüller et al., 2008). This could include artefactual DNA oxidation from ambient oxygen and transition metal ions during the extraction and/or analysis of samples (Finaud and Biologie, 2006; Nie et al., 2013).

Investigators have used other biomarkers (such as base lesions/AP sites, single- and double-strand breaks and PCR-based assays) to investigate oxidative damage to DNA (Gonzalez-Hunt et al., 2018). In a seminal study, Hartmann and colleagues (1994) used the comet assay to detect DNA migration in lymphocytes following an incremental treadmill test to volitional fatigue. They reported that DNA migration peaked at 24-hours following the oxidative insult; interestingly, this damage was still present for up to 72-hours before returning to baseline. This was followed by a series of studies by Niess which demonstrated a clear and significant increase in DNA migration 24-hours following exhaustive exercise (Niess et al., 1996) and a half marathon (Niess et al., 1998) in peripheral blood mononuclear cells. These results are supported by other similar work (Mars et al., 1998; Davison et al., 2005; Fogarty et al., 2013a,b; Williamson et al., 2018). The data corroborating exercise-induced DNA damage are not just confined to exhaustive running as there is evidence which suggests that maximal cycling (Zhang et al., 2004), marathon running (Tsai et al., 2001), Ironman triathlons (Neubauer et al., 2008; Reichhold et al., 2009) and rowing (Sardas et al., 2012), all provide a sufficient physiological challenge to damage nDNA.

The literature examining exercise-induced nDNA damage predominantly focuses on single-strand breaks through the use of the comet assay (Azqueta et al., 2014), with studies also including base-oxidation via enzyme incubation of ENDO III, FPG or hOGG1 (Soares et al., 2015). However, double-strand breaks (DSB) have greater biological significance with a single DSB able to induce cell apoptosis and genomic instability, yet the experimental evidence associated with exercise and DNA DSBs is lacking. To date, only three studies have investigated whether DNA DSBs occur as a consequent of exercise. Lippi and colleagues (2016) identified a dose-dependent response between exercise duration (5-, 10-, 21- and 42-km) and γ-H2AX detection, with maximum damage occurring following the 42-km run. A follow-up study demonstrated an increase in γ-H2AX following 21-km of running in healthy amateur athletes, a response which was exacerbated in diabetic runners (Lippi et al., 2018). Most recently, we demonstrated a 2.5- and 3.5-fold increase in γ-H2AX and 53BP1 foci respectively following 30-minutes of exercise at 80%–85% of $\dot{V}O_{2max}$ in normobaric hypoxia (12% FiO_2; ≈ 4,600 m) (Williamson et al., 2020a,b). Similar outcomes were demonstrable in the normoxic group (20.9% FiO_2), although not to the same extent. While the mechanisms of DNA damage have been briefly covered in this chapter (and certainly more comprehensively elsewhere; Chatterjee and Walker, 2017), it is worth highlighting additional mechanisms that may be responsible for potentiating DNA damage when exercising in hypoxia. For one, it is highly probable that mitochondria become inefficient and shift to a more reductive state; thus, causing the spontaneous and/or enzyme-catalysed generation of superoxide (Hernansanz-Agustín et al., 2014), perhaps due to a high proton motive force, increased ubiquinol/ubiquinone and increased $NADH/NAD^+$ (Loscalzo, 2016). It is also plausible that hypoxia increases vascular epinephrine, which is maximally stimulated during the initial exposure to acute hypoxia as arterial saturation declines (Mazzeo et al., 1998), and this may enhance primary radicals via an auto-oxidation mechanism.

mtDNA DAMAGE

Mitochondria contain a polyploid, 16,569 base pair circular genome which can be categorised as homo- or hetero-plasmy depending on the sequencing of the multiple copies within the cell. Maintaining the integrity of the mitochondrial genome is important to prevent irreversible loss or modification of its coded information, which is particularly detrimental in postmitotic tissue (Williamson and Davison, 2020). Comparative to nDNA, damage to mtDNA results in an array of oxidative modifications potentiated by the mitochondrial matrix microenvironment. Further, mtDNA is particularly susceptible to oxidative attack due to the close proximity of the genome (and abundance of cardiolipin and secondary peroxidative products) to multiple sources of mitochondrial superoxide, lack of protective histone proteins and limited capacity for repair (Cline, 2012; Brand, 2016). As a consequent of the lack of introns, genetic information within the mitochondrial genome is more tightly packed, resulting in higher relative levels of oxidative damage capable of harming a gene, thus amplifying the potential downstream detrimental and mutational effects (Pamplona, 2011). For one, correlations between mtROS and mtDNA strand breaks suggest that greater DNA damage has the potential to incite a greater formation of mtROS as impaired mtDNA translation leads to mitochondrial uncoupling with secondary increases in mtROS formation (Mikhed et al., 2015): i.e., 8-oxo-deoxyguanosine is a potential mutagenic lesion, and its accumulation is directly correlated with the development of pathological processes (Souza-Pinto et al., 1999; Barja and Herrero, 2000).

Currently, substantial experimental evidence linking the impact of exercise to mtDNA damage is hampered due to the lack of assay specificity and sensitivity as outlined. That said, in a series of experiments, Davison and colleagues determined the effect of continuous maximal concentric contractions on measures of mtDNA damage and oxidative stress. First, Fogarty et al. (2013a) reported an increase in muscle mitochondrial 8-OHdG with a concomitant decrease in total antioxidant capacity following exercise. Davison et al. (2014) then showed that increased oxidative damage to mtDNA is accompanied by an upregulation in haemeoxygenase-1, providing tentative evidence that cell oxidant stress is required to activate the signal transduction cascade related to haemeoxygenase-1. More recently, we have demonstrated an increase in global mtDNA damage in peripheral blood mononuclear cells and skeletal muscle (as quantified using LA-qPCR) following high-intensity intermittent exercise; a response that was attenuated by chronic supplementation (20 mg/day for 3 weeks) of the mitochondrial-targeted antioxidant, MitoQ (Williamson et al., 2020b). Taken together, these studies highlight a number of key concepts to the exercise redox field which require further clarity: (1) nuclear and mitochondrial DNA are damaged following high-intensity exercise, likely as a result of superoxide/hydrogen peroxide generation from both mitochondrial and non-mitochondrial sources; (2) the use of mitochondrial-targeted antioxidants may mitigate mtDNA damage potentially through (i) directly or indirectly metabolising key reactive species, (ii) altering the activities of respiratory chain complexes and/or (iii) exerting extramitochondrial effects; and (3) the ability of mitochondria to act as a sink by sequestering cytosolic hydrogen peroxide. At this stage, very little is known about the effect of exercise on the mitochondrial genome (from the perspective of DNA damage), and if anything, the findings of the presented research only raise further questions and ambiguity as to the role and implications of mtDNA damage.

CONTRIBUTING VARIABLES

Across various modalities and intensities, exercise (for the most part) induces damage to the nuclear and mitochondrial genomes (Fisher-Wellman and Bloomer, 2009); however, several confounding variables are worthy of brief discussion. For one, Fogarty et al. (2011) demonstrated a strong correlation between lipid-derived free radicals (i.e., alkoxyl free radicals) and exercise-induced DNA damage and lipid peroxidation. In addition to the ability of primary/secondary lipid radicals to directly damage DNA, the abundance of vulnerable mitochondrial lipids and the peroxidation of cell membranes by lipid-derived radicals can alter functionality, thus further increasing susceptibility of free radical-mediated DNA damage. Secondly, the ability of exercise to trigger an inflammatory response led to the hypothesis that DNA damage (and/or inhibition of DNA repair) may (in part) be explicated by the activation of immuno-inflammatory processes capable of propagating the oxidative stress response. Although Niess et al. (1998) demonstrated a correlation between endurance running and DNA damage in leukocytes, and they lacked specific markers to determine a specific mechanistic at play. Neubauer and colleagues (2008) later proposed that although exercise-induced muscle damage may activate an acute and/or delayed inflammatory response leading to DNA damage, they concluded more work is needed to elucidate this relationship; readers are directed towards the reviews of Kozakowska et al. (2015), Kawamura and Muraoka (2018) and Kay et al. (2019) for an appreciable understanding of the inflammation-induced DNA damage. Finally, training status, antioxidant capacity and/or nutritional (in)sufficiency, age and sex differences have all been shown to alter the exercise-mediated oxidative stress response (Radak et al., 2017).

CONCLUSION AND FUTURE PERSPECTIVES

The nuclear and mitochondrial genomes are subject to oxidative damage following high-intensity and/or prolonged exercise. While intensity and duration may be the two primary determinants controlling the extent of DNA damage, sex, age, nutritional and training status may also act as confounding variables that need to be considered. This builds on the recent review by Tryfidou et al. (2020) outlining that a one-dimensional cause-and-effect relationship doesn't fully encapsulate the complex network of variables involved in the interplay between exercise and DNA modification. Consideration as to the type of RONS and their biological significance, alongside an understanding of the type and magnitude of DNA damage is warranted. For example, very little is known about the effect of mtDNA damage and retrograde communication to the nuclear domain. The role of mitochondrial-targeted compounds on the

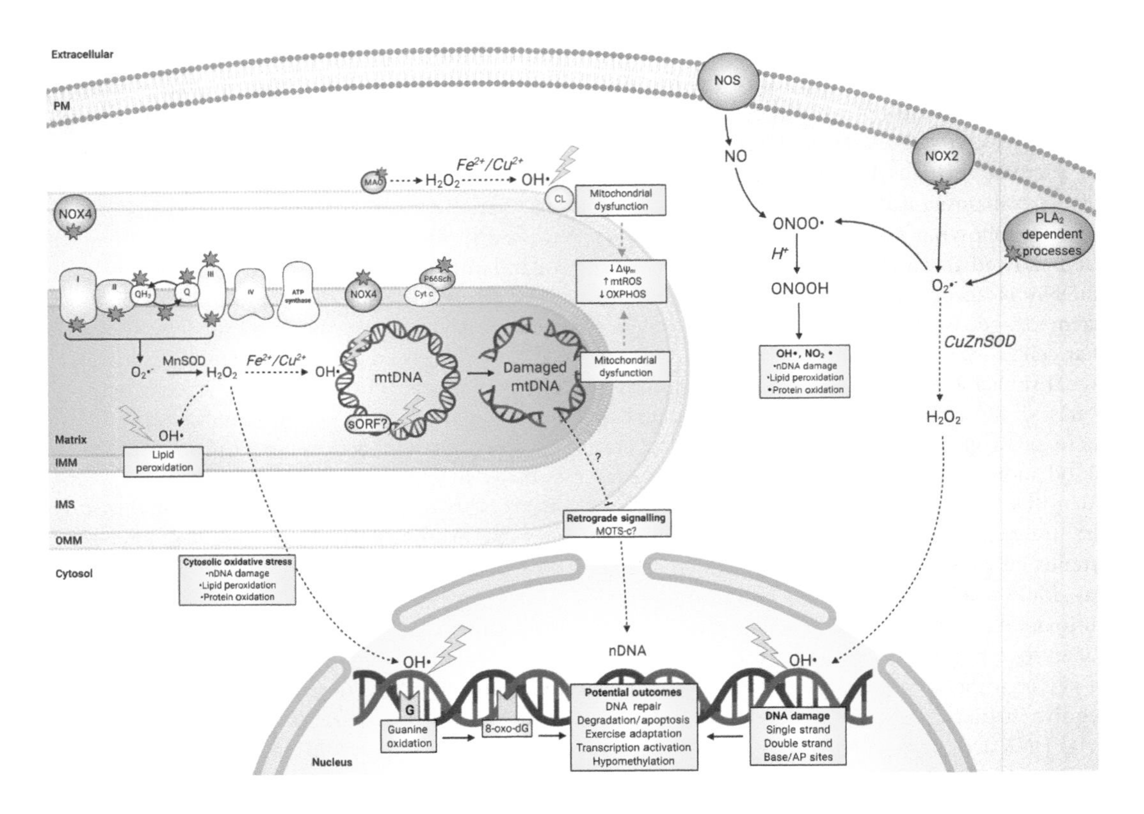

Figure 9.1 Overview of exercise-induced sources of RONS and potential consequences of mitochondrial and nuclear DNA damage. Red stars indicate potential sources of superoxide and hydrogen peroxide. Abbreviations: $OH^{•}$, hydroxyl free radical; CL, cardiolipin; CuZnSOD, copper-zinc superoxide dismutase; Cyt C, Cytochrome C; Fe^{2+}/Cu^{2+}, iron/copper; H_2O_2, hydrogen peroxide; IMM, inner mitochondrial membrane; IMS, intermembrane space; OMM, outer mitochondrial membrane; MAO, monoamine oxidase; MnSOD, manganese superoxide dismutase; mtDNA, mitochondrial DNA; NOX4, NADPH oxidase isoform 4; $NO_2^{•}$, nitrogen dioxide; nDNA, nuclear DNA; $O_2^{•-}$, superoxide; OMM, outer mitochondrial membrane; ONOOH, peroxynitrous acid, PM, plasma membrane; Q, Ubiquinone; QH2, Ubiquinol; sORF, short open-reading frame.

metabolism of key reactive species, the alteration of respiratory chain complex activity and/or exertion of extramitochondrial effects in the context of exercise is yet to be ascertained. We need to determine whether exercise damages a particular genomic region (i.e., promoter, exon, intron etc.), and the potential impact this may have on genome stability. Knowledge and understanding of exercise-induced damage in relation to epigenetics and transcriptional activation of oncogenic, pro-inflammatory and proangiogenic genes is also required. Traditionally, 13 protein-coding genes encompass the mitochondrial genome; however recently, short open reading frames (sORFs) have been identified, and it has been determined that the mitochondrial-derived peptide (MOTS-c) is produced from sORFs (Reynolds et al., 2021; Kim et al., 2018). It would be of interest to determine whether (1) oxidative damage to the mitochondria is an activator of MOTS-c expression and (2) if exercise-induced RONS damages a particular sORF which encodes for MOTS-c, is translocation to the nuclear genome and downstream nuclear expression halted until repair has taken place? The above future perspectives may be targeted at transient (i.e., exercise) and chronic (i.e., certain pathologies) mitochondrial oxidative stress scenarios, which have implications for age-related diseases and longevity.

REFERENCES

Azqueta, A., Slyskova, J., Langie, S.A.S., Gaivao, I. and Collins, A., (2014) Comet assay to measure DNA repair: Approach and applications. *Frontiers in Genetics*. doi: 10.3389/fgene.2014.00288.

Barja, G. and Herrero, A., (2000) Oxidative damage to mitochondrial DNA is inversely related to maximum life span in the heart and brain of mammals. *The FASEB Journal*, 142, pp. 312–318.

Booth, F.W., Roberts, C.K., Thyfault, J.P., Ruegsegger, G.N. and Toedebusch, R.G., (2017) Role of inactivity in chronic diseases: Evolutionary insight and pathophysiological mechanisms. *Physiological Reviews*, 97, pp. 1351–1402.

Boveris, A. and Chance, B., (1973) The mitochondrial generation of hydrogen peroxide. General properties and effect of hyperbaric oxygen. *Biochemical Journal*, 1343, pp. 707–716.

Brand, M.D., (2016) Mitochondrial generation of superoxide and hydrogen peroxide as the source of mitochondrial redox signaling. *Free Radical Biology and Medicine*, 100, pp. 14–31.

Cadet, J. and Davies, K.J.A., (2017) Oxidative DNA damage & repair: An introduction. *Free Radical Biology and Medicine*, 107, pp. 2–12.

Cadet, J., Davies, K.J.A., Medeiros, M.H., Di Mascio, P. and Wagner, J.R., (2017) Formation and repair of oxidatively generated damage in cellular DNA. *Free Radical Biology and Medicine*, 107, pp. 13–34.

Chatgilialoglu, C., D'Angelantonio, M., Kciuk, G. and Bobrowski, K., (2011) New insights into the reaction paths of hydroxyl radicals with 2′-deoxyguanosine. *Chemical Research in Toxicology*, 2412, pp. 2200–2206.

Chatterjee, N. and Walker, G.C., (2017) Mechanisms of DNA damage, repair, and mutagenesis. *Environmental and Molecular Mutagenesis*, 58, pp. 235–263.

Cline, S.D., (2012) Mitochondrial DNA damage and its consequences for mitochondrial gene expression. *Biochimica et Biophysica Acta: Gene Regulatory Mechanisms*, 18199–10, pp. 979–991.

Cobley, J.N., Margaritelis, N.V., Morton, J.P., Close, G.L., Nikolaidis, M.G. and Malone, J.K., (2015) The basic chemistry of exercise-induced DNA oxidation: Oxidative damage, redox signaling, and their interplay. *Frontiers in Physiology*. doi: 10.3389/fphys.2015.00182.

Collins, A., (2002) Comparative analysis of baseline 8-oxo-7, 8-dihydroguanine in mammalian cell DNA, by different methods in different laboratories : An approach to consensus ESCODD (European Standards Committee on Oxidative median value obtained with the enzymatic approach. *Carcinogenesis*, 2312, pp. 2129–2133.

Davison, G.W., (2016) Exercise and oxidative damage in nucleoid DNA quantified using single cell gel electrophoresis: Present and future application. *Frontiers in Physiology*. doi: 10.3389/fphys.2016.00249.

Davison, G.W., Hughes, C.M. and Bell, R.A., (2005) Exercise and mononuclear cell DNA damage: The effects of antioxidant supplementation. *International Journal of Sport Nutrition and Exercise Metabolism*, 155, pp. 480–492.

Davison, G.W., De Vito, G., Hughes, C.M., Burke, G., McEneny, J., Brown, D., McClean, C. and Fogarty, M.C., (2014) Mitochondrial DNA damage following isolated muscle exercise: A preliminary investigation into HO-1 gene expression. *Medicine and Science in Sports and Exercise*, 46(5S), p. 299.

Finaud, J. and Biologie, L., (2006) Oxidative stress: Relationship eith exercise and training. *Sports Medicine*, 364, pp. 327–358.

Fisher-Wellman, K. and Bloomer, R.J., (2009) Acute exercise and oxidative stress: A 30 year history. *Dynamic Medicine*, 81, pp. 1–25.

Fleming, A.M. and Burrows, C.J., (2020) On the irrelevancy of hydroxyl radical to DNA damage from oxidative stress and implications for epigenetics. *Chemical Society Reviews*, 49, pp. 6524–6528.

Fogarty, M.C., Hughes, C.M., Burke, G., Brown, J.C., Trinick, T.R., Duly, E., Bailey, D.M. and Davison, G.W., (2011) Exercise-induced lipid peroxidation: Implications for deoxyribonucleic acid damage and systemic free radical generation. *Environmental and Molecular Mutagenesis*, 521, pp. 35–42.

Fogarty, M.C., Devito, G., Hughes, C.M., Burke, G., Brown, J.C., McEneny, J., Brown, D., McClean, C. and Davison, G.W., (2013a) Effects of α-lipoic acid on mtDNA damage after isolated muscle contractions. *Medicine and Science in Sports and Exercise*, 458, pp. 1469–1477.

Fogarty, M.C., Hughes, C.M., Burke, G., Brown, J.C. and Davison, G.W., (2013b) Acute and chronic watercress supplementation attenuates exercise-induced peripheral mononuclear cell DNA damage and lipid peroxidation. *British Journal of Nutrition*, 1092, pp. 293–301.

Furda, A., Santos, J.H., Meyer, J.N. and Van Houten, B., (2014) Quantitative PCR-based measurement of nuclear and mitochondrial DNA damage and repair in mammalian cells. *Methods in Molecular Biology*, 1105, pp. 419–437.

Gonzalez-Hunt, C.P., Wadhwa, M. and Sanders, L.H., (2018) DNA damage by oxidative stress: Measurement strategies for two genomes. *Current Opinion in Toxicology*, 7, pp. 87–94.

Hartmann, A., Plappert, U., Raddata, K., Grunert-Fuchs, M. and Speit, G., (1994) Does physical activity induce DNA damage? *Mutagenesis*, 93, pp. 269–272.

Hernansanz-Agustín, P., Izquierdo-Álvarez, A., Sánchez-Gómez, F.J., Ramos, E., Villa-Piña, T., Lamas, S., Bogdanova, A. and Martínez-Ruiz, A., (2014) Acute hypoxia produces a superoxide burst in cells. *Free Radical Biology and Medicine*, 71, pp. 146–156.

Illés, E., Mizrahi, A., Marks, V. and Meyerstein, D., (2019) Carbonate-radical-anions, and not hydroxyl radicals, are the products of the Fenton reaction in neutral solutions containing bicarbonate. *Free Radical Biology and Medicine*, 131, pp. 1–6.

Illés, E., Patra, S.G., Marks, V., Mizrahi, A. and Meyerstein, D., (2020) The FeII(citrate) Fenton reaction under physiological conditions. *Journal of Inorganic Biochemistry*, 206, p. 111018.

Kawamura, T. and Muraoka, I., (2018) Exercise-induced oxidative stress and the effects of antioxidant intake from a physiological viewpoint. *Antioxidants*. doi: 10.3390/antiox7090119.

Kawanishi, S., Hiraku, Y. and Oikawa, S., (2001) Mechanism of guanine-specific DNA damage by oxidative stress and its role in carcinogenesis and aging. *Reviews in Mutation Research*, 4881, pp. 65–76.

Kay, J., Thadhani, E., Samson, L. and Engelward, B., (2019) Inflammation-induced DNA damage, mutations and cancer. *DNA Repair*, 83102673, pp. 1–53.

Kim, K.H., Son, J.M., Benayoun, B.A. and Lee, C., (2018) The mitochondrial-encoded peptide MOTS-c translocates to the nucleus to regulate nuclear gene expression in response to metabolic stress. *Cell Metabolism*, 283, pp. 516–524.

Knasmüller, S., Nersesyan, A., Mišík, M., Gerner, C., Mikulits, W., Ehrlich, V., Hoelzl, C., Szakmary, A. and Wagner, K.H., (2008) Use of conventional and -omics based methods for health claims of dietary antioxidants: A critical overview. *British Journal of Nutrition*, 99, ES3–52.

Kozakowska, M., Pietraszek-Gremplewicz, K., Jozkowicz, A. and Dulak, J., (2015) The role of oxidative stress in skeletal muscle injury and regeneration: Focus on antioxidant enzymes. *Journal of Muscle Research and Cell Motility*, 366, pp. 377–393.

Lippi, G., Buonocore, R., Tarperi, C., Montagnana, M., Festa, L., Danese, E., Benati, M., Salvagno, G.L., Bonaguri, C., Roggenbuck, D. and Schena, F., (2016) DNA injury is acutely enhanced in response to increasing bulks of aerobic physical exercise. *Clinica Chimica Acta*, 460, pp. 146–151.

Lippi, G., Tarperi, C., Danese, E., Montagnana, M., Festa, L., Benati, M., Salvagno, G.L., Bonaguri, C., Bacchi, E., Donà, S., Roggenbuck, D., Moghetti, P. and Schena, F., (2018) Middle-distance running and DNA damage in diabetics. *Journal of Laboratory and Precision Medicine*, 3, pp. 18–18.

Loscalzo, J., (2016) Adaptions to hypoxia and redox stress: Essential concepts confounded by misleading terminology. *Circulation Research*, 1194, pp. 511–513.

Mars, M., Govender, S., Weston, A., Naicker, V. and Chuturgoon, A., (1998) High intensity exercise: A cause of lymphocyte apoptosis? *Biochemical and Biophysical Research Communications*, 2492, pp. 366–370.

Mazzeo, R.S., Child, A., Butterfield, G.E., Mawson, J.T., Zamudio, S. and Moore, L.G., (1998) Catecholamine response during 12 days of high-altitude exposure (4,300 m) in women. *Journal of Applied Physiology*, 844, pp. 1151–1157.

Medinas, D.B., Cerchiaro, G., Trindade, D.F. and Augusto, O., (2007) The carbonate radical and related oxidants derived from bicarbonate buffer. *IUBMB Life*, 594–5, pp. 255–262.

Mikhed, Y., Daiber, A. and Steven, S., (2015) Mitochondrial oxidative stress, mitochondrial DNA damage and their role in age-related vascular dysfunction. *International Journal of Molecular Sciences*, 167, pp. 15918–15953.

Miyata, M., Kasai, H., Kawai, K., Yamada, N., Tokudome, M., Ichikawa, H., Goto, C., Tokudome, Y., Kuriki, K., Hoshino, H., Shibata, K., Suzuki, S., Kobayashi, M., Goto, H., Ikeda, M., Otsuka, T. and Tokudome, S., (2008) Changes of urinary 8-hydroxydeoxyguanosine levels during a two-day ultramarathon race period in Japanese non-professional runners. *International Journal of Sports Medicine*, 291, pp. 27–33.

Neubauer, O., Reichhold, S., Nersesyan, A., König, D. and Wagner, K.H., (2008) Exercise-induced DNA damage: Is there a relationship with inflammatory responses? *Exercise Immunology Review*, 14, pp. 51–72.

Nie, B., Gan, W., Shi, F., Hu, G.X., Chen, L.G., Hayakawa, H., Sekiguchi, M. and Cai, J.P., (2013) Age-dependent accumulation of 8-oxoguanine in the DNA and RNA in various rat tissues. *Oxidative Medicine and Cellular Longevity*, 2013, p. 303181.

Niess, A.M., Hartmann, A., Grünert-Fuchs, M., Poch, B. and Speit, G., (1996) DNA damage after exhaustive treadmill running in trained and untrained men. *International Journal of Sports Medicine*, 176, pp. 397–403.

Niess, A.M., Baumann, M., Roecker, K., Horstmann, T., Mayer, F. and Dickhuth, H.H., (1998) Effects of intensive endurance exercise on DNA damage in leucocytes. *The Journal of Sports Medicine and Physical Fitness*, 322, pp. 111–115.

Nikitaki, Z., Hellweg, C.E., Georgakilas, A.G. and Ravanat, J.L., (2015) Stress-induced DNA damage biomarkers: Applications and limitations. *Frontiers in Chemistry*. doi: 10.3389/fchem.2015.00035.

Okamura, K., Doi, T., Hamada, K., Sakurai, M., Yoshioka, Y., Mitsuzono, R., Migita, T., Sumida, S. and Sugawa-Katayama, Y., (1997) Effect of repeated exercise on urinary 8-hydroxy-deoxyguanosine excretion in humans. *Free Radical Research*, 266, pp. 507–514.

Pamplona, R., (2011) Mitochondrial DNA damage and animal longevity: Insights from comparative studies. *Journal of Aging Research*, 2011, 807108.

Pham-Huy, L.A., He, H. and Pham-Huy, C., (2008) Free radicals, antioxidants in disease and health. *International Journal of Biomedical Science : IJBS*, 42, pp. 89–96.

Poulsen, H.E., Loft, S. and Vistisen, K., (1996) Extreme exercise and oxidative DNA modification. *Journal of Sports Sciences*, 144, pp. 343–346.

Poulsen, H.E., Weimann, A., and Loft, S., (1999) Methods to detect DNA damage by free radicals: Relation to exercise. *The Proceedings of the Nutrition Society*, 584, pp. 1007–1014.

Powers, S.K., Deminice, R., Ozdemir, M., Yoshihara, T., Bomkamp, M.P. and Hyatt, H., (2020) Exercise-induced oxidative stress: Friend or foe? *Journal of Sport and Health Science*, 9, pp. 415–425.

Radák, Z., Pucsuk, J., Boros, S., Josfai, L. and Taylor, A.W., (2000) Changes in urine 8-hydroxydeoxyguanosine levels of super-marathon runners during a four-day race period. *Life Sciences*, 6618, pp. 1763–1767.

Radak, Z., Ishihara, K., Tekus, E., Varga, C., Posa, A., Balogh, L., Boldogh, I. and Koltai, E., (2017) Exercise, oxidants, and antioxidants change the shape of the bell-shaped hormesis curve. *Redox Biology*, 12, pp. 285–290.

Reichhold, S., Neubauer, O., Bulmer, A.C., Knasmüller, S. and Wagner, K.H., (2009) Endurance exercise and DNA stability: Is there a link to duration and intensity? *Reviews in Mutation Research*, 682, pp. 28–38.

Reynolds, J.C., Lai, R.W., Woodhead, J.S.T., Joly, J.H., Mitchell, C.J., Cameron-Smith, D., Lu, R., Cohen, P., Graham, N.A., Benayoun, B.A., Merry, T.L. and Lee, C., (2021) MOTS-c is an exercise-induced mitochondrial-encoded regulator of age-dependent physical decline and muscle homeostasis. *Nature Communications*, 121, pp. 1–11.

Sakellariou, G.K., Jackson, M.J. and Vasilaki, A., (2014) Redefining the major contributors to superoxide production in contracting skeletal muscle. The role of NAD(P)H oxidases. *Free Radical Research*, 481, pp. 12–29.

Sardas, S., Omurtag, G.Z., Monteiro, I.F.C., Beyogiu, D., Tozan-Beceren, A., Topesaki, N. and Cotuk, H.B., (2012) Assessment of DNA damage and protective role of vitamin E supplements after exhaustive exercise by comet assay in athletes. *Journal of Clinical Toxicology*, S501, pp. 1–4.

Soares, J.P., Silva, A.M., Oliveira, M.M., Peixoto, F., Gaivão, I. and Mota, M.P., (2015) Effects of combined physical exercise training on DNA damage and repair capacity: Role of oxidative stress changes. *Age*, 37(3), p. 9799.

Souza-Pinto, N.C., Croteau, D.L., Hudson, E.K., Hansford, R.G. and Bohr, V.A., (1999) Age-associated increase in 8-oxo-deoxyguanosine glycosylase/AP lyase activity in rat mitochondria. *Nucleic Acids Research*, 278, pp. 1935–1942.

St-Pierre, J., Buckingham, J.A., Roebuck, S.J. and Brand, M.D., (2002) Topology of superoxide production from different sites in the mitochondrial electron transport chain. *Journal of Biological Chemistry*, 27747, pp. 44784–44790.

Tryfidou, D.V., McClean, C., Nikolaidis, M.G. and Davison, G.W., (2020) DNA damage following acute aerobic exercise: A systematic review and meta-analysis. *Sports Medicine*, 501, pp. 103–127.

Tsai, K., Hsu, T.G., Hsu, K.M., Cheng, H., Y, L.T., Hsu, C.F. and Kong, C.W., (2001) Oxidative DNA damage in human peripheral leukocytes induced by massive aerobic exercise. *Free Radical Biology and Medicine*, 3111, pp. 1465–1472.

Valavanidis, A., Vlachogianni, T. and Fiotakis, C., (2009) 8-Hydroxy-2′ -deoxyguanosine (8-OHdG): A critical biomarker of oxidative stress and carcinogenesis. *Journal of Environmental Science and Health: Part C Environmental Carcinogenesis and Ecotoxicology Reviews*, 272, pp. 120–139.

von Sonntag, C. (1987) *The Chemical Basis of Radiation Biology*. London: Taylor & Francis.

Williamson, J. and Davison, G., (2020) Targeted antioxidants in exercise-induced mitochondrial oxidative stress: Emphasis on DNA damage. *Antioxidants*, 911, pp. 1–25.

Williamson, J., Hughes, C.M. and Davison, G.W., (2018) Exogenous plant-based nutraceutical supplementation and peripheral cell mononuclear DNA damage following high intensity exercise. *Antioxidants*, 7, 70.

Williamson, J., Hughes, C.M., Burke, G. and Davison, G.W., (2020a) A combined γ-H2AX and 53BP1 approach to determine the DNA damage-repair response to exercise in hypoxia. *Free Radical Biology and Medicine*, 154, pp. 9–17.

Williamson, J., Hughes, C.M., Cobley, J.N. and Davison, G.W., (2020b) The mitochondria-targeted antioxidant MitoQ, attenuates exercise-induced mitochondrial DNA damage. *Redox Biology*, 36, p. 101673.

Zhang, M., Izumi, I., Kagamimori, S., Sokejima, S., Yamagami, T., Liu, Z. and Qi, B., (2004) Role of taurine supplementation to prevent exercise-induced oxidative stress in healthy young men. *Amino Acids*, 262, pp. 203–207.

CHAPTER TEN

Nutritional Antioxidants for Sports Performance

Jamie N. Pugh and Graeme L. Close

CONTENTS

Introduction / 115
Heterogeneity of Antioxidant Supplements / 117
Methodologies Used / 117
Context / 118
Conclusion / 120
References / 120

INTRODUCTION

Following the first robust evidence that endurance exercise leads to increases in free radical production (Davies et al., 1982), there has long been a perception of negativity regarding their role in athletic performance. That is, increased free radical production, often termed 'oxidative stress', was considered a deleterious consequence of exercise that could negatively affect different facets of athletic performance, recovery and health. The production of free radicals can also be triggered by several other endogenous and exogenous factors including exposure to radiation, excessive heat, inflammation, infection and trauma (Halliwell, 2009). Given that athletes may face many of these factors on a daily basis, it can be seen why nutritional antioxidant sources have been of interest to athletes to prevent or attenuate such damage. However, knowledge around the biological implications of exercise-induced free radical production has expanded rapidly. It is now appreciated that while high levels of free radicals can damage cellular components, physiological production of low-to-moderate levels may play regulatory roles in cells such as the control of gene expression, regulation of cell signalling pathways and modulation of skeletal muscle force production (Powers and Jackson, 2008). Indeed, the definition of 'oxidative stress' has been updated to emphasise this with the current definition being 'an imbalance between oxidants and antioxidants in favour of the oxidants, leading to a disruption of redox signalling and control and/or molecular damage' (Sies, 2007). Despite this updated definition, the original misnomer that oxidative stress is always a deleterious consequence of exercise remains strong in both the exercise science literature and applied sports practice, with dietary antioxidant supplements often prompted as *sine qua non* for athletes. Moreover, outdated techniques that lack validity and have been widely criticised in the mainstream redox biology literature beset the sport science literature which has, without doubt, led to erroneous conclusions being drawn on the efficacy of dietary antioxidant supplements during exercise. Given the complexities of the biology at play and the multitude of scenarios that athletes may encounter, a nuanced approach should be taken with respect to dietary antioxidant advice with a critical appraisal of pertinent research

DOI: 10.1201/9781003051619-10

studies to evaluate their immediate translational potential to applied practice.

Although relatively simple in concept, the translation of research to practice is rarely a straightforward process. Elite sport is dynamic, unpredictable, chaotic and often contextual, with the factors affecting performance being vast. It is therefore imperative that a critical framework is applied to the interpretation of antioxidant research for athletes. The Paper-to-Podium (P-2-P) framework (Close et al., 2019) is a recent attempt to evaluate the translational potential of nutritional research (Figure 10.1). It is important to stress that the P-2-P matrix is not designed to assess the quality of the research *per se* as clearly, while a study may lack immediate translational potential, the generation of new knowledge or the stepwise increment in our understanding clearly requires an early stage, often purely mechanistic research.

Through this lens, redox biology, and antioxidant supplementation, may be one of the most difficult and confusing fields for athletes (and nutritional practitioners) to navigate. Beyond the complexity of the biology and chemistry involved, there are a multitude of factors that have made interpretation of the research challenging. Throughout this chapter, we will attempt to highlight some of the major considerations and limitations in sport science-related free radical research framed around the P-2-P matrix.

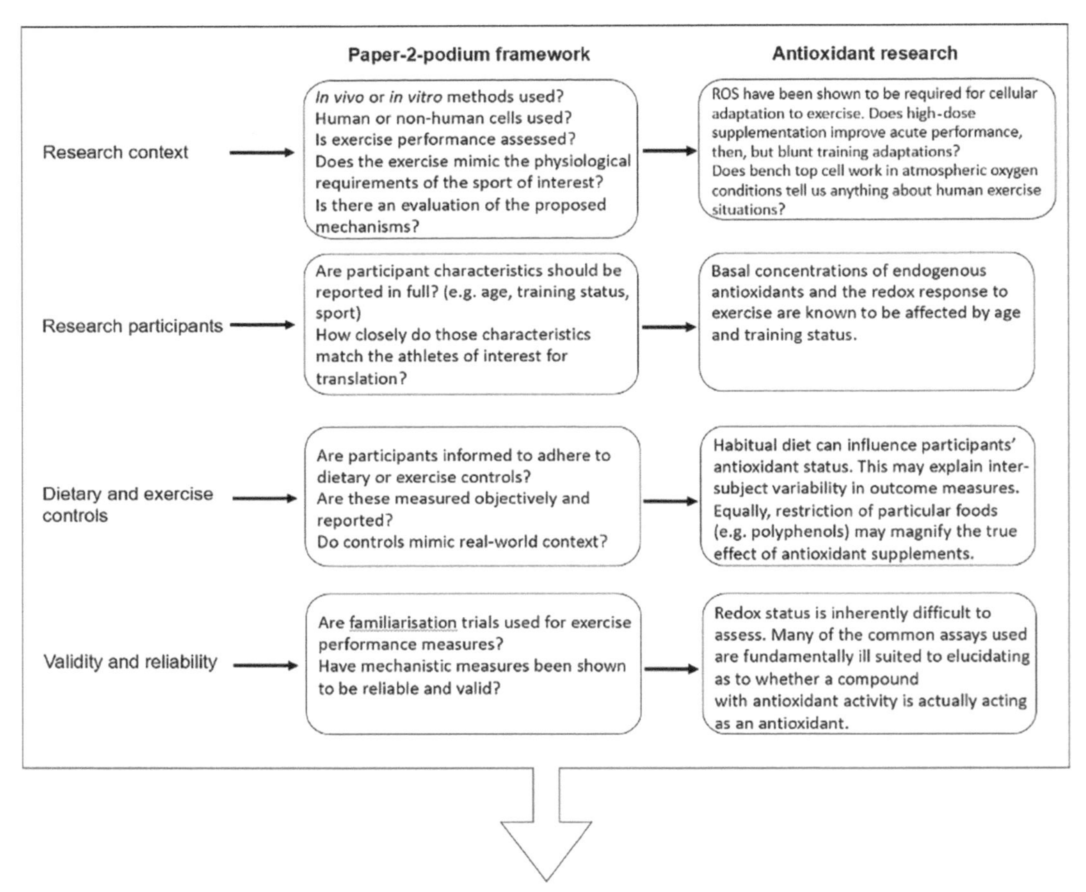

Figure 10.1 Adaptation of the Paper-2-Podium (P-2-P) matrix: an operational framework to evaluate the translational potential of performance nutrition research including specific considerations for antioxidant research (Close et al., 2019).

HETEROGENEITY OF ANTIOXIDANT SUPPLEMENTS

The term antioxidant has been defined in several ways and may be defined as any substance that delays or prevents the oxidation of a substrate (Powers and Jackson, 2008). Such a broad definition, therefore, covers both substances that directly scavenges ROS (e.g., vitamin E's ability to scavenge lipid peroxyl radicals) and those that inhibit the initiation and propagation step leading to the termination of the reaction and delay the oxidation process. When ROS are generated in vivo, many antioxidants come into play. Their relative importance depends upon which ROS is generated, how, where and what target of damage is measured (Halliwell, 1995). Many substances can therefore have both pro- and antioxidant properties. For example, nitric oxide, a free radical, could be considered to have antioxidant properties via its ability to sequester cellular iron (Sahni et al., 2018). The antioxidant defence system of the body consists of antioxidant enzymes (including superoxide dismutases, catalase and glutathione peroxidase) and non-enzymatic antioxidants (including vitamins A, C and E, coenzyme Q10 and glutathione). The number of nutritional antioxidants presented and discussed in the scientific literature is seemingly endless and includes products such as colostrum, caffeine, selenium, carotenoids such as β-carotene, lycopene, lutein, probiotics, coenzyme Q, quercetin, resveratrol, pterostilbene, pycnogenol and astaxanthine. This broad range of antioxidant supplements makes generalisations difficult because antioxidants are chemically heterogeneous, i.e., they all work in distinct ways. Added to this, the various duration and doses of supplementation protocols means extreme caution is needed when interpreting research findings. It is beyond the scope of the current chapter to discuss each of these supplements individually, but we point the reader to broad reviews that have discussed many of these supplements and athletic performance (Mason et al., 2020; Braakhuis and Hopkins, 2015; Belviranli and Okudan, 2015). What is clear, however, is that the major contributor to antioxidant defence in exercising humans will be the enzymatic antioxidant defence enzymes. It is therefore somewhat confusing that the sport science literature has tended to focus on low molecular weight antioxidant compounds like vitamin C whose specific role in defence in these exercise situations is likely to be trivial.

METHODOLOGIES USED

As with all scientific studies, the methods used should be appropriate for the research question. Many of these aspects are highlighted in the P-2-P framework including research participants, research design and dietary and exercise controls. For example, in considering the research participants, it is clear that exercise training increases the resistance of skeletal muscle against oxidative stress and provides enhanced protection through the upregulation of its own antioxidant enzymes (Radak et al., 2008). The use of untrained or recreationally active participants is likely then to have only limited translation to elite athletes. Equally, even within defined participant groups, care should be taken to consider participants' dietary intake. Large heterogeneity of study outcomes maybe a direct result of differences in the baseline antioxidant status of the research participants. Dietary intake has been shown to relate to blood concentrations of various antioxidants (Margaritelis et al., 2020), particularly the habitual intake of fruits and vegetables (Esfahani et al., 2011). It follows then that antioxidant supplementation has been shown to increase blood concentrations of vitamin C (Paschalis et al., 2016) and glutathione (Paschalis et al., 2018) only in those with the lowest baseline concentrations and that only these 'deficient' participants improved their exercise performance following supplementation. As such, personalized antioxidant interventions to enhance exercise performance have been proposed (Margaritelis et al., 2018) although this hypothesis certainly needs further exploration.

One of the fundamental difficulties in sports performance research is the quantification of oxidative stress or damage caused by a single bout, or multiple bouts, of exercise. In theory, a blood-borne marker that could be measured before and after a damaging bout of exercise, with and without an antioxidant supplement, could give insights into the efficacy of such supplement, and there are many studies that have utilised this approach. However, extreme caution should be used when evaluating such studies as many of the assays used are fundamentally flawed and will without doubt generate spurious results (Cobley et al., 2017). For example, the thiobarbituric acid-reactive substance (TBARS) assay is a purported marker of lipid peroxidation and appears regularly in sport science–related antioxidant studies. However, TBARS are also generated by several non-redox regulated means, and artificial lipid peroxidation is generated by the assay itself (Halliwell and Whiteman, 2004). Perhaps more concerning, however, is the persistence with the total antioxidant capacity (TAC) assay (and other similar methods) that has been used extensively in sports science research despite its criticism in mainstream redox biology (Pompella et al., 2014; Sies, 2007; Arts et al., 2004). The TAC assay is particularly misleading because the assay is orthogonal to antioxidant enzymes and as such there is nothing 'Total' about it! Indeed, in the sport sciences, the TAC assay is still used regularly to define exercise-induced "oxidative stress" despite many highly respected redox biologists clearly establishing that this assay cannot give useful information on the state of the organism and its use should therefore be discouraged. Whilst the idea of a one stop, all-encompassing plasma marker of oxidative damage is fanciful, the studies that have used them are likely drawing inappropriate conclusions.

Another key limitation in understanding the roles of oxidants and antioxidants in sport science research involves access to the tissue of interest. It is not uncommon to see blood markers of oxidative stress used to explain what is occurring in skeletal muscle. Given the extremely short half-lives of free radicals (i.e., in the nanosecond range), it is questionable what a marker in blood plasma that may have been left exposed to atmospheric ground-state molecular dioxygen prior to the assay can tell us about the production of specific free radicals arising from skeletal muscle. Whilst muscle biopsies are now routinely taken in sport science research, many of the more specific techniques used to measure free radical production following muscle contraction are not yet translatable to human applied sport science research. There are, however, assays in development that have the potential to change this, and as such the future studies in this space are particularly exciting. For example, novel immunological assays, like catalyst-free Click-PEG (Cobley and Husi, 2020; Cobley et al., 2019), have the potential to transform how oxidative stress is measured (reviewed in Cobley and Husi, 2020). Specifically, it will be possible to measure the redox state of protein thiols to shed new light on exercise-induced redox signalling, as well as developing new valid and reliable biomarkers of oxidative stress.

CONTEXT

When considering the efficacy of any intervention for sports performance, context is perhaps one of the most important factors. In the P-2-P framework, it is suggested that the greatest translational potential comes from those studies that include participants of similar characteristics as the target population, using the appropriate exercise performance measures, and attempts to evaluate the potential mechanisms of action. In antioxidant supplement research, this has perhaps been best exemplified by the contrasting aims of sports competition (performance) and training (adaptation to exercise).

There is abundant evidence that the increase in free radical production in contracting skeletal muscle contributes to muscle fatigue induced by prolonged muscular contractions (Powers and Jackson, 2008). Despite this, data pertaining to the ergogenic effects of antioxidant supplementation is, at best, equivocal. For example, studies that have shown improvements in acute exercise performance following N-acetyl cysteine (NAC) supplementation have mainly been found following intravenous administration immediately before and during exercise (Medved et al., 2004) and in untrained muscle groups (Matuszczak et al., 2005), which has little translational potential for use with athletes, considering IV infusions are not only impractical but are also prohibited by the World Anti-Doping Agency (WADA), unless the rate is less than 100 mL per 12 hours. Even where oral NAC supplementation has been shown to attenuate muscle fatigue following intermittent running exercise, the dose required led to mild gastrointestinal side effects that could ultimately negate any potential benefits (Cobley et al., 2011). Equally, supplementation of co-enzyme Q10 has been shown to improve exercise performance in patients with coronary artery disease (Tiano et al., 2007) and middle-aged participants (Bonetti et al., 2000) but not in trained individuals (Laaksonen et al., 1995), despite its poor bio-availability and limited accumulation in the mitochondria (Williamson et al., 2020). While the majority of studies have investigated the effects of a single antioxidant, one study showed a 3.1% improvement in 30 km time trial performance in elite cyclists following supplementation containing lower doses of multiple antioxidants (MacRae and Mefferd, 2006). While such work is clearly promising, it is difficult to ascertain the underlying mechanisms. To truly link antioxidant supplements to enhanced performance via redox related mechanisms, the nutritional antioxidant must competitively react with the relevant species with spatiotemporal fidelity, at the relevant place and time (Murphy, 2014; Day, 2014), something that to date has not been established in athletic populations (Cobley et al., 2015).

As well as considering athletic performance during a single exercise bout, many athletes are required to compete multiple times with only hours between, particularly during championship events. Therefore, there is a strong interest in facilitating or enhancing physiological recovery and subsequent performance following an acute exercise bout. Studies have utilised methods that have assessed non-specific post-exercise markers of muscle damage such as creatine kinase and lactate dehydrogenase, recovery of muscle function following a damaging bout of exercise and subjective markers of delayed onset of muscle soreness (DOMS). In regard to the latter, a Cochrane review concluded that there is 'moderate to low-quality evidence that high dose antioxidant supplementation does not result in a clinically relevant reduction of muscle soreness after exercise up to 6 hours or at 24, 48, 72 and 96 hours after exercise' (Ranchordas et al., 2020). However, some studies have shown greater recovery of muscle force following antioxidant supplementation, often without attenuation of subjective markers of DOMS. For example, Montmorency tart cherry juice have been extensively studied due to their high phytochemical concentrations with some data suggesting positive outcomes. However, often these studies have been performed following the restriction of polyphenols for several days prior to the intervention (Bell et al., 2015, 2016) and therefore do not necessarily reflect nutritional intakes that are typically consumed from team sport athletes. Moreover, given the lack of measures of free radicals and antioxidant in these studies, it is entirely feasible that if there is an actual improvement in recovery following polyphenol supplementation, the mechanisms responsible may not necessarily be redox related. Of those studies that have not restricted habitual diet, benefits have been shown in recreational athletes (Howatson et al., 2010; Quinlan and Hill, 2019) but not well-trained or elite athletes (Morehen et al., 2020; Kupusarevic et al., 2019). Given the equivocal findings for antioxidants to improve recovery from damaging exercise, other nutritional strategies (e.g., glycogen replenishment, optimal protein intake) are likely to be of greater value to athletes.

Before an athlete competes, they first undertake extensive technical and physiological training, with the explicit goal of inducing positive adaptations, including those at a cellular level within skeletal muscle. It has become increasingly apparent that free radicals play a key role within this process and can act through several different pathways of signalling transduction, both directly and indirectly (Cobley et al., 2015). In vitro and animal models have elegantly shown a multitude of adaptations following free radical exposure that are blunted by antioxidants (Gomez-Cabrera et al., 2015; Radak et al., 2017). However, the question still remains if these theoretical attenuations in training adaptations translate as being detrimental for exercise training adaptations in highly trained athletes. Work from as early as 1971 suggested that vitamin E supplementation had unfavourable results for trained swimmers (Sharman et al., 1971). Since then there has been fierce debate as to whether high dose antioxidant supplementation blunts the cellular, and thus phenotypical, adaptation to exercise training (Holloszy et al., 2012). Studies have shown that improvements in exercise performance are blunted when training is accompanied by chronic supplementation of several antioxidants (Malm et al., 1997; Gomez-Cabrera et al., 2008). In contrast, no such effect was seen in elite cyclists (Rokitzki et al., 1994) and in moderately trained cyclists after 12 weeks of supplementation (Yfanti et al., 2010). Moreover, it has also been shown that, despite attenuations in markers of mitochondrial biogenesis, there was no difference in performance improvements following 11 weeks of exercise training with vitamin C and E supplementation, compared with a placebo (Paulsen et al., 2014). High-dose antioxidant supplements may also interfere with the uptake of other nutrients, while low doses of multiple antioxidants have been shown in numerous clinical cases to be more effective than high doses of a single one (Halliwell, 2009). Whilst the data is, at best, equivocal with regards to enhanced recovery with antioxidant supplements, combined with emerging evidence that antioxidants may blunt training adaptations for now the best approach may be to abstain from high-dose supplements during training. Although not explicitly investigated to date, there have been no data to suggest that eating high-quality fruit and vegetables attenuates adaptations to exercise. Perhaps our recommendations to athletes for now may not be any more complicated than to ensure they regularly consume a wide variety of fruits and vegetables.

CONCLUSION

Intake of dietary supplements is prevalent among athletes at all competitive levels (Garthe and Maughan, 2018) with antioxidants being particularly popular. Worryingly, a significant number of commercially available dietary supplements, including individual vitamins, can be contaminated with WADA prohibited substances (Martínez-Sanz et al., 2017). Many sports, therefore, recommend a 'food first' philosophy whereby wherever possible, dietary deficiencies should be corrected with food rather than immediately looking to dietary supplements. Moreover, it is often recommended that dietary supplements should only be given when there is a clear rationale for their use, and, given the lack of consensus on the benefits of dietary antioxidants, it is difficult to justify their inclusion in an athlete's dietary supplement strategy. Whilst research in this field is vast, the lack of appropriate assays on appropriate tissues in appropriate participants besets this field of research. We, therefore, believe that future studies must now try to add some clarity to this research field which will require collaboration between sport scientists and redox biologists using state-of-the-art techniques to progress a fascinating area.

REFERENCES

Arts, M. J., Haenen, G. R., Voss, H. P. & Bast, A. 2004. Antioxidant capacity of reaction products limits the applicability of the Trolox Equivalent Antioxidant Capacity (TEAC) assay. *Food Chem Toxicol*, 42, 45–9.

Bell, P. G., Walshe, I. H., Davison, G. W., Stevenson, E. J. & Howatson, G. 2015. Recovery facilitation with montmorency cherries following high-intensity, metabolically challenging exercise. *Appl Physiol Nutr Metab*, 40, 414–23.

Bell, P. G., Stevenson, E., Davison, G. W. & Howatson, G. 2016. The Effects of Montmorency Tart Cherry Concentrate Supplementation on Recovery Following Prolonged, Intermittent Exercise. *Nutrients*, 8, 441.

Belviranli, M. & Okudan, N. 2015. Well-known antioxidants and newcomers in sport nutrition: Coenzyme Q10, quercetin, resveratrol, pterostilbene, pycnogenol and astaxanthin. In: Lamprecht, M. (ed.) *Antioxidants in Sport Nutrition*. Boca Raton (FL): CRC Press/Taylor & Francis, 79–102.

Bonetti, A., Solito, F., Carmosino, G., Bargossi, A. M. & Fiorella, P. L. 2000. Effect of ubidecarenone oral treatment on aerobic power in middle-aged trained subjects. *J Sports Med Phys Fitness*, 40, 51–7.

Braakhuis, A. J. & Hopkins, W. G. 2015. Impact of dietary antioxidants on sport performance: A review. *Sports Med*, 45, 939–55.

Close, G. L., Kasper, A. M. & Morton, J. P. 2019. From paper to podium: Quantifying the translational potential of performance nutrition research. *Sports Med*, 49, 25–37.

Cobley, J. N. & Husi, H. 2020. Immunological techniques to assess protein thiol redox state: Opportunities, challenges and solutions. *Antioxidants (Basel)*, 9, 315.

Cobley, J. N., Mcglory, C., Morton, J. P. & Close, G. L. 2011. N-Acetylcysteine's attenuation of fatigue after repeated bouts of intermittent exercise: Practical implications for tournament situations. *Int J Sport Nutr Exerc Metab*, 21, 451–61.

Cobley, J. N., Mchardy, H., Morton, J. P., Nikolaidis, M. G. & Close, G. L. 2015. Influence of vitamin C and vitamin E on redox signaling: Implications for exercise adaptations. *Free Radic Biol Med*, 84, 65–76.

Cobley, J. N., Close, G. L., Bailey, D. M. & Davison, G. W. 2017. Exercise redox biochemistry: Conceptual, methodological and technical recommendations. *Redox Biol*, 12, 540–8.

Cobley, J. N., Noble, A., Jimenez-Fernandez, E., Valdivia Moya, M. T., Guille, M. & Husi, H. 2019. Catalyst-free click pegylation reveals substantial mitochondrial ATP synthase sub-unit alpha oxidation before and after fertilisation. *Redox Biol*, 26, 101258.

Davies, K. J., Quintanilha, A. T., Brooks, G. A. & Packer, L. 1982. Free radicals and tissue damage produced by exercise. *Biochem Biophys Res Commun*, 107, 1198–205.

Day, B. J. 2014. Antioxidant therapeutics: Pandora's box. *Free Radic Biol Med*, 66, 58–64.

Esfahani, A., Wong, J. M., Truan, J., Villa, C. R., Mirrahimi, A., Srichaikul, K. & Kendall, C. W. 2011. Health effects of mixed fruit and vegetable concentrates: A systematic review of the clinical interventions. *J Am Coll Nutr*, 30, 285–94.

Garthe, I. & Maughan, R. J. 2018. Athletes and supplements: Prevalence and perspectives. *Int J Sport Nutr Exerc Metab*, 28, 126–38.

Gomez-Cabrera, M. C., Domenech, E., Romagnoli, M., Arduini, A., Borras, C., Pallardo, F. V., Sastre, J. & Viña, J. 2008. Oral administration of vitamin C decreases muscle mitochondrial biogenesis and hampers training-induced adaptations in endurance performance. *Am J Clin Nutr*, 87, 142–9.

Gomez-Cabrera, M. C., Salvador-Pascual, A., Cabo, H., Ferrando, B. & Viña, J. 2015. Redox modulation of mitochondriogenesis in exercise. Does antioxidant supplementation blunt the benefits of exercise training? *Free Radic Biol Med*, 86, 37–46.

Halliwell, B. 1995. Antioxidant characterization. Methodology and mechanism. *Biochem Pharmacol*, 49, 1341–8.

Halliwell, B. 2009. The wanderings of a free radical. *Free Radic Biol Med*, 46, 531–42.

Halliwell, B. & Whiteman, M. 2004. Measuring reactive species and oxidative damage in vivo and in cell culture: How should you do it and what do the results mean? *Br J Pharmacol*, 142, 231–55.

Holloszy, J. O., Higashida, K., Kim, S. H., Higuchi, M. & Han, D.-H. 2012. Response to letter to the editor by Gomez-Cabrera et al. *Am J Physiol-Endocrinol Metab*, 302, E478–79.

Howatson, G., Mchugh, M. P., Hill, J. A., Brouner, J., Jewell, A. P., Van Someren, K. A., Shave, R. E. & Howatson, S. A. 2010. Influence of tart cherry juice on indices of recovery following marathon running. *Scand J Med Sci Sports*, 20, 843–52.

Kupusarevic, J., Mcshane, K. & Clifford, T. 2019. Cherry gel supplementation does not attenuate subjective muscle soreness or alter wellbeing following a match in a team of professional rugby union players: A pilot study. *Sports (Basel)*, 7, 84.

Laaksonen, R., Fogelholm, M., Himberg, J. J., Laakso, J. & Salorinne, Y. 1995. Ubiquinone supplementation and exercise capacity in trained young and older men. *Eur J Appl Physiol Occup Physiol*, 72, 95–100.

Macrae, H. S. & Mefferd, K. M. 2006. Dietary antioxidant supplementation combined with quercetin improves cycling time trial performance. *Int J Sport Nutr Exerc Metab*, 16, 405–19.

Malm, C., Svensson, M., Ekblom, B. & Sjödin, B. 1997. Effects of ubiquinone-10 supplementation and high intensity training on physical performance in humans. *Acta Physiol Scand*, 161, 379–84.

Margaritelis, N. V., Paschalis, V., Theodorou, A. A., Kyparos, A. & Nikolaidis, M. G. 2018. Antioxidants in personalized nutrition and exercise. *Adv Nutr*, 9, 813–23.

Margaritelis, N. V., Chatzinikolaou, P. N., Bousiou, F. V., Malliou, V. J., Papadopoulou, S. K., Potsaki, P., Theodorou, A. A., Kyparos, A., Geladas, N. D., Nikolaidis, M. G. & Paschalis, V. 2020. Dietary cysteine intake is associated with blood glutathione levels and isometric strength. *Int J Sports Med*, 42, 441–7.

Martínez-Sanz, J. M., Sospedra, I., Ortiz, C. M., Baladía, E., Gil-Izquierdo, A. & Ortiz-Moncada, R. 2017. Intended or unintended doping? A review of the presence of doping substances in dietary supplements used in sports. *Nutrients*, 9, 1093.

Mason, S. A., Trewin, A. J., Parker, L. & Wadley, G. D. 2020. Antioxidant supplements and endurance exercise: Current evidence and mechanistic insights. *Redox Biol*, 35, 101471.

Matuszczak, Y., Farid, M., Jones, J., Lansdowne, S., Smith, M. A., Taylor, A. A. & Reid, M. B. 2005. Effects of N-acetylcysteine on glutathione oxidation and fatigue during handgrip exercise. *Muscle Nerve*, 32, 633–8.

Medved, I., Brown, M. J., Bjorksten, A. R., Murphy, K. T., Petersen, A. C., Sostaric, S., Gong, X. & Mckenna, M. J. 2004. N-acetylcysteine enhances muscle cysteine and glutathione availability and attenuates fatigue during prolonged exercise in endurance-trained individuals. *J Appl Physiol*, 97, 1477–85.

Morehen, J. C., Clarke, J., Batsford, J., Barrow, S., Brown, A. D., Stewart, C. E., Morton, J. P. & Close, G. L. 2020. Montmorency tart cherry juice does not reduce markers of muscle soreness, function and inflammation following professional male rugby League match-play. *Eur J Sport Sci*, 21(2), 1–10.

Murphy, M. P. 2014. Antioxidants as therapies: Can we improve on nature? *Free Radic Biol Med*, 66, 20–3.

Paschalis, V., Theodorou, A. A., Kyparos, A., Dipla, K., Zafeiridis, A., Panayiotou, G., Vrabas, I. S. & Nikolaidis, M. G. 2016. Low vitamin C values are linked with decreased physical performance and increased oxidative stress: Reversal by vitamin C supplementation. *Eur J Nutr*, 55, 45–53.

Paschalis, V., Theodorou, A. A., Margaritelis, N. V., Kyparos, A. & Nikolaidis, M. G. 2018. N-acetylcysteine supplementation increases exercise performance and reduces oxidative stress only in individuals with low levels of glutathione. *Free Radic Biol Med*, 115, 288–97.

Paulsen, G., Cumming, K. T., Holden, G., Hallén, J., Rønnestad, B. R., Sveen, O., Skaug, A., Paur, I., Bastani, N. E., Østgaard, H. N., Buer, C., Midttun, M., Freuchen, F., Wiig, H., Ulseth, E. T., Garthe, I., Blomhoff, R., Benestad, H. B. & Raastad, T. 2014. Vitamin C and E supplementation hampers cellular adaptation to endurance training in humans: A double-blind, randomised, controlled trial. *J Physiol*, 592, 1887–901.

Pompella, A., Sies, H., Wacker, R., Brouns, F., Grune, T., Biesalski, H. K. & Frank, J. 2014. The use of total antioxidant capacity as surrogate marker for food quality and its effect on health is to be discouraged. *Nutrition*, 30, 791–3.

Powers, S. K. & Jackson, M. J. 2008. Exercise-induced oxidative stress: Cellular mechanisms and impact on muscle force production. *Physiol Rev*, 88, 1243–76.

Quinlan, R. & Hill, J. A. 2019. The efficacy of tart cherry juice in aiding recovery after intermittent exercise. *Int J Sports Physiol Perform*, 1–7. doi: 10.1123/ijspp.2019-0101.

Radak, Z., Chung, H. Y. & Goto, S. 2008. Systemic adaptation to oxidative challenge induced by regular exercise. *Free Radic Biol Med*, 44, 153–9.

Radak, Z., Ishihara, K., Tekus, E., Varga, C., Posa, A., Balogh, L., Boldogh, I. & Koltai, E. 2017. Exercise, oxidants, and antioxidants change the shape of the bell-shaped hormesis curve. *Redox Biol*, 12, 285–90.

Ranchordas, M. K., Rogerson, D., Soltani, H. & Costello, J. T. 2020. Antioxidants for preventing and reducing muscle soreness after exercise: A Cochrane systematic review. *Br J Sports Med*, 54, 74–8.

Rokitzki, L., Logemann, E., Huber, G., Keck, E. & Keul, J. 1994. alpha-Tocopherol supplementation in racing cyclists during extreme endurance training. *Int J Sport Nutr*, 4, 253–64.

Sahni, S., Hickok, J. R. & Thomas, D. D. 2018. Nitric oxide reduces oxidative stress in cancer cells by forming dinitrosyliron complexes. *Nitric Oxide*, 76, 37–44.

Sharman, I. M., Down, M. G. & Sen, R. N. 1971. The effects of vitamin E and training on physiological function and athletic performance in adolescent swimmers. *Br J Nutr*, 26, 265–76.

Sies, H. 2007. Total antioxidant capacity: Appraisal of a concept. *J Nutr*, 137, 1493–5.

Tiano, L., Belardinelli, R., Carnevali, P., Principi, F., Seddaiu, G. & Littarru, G. P. 2007. Effect of coenzyme Q10 administration on endothelial function and extracellular superoxide dismutase in patients with ischaemic heart disease: A double-blind, randomized controlled study. *Eur Heart J*, 28, 2249–55.

Williamson, J., Hughes, C. M., Cobley, J. N. & Davison, G. W. 2020. The mitochondria-targeted antioxidant MitoQ, attenuates exercise-induced mitochondrial DNA damage. *Redox Biol*, 36, 101673.

Yfanti, C., Akerström, T., Nielsen, S., Nielsen, A. R., Mounier, R., Mortensen, O. H., Lykkesfeldt, J., Rose, A. J., Fischer, C. P. & Pedersen, B. K. 2010. Antioxidant supplementation does not alter endurance training adaptation. *Med Sci Sports Exerc*, 42, 1388–95.

CHAPTER ELEVEN

Antioxidant Supplements and Exercise Adaptations

Shaun A. Mason, Lewan Parker, Adam J. Trewin and Glenn D. Wadley

CONTENTS

Introduction / 123
Antioxidant Defenses / 124
Mitochondrial Biogenesis / 125
Muscle Hypertrophy/Strength / 126
Substrate Metabolism / 127
Oxidative Stress and Vascular Function / 128
Performance / 130
  Endurance Performance / 130
  Muscle Recovery / 130
  VO_2 max / 130
  Personalized Supplementation / 130
Conclusion / 131
References / 131

INTRODUCTION

Antioxidant supplements (e.g., vitamin C) are commonly consumed to limit the presumably deleterious effects of exercise-induced reactive oxygen species (ROS), thus reducing fatigue and improving recovery and performance (Knapik et al., 2016). In this chapter, we will discuss the evidence from studies of antioxidant supplementation and its effects on physiological responses and adaptations to exercise. We highlight certain compounds that may ameliorate muscle damage associated with delayed onset muscle soreness (DOMS) (Bryer and Goldfarb, 2006), delay fatigue during prolonged submaximal exercise (Corn and Barstow, 2011), and improve recovery of muscle function following damaging eccentric-based exercise (Jakeman and Maxwell, 1993). However, in general, there is insufficient evidence to recommend antioxidant compounds for use in athletic and recreational exercise scenarios (Mason et al., 2020). Antioxidants may even impair some acute and chronic responses to exercise (Ristow et al., 2009, Gomez-Cabrera et al., 2008a, Morrison et al., 2015) because redox signaling might regulate certain exercise adaptations (Mason et al., 2016) (Figure 11.1). Effects of supplementation will depend on the specific antioxidant compound and its mechanism(s) of action in vivo, and the dose, bioavailability, and route of administration. Some antioxidants can act, at least in part, as free radical scavengers (e.g., vitamin C), while other compounds with antioxidant properties exert their effects indirectly through induction of endogenous antioxidant enzymes and/or via inhibition of ROS-producing enzymes (e.g., polyphenols act as pro-oxidants to induce Nrf-2 signaling). In this chapter, we will explore the impact of exogenous antioxidants on exercise training adaptations including mitochondrial biogenesis,

DOI: 10.1201/9781003051619-11

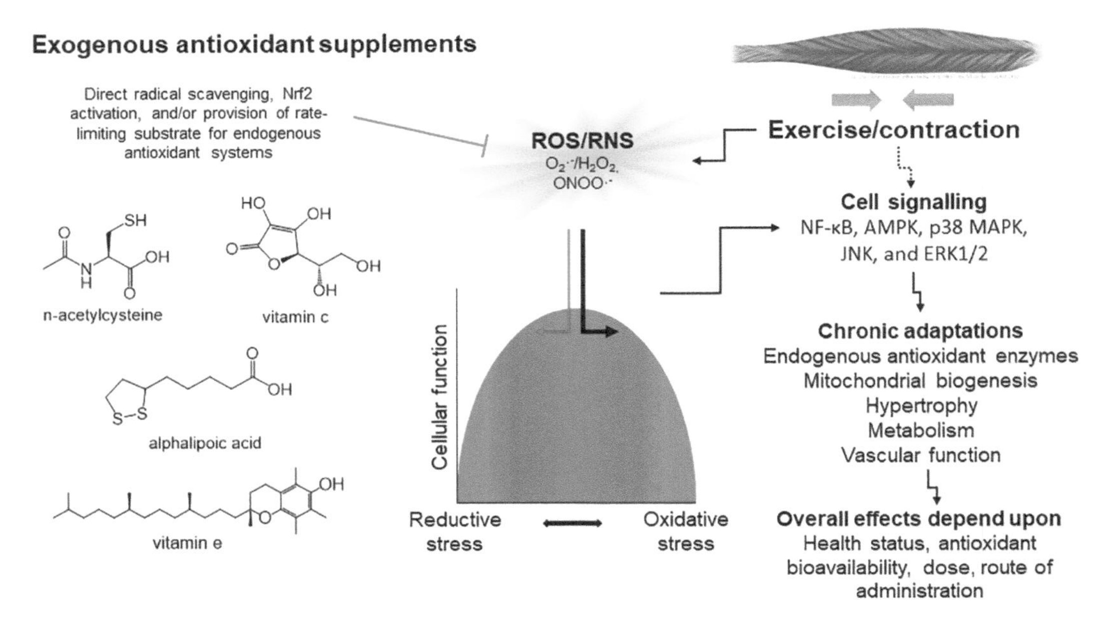

Figure 11.1 Antioxidant compounds are often consumed in an attempt to decrease the presumed deleterious exercise-induced oxidative stress. Depending on the complex combination of conditions, this may attenuate or enhance the normal activation of redox-sensitive cell signaling pathways that contribute to long-term beneficial physiological adaptations to exercise (Nrf2 – nuclear factor erythroid 2–related factor 2; ROS – reactive oxygen species; RNS – reactive nitrogen species; NF-κB – nuclear factor κB; AMPK – adenosine 5′ monophosphate-activated protein kinase; p38 MAPK – p38 mitogen-activated protein kinases; JNK – c-Jun N-terminal kinase; ERK1/2 – Extracellular signal-regulated protein kinase 1/2). Image of skeletal muscle within the figure was copied under Creative Commons Attribution-Share Alike 4.0 International license from the site https://commons.wikimedia.org/wiki/File:Muscle_Types.png.

antioxidant enzyme induction, vascular function, muscle hypertrophy and strength, substrate metabolism, and exercise performance enhancement.

ANTIOXIDANT DEFENSES

Endurance exercise itself has been referred to as an "antioxidant" (Gomez-Cabrera et al., 2008b) due to the increased endogenous antioxidant defenses observed following training. These adaptations improve the capacity of skeletal muscles to tolerate ROS that might be induced by future bouts of exhaustive or fatiguing exercise. The primary antioxidant enzyme defenses within cells consist of superoxide dismutase (SOD), glutathione peroxidase (GP_X), catalase (CAT), and peroxiredoxin/thioredoxin (PRDX/TRDX). In human skeletal muscle, endurance exercise training increases many of these endogenous antioxidant defenses including increased SOD activity, SOD2 protein abundance, GPx-1 mRNA levels, and reduced glutathione (GSH) levels (Morrison et al., 2015). However, there is little evidence that exercise training increases skeletal muscle CAT activity in humans.

A systematic review and meta-analysis of 19 human exercise studies found that antioxidant parameters increase, while pro-oxidant parameters such as thiobarbituric acid-reactive substances (TBARS) and F2-isoprostanes decrease after exercise training (de Sousa et al., 2017). The vast majority of the included studies were of endurance exercise training and measured redox parameters in blood, for which there was strong evidence regardless of training intensity or duration of a positive effect on redox balance (de Sousa et al., 2017).

ROS produced within skeletal muscles during contraction drives the induction of antioxidant enzymes to increase antioxidant defenses. The initial signaling events in skeletal muscle occur by the activation of redox-sensitive proteins such as the transcription factor nuclear factor κB (NF-κB), adenosine 5′ monophosphate-activated protein kinase (AMPK), and the MAP kinases p38 mitogen-activated protein kinase (p38 MAPK), c-Jun

N-terminal kinase (JNK), and extracellular signal-regulated protein kinase 1/2 (ERK1/2) that subsequently stimulate the Kelch-like ECH-associated protein 1-nuclear factor erythroid 2–related factor 2 (Keap1-Nrf2) pathway to initiate transcription of antioxidant enzymes through its binding to antioxidant response elements (ARE) (Merry and Ristow, 2016). However, much of these aforementioned pathways have been established in rodent and in vitro models, and evidence of direct oxidation of these proteins in response to exercise is still lacking in humans. Nevertheless, given the central role of ROS as a signal to increase adaptations to exercise such as antioxidant defenses, this raises the possibility that excess exogenous antioxidant supplementation may blunt beneficial redox signaling dependent adaptations to training.

Vitamins C and E are the most widely used antioxidant supplements with up to 20% of the population consuming them (Rock et al., 2004). However, despite some supporting evidence from rodent exercise studies, there is a lack of evidence in humans that supplementation with vitamin C or E has much of an impact on the adaptation of increased skeletal muscle antioxidant defenses following exercise training (Mason et al., 2020). Furthermore, findings from rodent research should be interpreted with caution, particularly for vitamin C since unlike rodents, humans lack the enzyme L-gulono-gamma-lactone oxidase required to synthesize vitamin C and are entirely dependent on vitamin C from the diet. When taken in combination, there appear to be no negative effects using doses of vitamin C (500 mg/day) and E (400 IU/day) in humans on the adaptive antioxidant responses of skeletal muscle to endurance training (Yfanti et al., 2010). However, human studies at higher doses of 1 g/day vitamin C in combination with 400 IU/day vitamin E have reported blunting of skeletal muscle antioxidant responses (Morrison et al., 2015, Ristow et al., 2009) to endurance training, particularly SOD activity and protein abundance. This suggests people taking 1 g/day vitamin C in combination with 400 IU/day vitamin E may miss out on some improvements in antioxidant defenses with endurance training, although there is little evidence that this is actually harmful.

N-acetylcysteine (NAC) is a widely used antioxidant compound in research that provides cysteine as a precursor for endogenous glutathione synthesis, although how it acts as an antioxidant is still unclear. NAC infusion in humans during an acute bout of endurance exercise attenuates skeletal muscle JNK phosphorylation and the increase in mRNA levels of SOD2 during acute exercise (Petersen et al., 2012). However, NAC is not well tolerated at high doses, with oral doses above 50–70 mg/kg reported to produce gastrointestinal problems (Ferreira et al., 2011, Cobley et al., 2011), and its effects on antioxidant defenses following endurance training have not been investigated.

Resveratrol is a polyphenol that has antioxidant properties at least in vitro (Baur, 2010) and is also known to activate the redox-sensitive kinases AMPK and sirtuin1 (SIRT1) in vitro (Baur, 2010). However, oral doses in humans of 250–1,500 mg/day do not appear to activate skeletal muscle AMPK or other redox-sensitive kinases such as p38 MAPK or JNK (Poulsen et al., 2013, Olesen et al., 2014). Nevertheless, in rodents, there is considerable evidence resveratrol reduces skeletal muscle oxidative stress and increases the expression and activity of endogenous antioxidant enzymes (reviewed in Mason et al., 2020). In contrast, the evidence is limited and unclear in humans. For example, in young (22-year-old) (Scribbans et al., 2014), but not older (65-year-old) males (Olesen et al., 2014), 150–250 mg/day resveratrol prevents increases in gene expression of sirtuin 1 (SIRT1) and SOD2 following 4 weeks of endurance training. However, this does not translate to blunted antioxidant enzymes activity, so the functional effects on antioxidant defenses are likely minimal (Olesen et al., 2014, Scribbans et al., 2014). Doses above 2 g/day may impact the adaptation of antioxidant defenses in humans, but with an increased risk of adverse effects such as gastrointestinal disturbance.

MITOCHONDRIAL BIOGENESIS

Mitochondria are the primary hub of cellular metabolism and highly abundant in tissues such as skeletal muscle. Indeed, the adaptation of increased mitochondrial content in skeletal muscle following endurance training is thought to be mostly due to the cumulative effects of each acute bout of exercise. During contraction, elevated levels of ROS, but also increased cytosolic calcium, AMP and possibly NAD^+ activate stress signaling kinases such as the MAPKs, AMPK, and Ca^{2+}/calmodulin-dependent protein kinase (CAMK) and possibly SIRT3 to stimulate the mitochondrial biogenesis pathway (Mason et al., 2016). Given increased ROS levels during contraction have a regulatory role in the induction of skeletal muscle mitochondrial biogenesis, there has been considerable research into whether supplementation with exogenous antioxidants can blunt mitochondrial biogenesis following endurance training.

Despite several studies in rodents establishing that vitamin C or vitamin E can blunt some of the markers of mitochondrial biogenesis following endurance training, there is little research in humans. The impact of combined vitamin C and E supplementation on skeletal muscle mitochondrial biogenesis in humans may be dose dependent. For example, a combination of vitamin C (500 mg/day) and E (400 IU/day) does not impact the mitochondrial biogenesis response in skeletal muscle to endurance training (Yfanti et al., 2010). However, at higher doses of 1 g/day vitamin C in combination with 240–400 IU/day vitamin E have reported blunting of skeletal muscle mRNA and protein responses in several markers of mitochondrial biogenesis following endurance training (Morrison et al., 2015, Paulsen et al., 2014, Ristow et al., 2009). Specific markers include blunting of the training-induced increase in protein abundance of transcription factor A mitochondrial (Tfam) and cytosolic (but not whole cell) levels of a key regulator of the mitochondrial biogenesis pathway, peroxisome proliferator-activated receptor gamma coactivator 1-alpha (PGC-1α). There is also some evidence that protein abundance of some electron transport chain enzymes such as cytochrome C oxidase subunit IV (COXIV) are attenuated with supplementation following training (Morrison et al., 2015). Despite some of these cellular adaptations being impaired with combined vitamin C and E supplementation, it appears this is unlikely to translate to a blunting of the increase in mitochondrial content since skeletal muscle citrate synthase activity (an indicator of mitochondrial content) increases normally (Morrison et al., 2015). Collectively, there is good evidence that 1 g/day of vitamin C in combination with 400 IU/d of vitamin E in humans hampers some, but not all of the skeletal muscle mitochondrial adaptations to endurance training.

NAC can scavenge certain ROS, and in skeletal muscle in vitro, it prevents PGC-1α promoter and coactivation activity following contractile activity. However, there is limited evidence in humans that NAC hampers skeletal muscle mitochondrial adaptations to endurance training. Indeed, infusion of NAC during acute endurance exercise in humans does not alter the increase of PGC-1α mRNA levels in skeletal muscle (Petersen et al., 2012). The impact of NAC on skeletal muscle mitochondrial training adaptations, particularly using oral supplementation has yet to be examined and requires further research but may be limited given that the chronic high doses needed to achieve sufficient bioavailability lead to increased risk of adverse effects.

The evidence of resveratrol affecting exercise training adaptations such as increased mitochondrial biogenesis is mixed in rodent studies (for review see Mankowski et al., 2015). In young men, the impact is relatively minor, with the mRNA response to endurance training of some mitochondrial biogenesis markers such as PGC-1α being blunted by resveratrol supplementation (150 mg/day), but no effect on the increased skeletal muscle abundance or enzyme activity of mitochondrial proteins (Scribbans et al., 2014). Skeletal muscle mitochondrial adaptations to training in older men using higher doses of 250 mg/day resveratrol are unaffected (Olesen et al., 2014). Given that none of these findings translate to attenuated muscle oxidative capacity, the effects at the whole-muscle level are probably minor. Higher doses (>2 g per day) yield higher plasma concentrations, although effects of these doses on outcomes have not been explored in humans.

MUSCLE HYPERTROPHY/STRENGTH

Exercise-induced skeletal muscle hypertrophy represents a net positive balance between rates of skeletal muscle protein synthesis and degradation in response to overload-induced muscle remodeling. Numerous anabolic and catabolic signaling pathways are involved, with some known to be redox-sensitive. Among these pathways are the insulin/IGF-1-PI3K (insulin-like growth factor 1-phosphoinositide 3-kinase) pathway, which involves activation of kinases Akt and mammalian target of rapamycin (mTOR) and their downstream effectors ribosomal protein S6 kinase (p70S6k) and eukaryotic translation initiation factor 4E (eIF4E). IGF-1 induces ROS production in mouse C2C12 myocytes (Handayaningsih et al., 2011), and experimentally elevated ROS was shown to increase IGF-1-induced tyrosine phosphorylation of the IGF-I receptor in C2C12 myocytes (Handayaningsih et al., 2011). Increased IGF-I promoted increased myocyte hypertrophy and phosphorylation of Akt, mTOR, and p70S6K and extracellular signal-regulated kinase 1/2 (ERK1/2), while antioxidant NAC treatment prevented these changes (Handayaningsih et al., 2011). Thus, in vitro evidence suggests that ROS are important for skeletal muscle hypertrophy mediated via IGF-1 signaling. NO and RNS are also important in redox signaling pathways involved in muscle hypertrophy and Akt-mTOR activation in studies conducted in rodents that induced compensatory hypertrophy through synergistic ablation (Ito et al., 2013). Findings in rodents have also reported that antioxidants such as vitamin C may attenuate normal hypertrophic responses and activation of anabolic and catabolic signaling pathways in skeletal muscle following synergistic ablation (Makanae et al., 2013).

A few human studies have investigated antioxidant supplementation and muscle strength/ hypertrophy following training. A recent systematic review of seven randomized controlled trials (Dutra et al., 2020) found no effect of vitamin C and E supplementation on muscle strength after resistance training. Bobeuf and colleagues observed a significant increase in lean mass in elderly individuals after 6 months of resistance training only when training was combined with vitamin C (1 g/day) and vitamin E (400 IU/day) supplementation (Bobeuf et al., 2011). In contrast, Bjørnsen et al. found attenuated gains in lean muscle mass and muscle thickness in elderly males with concomitant vitamin C (500 mg/ day) and vitamin E (175 IU/day) supplementation during 12 weeks of resistance exercise (Bjørnsen et al., 2016). A recent study in adult females (Dutra et al., 2018) found no effect of a combination of vitamin C (1 g/day) and vitamin E (400IU/day) on quadriceps muscle thickness during 10 weeks of strength training. Paulsen et al. reported no significant difference in the improvement in lean body mass accretion, muscle group cross-sectional areas, or fractional protein synthetic rate in young healthy adults in response to supplementation with vitamin C (1 g/day) and vitamin E (350 IU/day) during 10 weeks of resistance exercise training (Paulsen et al., 2014). However, antioxidant supplementation attenuated the acute exercise-induced activation of p70S6k and MAP kinases p38 MAPK and ERK1/2, and post-exercise activation of the ubiquitin-proteasome pathway. This latter finding lends support to the findings of Makanae et al., in that antioxidants might attenuate not only anabolic signalling pathways but also catabolic signaling pathways in muscle mass regulation (Makanae et al., 2013).

SUBSTRATE METABOLISM

Elevated cellular concentrations of ROS/RNS may impair selective redox-sensitive pathways of energy metabolism. For instance, hydrogen peroxide (H_2O_2) and peroxynitrite ($ONOO^-$) can inhibit glyceraldehyde 3-phosphate dehydrogenase (GAPDH) activity by reacting directly with the active site thiol, thus potentially impairing glycolysis (Quijano et al., 2016). Further, ROS can impair activity of the enzyme aconitase by releasing an Fe atom from an Fe-S cluster that functions as a Lewis acid during catalysis in the tricarboxylic acid cycle, potentially diminishing the supply of reducing agents to the electron transport chain and thus diminishing the rate of ROS production (Quijano et al., 2016). Elevated ROS and RNS may also reduce beta-oxidation efficiency through the generation of nitro-fatty acids that can undergo beta-oxidation in the mitochondria (Quijano et al., 2016). Despite the known interplay between oxidants and energy metabolism, effects of elevated ROS/RNS on these pathways of energy metabolism during exercise are unclear. Moreover, effects of antioxidants on energy metabolism pathways have been scarcely explored in the context of acute exercise or exercise training adaptations. Effects of exogenous antioxidants on substrate metabolism are likely complex and will depend on the specific antioxidant compound, its bioavailability, and dosing regimen administered.

The carotenoid astaxanthin is a free radical scavenger of singlet oxygen and peroxyl radical intermediates that may provide protection against lipid peroxidation in vivo, although evidence is limited in humans (Goto et al. 2001). In rodents, chronic supplementation with astaxanthin decreased respiratory exchange ratio (RER) during exercise and increased fat utilization while reducing carbohydrate oxidation and sparing muscle glycogen during exercise (Aoi et al., 2018). Enhanced fat oxidation and preservation of carbohydrate stores might contribute to enhanced exercise time to exhaustion observed with astaxanthin in some rodent studies (Aoi et al., 2018, Polotow et al., 2014, Ikeuchi et al., 2006). However, these findings need to be considered along with recent findings in rodents (Polotow et al., 2014, Zhou et al., 2019) indicating a hampering of endurance training-related induction of antioxidant enzymes in skeletal muscle. In contrast to rodent findings, chronic astaxanthin supplementation did not result in a change in whole-body fat oxidation rate during submaximal exercise, nor improve time trial performance in well-trained male athletes (Res et al., 2013).

Catechins are polyphenols that probably act as indirect antioxidants in vivo via activation of endogenous antioxidant enzymes. The effects of catechin supplementation on substrate oxidation are varied. Chronic supplementation with the catechin (-)-epigallocatechin gallate (EGCG) was found to increase beta-oxidation activity in gastrocnemius muscle and upregulate genes involved in beta-oxidation such as fatty acid translocase (FAT/CD36), in addition to increasing swim time to fatigue in BALB/c mice (Murase et al., 2005). Ten weeks of supplementation with catechins (572.8 mg/day) decreased RER and carbohydrate oxidation during submaximal exercise in healthy young males (Ichinose et al., 2011). In contrast, no effects on whole-body substrate oxidation were found during a cycling time trial after 2 days supplementation with a green tea supplement in recreationally active males (Martin et al., 2014); or during a 40 km time trial after 6 days of supplementation with green tea extract (TEAVIGO) in trained competitive cyclists (Dean et al., 2009). A recent meta-analysis of 8 randomized controlled trials found significantly decreased RER and increased fasting/postprandial energy expenditure with EGCG supplementation ranging from 300 to 600 mg per day over 2 days to 12 weeks (Kapoor et al., 2017).

Performance-related implications of changes in substrate metabolism with antioxidants are unclear in humans. Trewin et al. found an increased fat oxidation rate and higher blood glucose concentrations during high-intensity interval exercise in well-trained cyclists after acute NAC infusion and impaired mean power output during a subsequent time trial (Trewin et al., 2013). In this study, a greater reliance on fat oxidation might have been counterproductive to performance given the reliance on carbohydrates for fuel during high-intensity exercise.

OXIDATIVE STRESS AND VASCULAR FUNCTION

The vascular network is vital for the distribution and exchange of blood gases, nutrients, and hormones within and between cells and organs. Oxidative stress is well-known to contribute to vascular dysfunction and vascular-related chronic disease (Akoumianakis and Antoniades, 2019); however, ROS in low quantities are also necessary for vascular health, blood flow regulation, and vascular adaptations to acute and chronic exercise training (Trinity et al., 2016).

Oxidative stress-induced vascular dysfunction transpires predominantly through superoxide reacting with the potent vasodilator nitric oxide (NO) (Sena et al., 2018, Jackson et al., 1998, Taddei et al., 1998). This process not only directly decreases NO bioavailability but also leads to the formation of ROS including peroxynitrate and the hydroxyl radical which cause endothelial NO synthase (eNOS) uncoupling through oxidation of the eNOS cofactor tetrahydrobiopterin (BH4) and substrate l-arginine (Crabtree and Channon, 2011, Alkaitis and Crabtree, 2012, Chen et al., 2010, Li and Forstermann, 2013). eNOS uncoupling favors superoxide production over NO further perpetuating decreased NO bioavailability, increased ROS production, and subsequent endothelial dysfunction (Li and Forstermann, 2013). ROS can also lead to oxidation of low-density lipoprotein (ox-LDL) and the activation of redox-sensitive protein kinases, phosphatases, mitogen-activated protein kinases, and transcription factors, which lead to aberrant vascular remodeling, eNOS uncoupling, and release of vasoconstricting agents including thromboxane A2, endothelin-1, and prostaglandin H2, and chemokines such as monocyte chemotactic protein-1 and intercellular and vascular cell adhesion molecule-1 (Bhatt et al., 2014, Sena et al., 2018, Forstermann et al., 2017, Stepp et al., 2002, Wu et al., 2017, Steinberg et al., 1989). In some cases, such as with hydrogen peroxide, ROS may also directly regulate vascular tone and promote vasoconstriction (Lucchesi et al., 2005).

Oxidative stress can acutely and chronically lead to cardiac, macro, and microvascular dysfunction, likely through the direct and indirect involvement in endothelial dysfunction, aberrant vascular remodeling, inflammation, and fibrosis (Trinity et al., 2016, Akoumianakis and Antoniades, 2019). Although antioxidant supplementation appears to be effective at improving or restoring vascular function in various clinical populations (Taddei et al., 1998, Schneider et al., 2005, Wray et al., 2012, Trinity et al., 2016), its capacity to influence exercise-mediated vascular function and training adaptations are unclear. In humans, targeting mitochondrial ROS with acute mitochondrial-targeted antioxidant treatment (80 mg of MitoQ) can improve brachial and popliteal endothelial function, claudication pain, and exercise capacity in peripheral artery disease patients (Park et al., 2020). These findings reflect rodent studies that have reported mitochondrial antioxidants and Nrf2 activators to restore cardiac function and exercise capacity in cardiomyocyte-specific NADPH oxidase 4 (Nox4)- and Nrf2-deficient mice (Hancock et al., 2018). However, the findings are not always consistent. For example, intravenous vitamin C infusion (7.5 g total infusion at 1.7–5 ml/min) improves diastolic myocardial function in patients with type 2 diabetes (T2D) and healthy controls, whereas brachial artery endothelial function and exercise capacity are not significantly impacted (Scalzo et al., 2018). Similarly, a recent meta-analysis concluded that despite cocoa flavanol supplementation improving vascular function, it fails to translate to improved exercise performance in most studies that investigated trained athletes (Decroix et al., 2018).

Antioxidant treatment appears to be less effective in the absence of pathology where redox homeostasis and vascular function are normal. For example, acute pretreatment with an antioxidant cocktail containing vitamins C (1,000 mg) and E (600 IU) and α-lipoic acid (600 mg) prior to knee-extension exercise improve exercise-induced total leg blood flow, leg oxygen utilization, and arterial oxygen saturation in chronic obstructive pulmonary disease patients (Rossman et al., 2015). However, these vascular benefits were not observed in healthy individuals undergoing the same intervention (Rossman et al., 2015). Likewise, intra-arterial vitamin C infusion (18 mg/min or 0.8–8 mg/100 mL forearm tissue per minute) improves endothelial function in hypertensive patients, yet the same vascular benefits are not observed in normotensive individuals (Schneider et al., 2005, Taddei et al., 1998). A few studies have investigated the effects of antioxidant treatment on exercise training-induced vascular adaptations in healthy individuals. In young healthy men, daily oral ingestion of MitoQ (10 mg) did not significantly affect training-induced increases in exercise capacity, muscle mitochondrial capacity, and circulating angiogenic cells following 3 weeks of endurance training (Shill et al., 2016). Similar MitoQ treatment did not improve cardiovascular function and exercise capacity in healthy control mice (Hancock et al., 2018). Collectively, antioxidants do not appear to improve vascular function and exercise capacity in healthy individuals, at least with the compounds used to date.

ROS play a pivotal role in maintaining physiological vascular function. This includes endothelial cell and vascular smooth muscle cell proliferation, differentiation and migration, the upregulation of eNOS expression in endothelial cells, and the direct regulation of blood flow during exercise and muscle contraction (Trinity et al., 2016). This may explain why in some cases acute treatment with an antioxidant cocktail (vitamin C, 1,000 mg; vitamin E, 600 IU; α-lipoic acid, 600 mg) leads to impaired basal (Wray et al., 2012) and contraction-mediated (Richardson et al., 2007) vascular function in healthy individuals. Similarly, 6 weeks of knee-extensor exercise training in older individuals leads to decreased systemic oxidative stress (oxygen-centered free radicals measured via spin trapping) and improvements in blood pressure and flow-mediated and contraction-mediated arterial dilation (Wray et al., 2009, Donato et al., 2010). However, these training-induced vascular adaptations are prevented when an antioxidant cocktail (vitamin C, 1,000 mg; vitamin E, 600 IU; α-lipoic acid, 600 mg) is ingested prior to the post-training measurement of vascular function (Wray et al., 2009, Donato et al., 2010). Similarly, resveratrol treatment (250 mg/day) in older men during 8 weeks of high-intensity exercise training led to the attenuation of training-induced improvements in exercise capacity and mean arterial pressure, and several markers of vascular function including muscle thromboxane synthase protein content and muscle interstitial prostacyclin (Gliemann et al., 2013). Research in rodents has provided further support that exercise-induced oxidative stress is in many cases beneficial and linked to cardiovascular function and exercise capacity. For example, exercise-induced myocardial oxidative stress in rodents is required for the induction of Nrf2 and its downstream antioxidant response element (ARE) targets, which leads to protection against oxidative-stress-induced cardiac dysfunction and exercise intolerance (Hancock et al., 2018). Furthermore, Nrf2 activators or mitochondrial-targeted antioxidant treatment restores cardiac function and exercise capacity in cardiomyocyte-specific Nox4- and Nrf2-deficient mice, whereas no additional benefits are found in control mice (Hancock et al., 2018). Taken together, antioxidant treatment may only be effective in cases where redox homeostasis and vascular function are perturbed.

PERFORMANCE

Endurance Performance

Endurance exercise is an exercise that involves contractions of large muscle groups that are sustained over an extended period. Evidence from rodent studies supports endurance performance-enhancing effects of certain antioxidant compounds, including astaxanthin, curcumin, catechins, resveratrol, and melatonin (Mason et al., 2020); although impairments were found in one study after vitamin C supplementation (Gomez-Cabrera et al., 2008a). In humans, some studies of acute or chronic supplementation with NAC, anthocyanins, coenzyme Q10, and vitamin E have shown enhanced endurance performance and/or capacity (Mason et al., 2020). However, findings are varied for these antioxidants (Mason et al., 2020), and other studies have reported impairments following vitamin C (Braakhuis et al., 2014) and NAC (Trewin et al., 2013) supplementation. Chronic quercetin supplementation may enhance endurance performance in previously untrained individuals (Pelletier et al., 2013). However, much of the evidence for performance outcomes stems from a limited number of studies involving a small number of male participants. For most antioxidants, there is a lack of evidence demonstrating ergogenic effects in healthy humans (Mason et al., 2020).

Muscle Recovery

Antioxidants have also been investigated with respect to their effects on muscle recovery from intense bouts of eccentric exercise. Some human studies have shown enhanced force recovery or reduced soreness following supplementation with quercetin, curcumin, anthocyanins, EGCG, and vitamin C (Mason et al., 2020). However, there is no convincing evidence for benefits of antioxidant supplements for recovery of muscle strength, amelioration of muscle damage, or reduction in DOMS following intense damaging exercise (Ranchordas et al., 2020, Martinez-Ferran et al., 2020, Mason et al., 2020).

VO_2 max

Effects of antioxidant supplementation on exercise-induced changes in VO_2 max are equivocal. Catechin supplementation increased VO_2 max in sedentary middle-aged participants (Taub et al., 2016), but attenuated training-induced increases in VO_2 max in recreationally active males. (Schwarz et al., 2018). Resveratrol supplementation augmented the improvement in VO_2 max and fatigue resistance following exercise training in older individuals in one study (Alway et al., 2017), but blunted this in another study in older individuals (Gliemann et al., 2013). In young males, there was no effect of resveratrol supplementation combined with exercise training on VO_2 max (Scribbans et al., 2014). Three weeks of supplementation with the mitochondria-targeted antioxidant Mito-Q (10 mg/day) had no effect on exercise-training-induced VO_2 max in young healthy males (Shill et al., 2016). Acute vitamin C infusion (7.5 g) also had no effect on VO_2 peak in adults with type 2 diabetes or age and weight-matched healthy control individuals (Scalzo et al., 2018). Finally, chronic supplementation with zinc in trained female Futsal players (Saeedy et al., 2016) and selenium in male students (Margaritis et al., 1997) did not affect training-related changes in VO_2 max.

Personalized Supplementation

Recent research indicates that 30 days of NAC supplementation in healthy individuals improve VO_2 max, time trial, and Wingate cycling performance only in individuals with low GSH levels (Paschalis et al., 2018). Likewise, VO_2 max is only consistently improved by antioxidant supplementation in healthy individuals when vitamin C or GSH deficiencies are present and restored by the respective supplementation of vitamin C or NAC (Margaritelis et al., 2020). These findings suggest that a personalized targeted approach to antioxidant treatment may be required to improve exercise performance. However, to pursue a comprehensive personalized supplementation approach, further knowledge about an individual's genotype including single nucleotide polymorphisms relevant to redox regulation is required.

CONCLUSION

Overall, it is clear that antioxidant supplements are not a one-stop-shop solution for improving exercise training adaptations such as mitochondrial biogenesis, antioxidant enzyme induction, vascular function, muscle hypertrophy and strength, substrate metabolism, and exercise performance. The effects of antioxidant treatment on exercise-related outcomes are likely influenced by a number of factors. These include whether an antioxidant deficiency exists, whether the antioxidant treatment is designed to target that deficiency, how it interacts with the multiple endogenous antioxidant systems, and whether the treatment dose, duration, method of administration (oral versus intravenous), and bioavailability of the antioxidant lead to biologically relevant concentrations in the target tissue. Furthermore, given the evidence suggesting that antioxidant supplementation can interfere with the beneficial effects of exercise, an individualized approach to antioxidant supplementation that carefully considers these factors will be required for an antioxidant treatment strategy to be effective.

REFERENCES

Akoumianakis, I. & Antoniades, C. 2019. Impaired vascular redox signaling in the vascular complications of obesity and diabetes mellitus. *Antioxid Redox Signal*, 30, 333–53.

Alkaitis, M. S. & Crabtree, M. J. 2012. Recoupling the cardiac nitric oxide synthases: Tetrahydrobiopterin synthesis and recycling. *Curr Heart Fail Rep*, 9, 200–10.

Alway, S. E., Mccrory, J. L., Kearcher, K., Vickers, A., Frear, B., Gilleland, D. L., Bonner, D. E., Thomas, J. M., Donley, D. A., Lively, M. W. & Mohamed, J. S. 2017. Resveratrol enhances exercise-induced cellular and functional adaptations of skeletal muscle in older men and women. *J Gerontol A Biol Sci Med Sci*, 72, 1595–606.

Aoi, W., Maoka, T., Abe, R., Fujishita, M. & Tominaga, K. 2018. Comparison of the effect of non-esterified and esterified astaxanthins on endurance performance in mice. *J Clin Biochem Nutr*, 62, 161–6.

Baur, J. A. 2010. Biochemical effects of SIRT1 activators. *Biochim Biophys Acta*, 1804, 1626–34.

Bhatt, S. R., Lokhandwala, M. F. & Banday, A. A. 2014. Vascular oxidative stress upregulates angiotensin II type I receptors via mechanisms involving nuclear factor kappa B. *Clin Exp Hypertens*, 36, 367–73.

Bjørnsen, T., Salvesen, S., Berntsen, S., Hetlelid, K. J., Stea, T. H., Lohne-Seiler, H., Rohde, G., Haraldstad, K., Raastad, T., Køpp, U., Haugeberg, G., Mansoor, M. A., Bastani, N. E., Blomhoff, R., Stølevik, S. B., Seynnes, O. R. & Paulsen, G. 2016. Vitamin C and E supplementation blunts increases in total lean body mass in elderly men after strength training. *Scand J Med Sci Sports*, 26, 755–63.

Bobeuf, F., Labonte, M., Dionne, I. J. & Khalil, A. 2011. Combined effect of antioxidant supplementation and resistance training on oxidative stress markers, muscle and body composition in an elderly population. *J Nutr Health Aging*, 15, 883–9.

Braakhuis, A. J., Hopkins, W. G. & Lowe, T. E. 2014. Effects of dietary antioxidants on training and performance in female runners. *Eur J Sport Sci*, 14, 160–8.

Bryer, S. C. & Goldfarb, A. H. 2006. Effect of high dose vitamin C supplementation on muscle soreness, damage, function, and oxidative stress to eccentric exercise. *Int J Sport Nutr Exerc Metab*, 16, 270–80.

Cobley, J. N., Mcglory, C., Morton, J. P. & Close, G. L. 2011. N-acetylcysteine's attenuation of fatigue after repeated bouts of intermittent exercise: Practical implications for tournament situations. *Int J Sport Nutr Exerc Metab*, 21, 451–61.

Corn, S. D. & Barstow, T. J. 2011. Effects of oral N-acetylcysteine on fatigue, critical power, and W' in exercising humans. *Respir Physiol Neurobiol*, 178, 261–8.

Chen, W., Druhan, L. J., Chen, C.-A., Hemann, C., Chen, Y.-R., Berka, V., Tsai, A.-L. & Zweier, J. L. 2010. Peroxynitrite induces destruction of the tetrahydrobiopterin and heme in endothelial nitric oxide synthase: Transition from reversible to irreversible enzyme inhibition. *Biochemistry*, 49, 3129–37.

Crabtree, M. J. & Channon, K. M. 2011. Synthesis and recycling of tetrahydrobiopterin in endothelial function and vascular disease. *Nitric Oxide*, 25, 81–8.

De Sousa, C. V., Sales, M. M., Rosa, T. S., Lewis, J. E., De Andrade, R. V. & Simões, H. G. 2017. The antioxidant effect of exercise: A systematic review and meta-analysis. *Sports Med*, 47, 277–93.

Dean, S., Braakhuis, A. & Paton, C. 2009. The effects of Egcg on fat oxidation and endurance performance in male cyclists. *Int J Sport Nutr Exerc Metab*, 19, 624–44.

Decroix, L., Soares, D. D., Meeusen, R., Heyman, E. & Tonoli, C. 2018. Cocoa flavanol supplementation and exercise: A systematic review. *Sports Med*, 48, 867–92.

Donato, A. J., Uberoi, A., Bailey, D. M., Wray, D. W. & Richardson, R. S. 2010. Exercise-induced brachial artery vasodilation: Effects of antioxidants and exercise training in elderly men. *Am J Physiol Heart Circ Physiol*, 298, H671–8.

Dutra, M. T., Alex, S., Mota, M. R., Sales, N. B., Brown, L. E. & Bottaro, M. 2018. Effect of strength training combined with antioxidant supplementation on muscular performance. *Appl Physiol Nutr Metab*, 43, 775–81.

Dutra, M. T., Martins, W. R., Ribeiro, A. L. A. & Bottaro, M. 2020. The effects of strength training combined with Vitamin C and E supplementation on skeletal muscle mass and strength: a systematic review and meta-analysis. *J Sports Med (Hindawi Publ Corp)*, 2020, 3505209.

Ferreira, L. F., Campbell, S. & Reid, M. B. 2011. N-acetylcysteine in handgrip exercise: Plasma thiols and adverse reactions. *Int J Sport Nutr Exerc Metab*, 21, 146–54.

Forstermann, U., Xia, N. & Li, H. 2017. Roles of vascular oxidative stress and nitric oxide in the pathogenesis of atherosclerosis. *Circ Res*, 120, 713–35.

Gliemann, L., Schmidt, J. F., Olesen, J., Bienso, R. S., Peronard, S. L., Grandjean, S. U., Mortensen, S. P., Nyberg, M., Bangsbo, J., Pilegaard, H. & Hellsten, Y. 2013. Resveratrol blunts the positive effects of exercise training on cardiovascular health in aged men. *J Physiol*, 591, 5047–59.

Gomez-Cabrera, M. C., Domenech, E., Romagnoli, M., Arduini, A., Borras, C., Pallardo, F. V., Sastre, J. & Vina, J. 2008a. Oral administration of vitamin C decreases muscle mitochondrial biogenesis and hampers training-induced adaptations in endurance performance. *Am J Clin Nutr*, 87, 142–9.

Gomez-Cabrera, M. C., Domenech, E. & Viña, J. 2008b. Moderate exercise is an antioxidant: Upregulation of antioxidant genes by training. *Free Radic Biol Med*, 44, 126–31.

Goto, S., Kogure, K., Abe, K., Kimata, Y., Kitahama, K., Yamashita, E. & Terada, H. 2001. Efficient radical trapping at the surface and inside the phospholipid membrane is responsible for highly potent antiperoxidative activity of the carotenoid astaxanthin. *Biochim Biophys Acta*, 1512, 251–8.

Hancock, M., Hafstad, A. D., Nabeebaccus, A. A., Catibog, N., Logan, A., Smyrnias, I., Hansen, S. S., Lanner, J., Schroder, K., Murphy, M. P., Shah, A. M. & Zhang, M. 2018. Myocardial Nadph oxidase-4 regulates the physiological response to acute exercise. *Elife*, 7, e41044.

Handayaningsih, A.-E., Iguchi, G., Fukuoka, H., Nishizawa, H., Takahashi, M., Yamamoto, M., Herningtyas, E.-H., Okimura, Y., Kaji, H., Chihara, K., Seino, S. & Takahashi, Y. 2011. Reactive oxygen species play an essential role in IGF-I signaling and IGF-I-induced myocyte hypertrophy in C2C12 myocytes. *Endocrinology*, 152, 912–21.

Ichinose, T., Nomura, S., Someya, Y., Akimoto, S., Tachiyashiki, K. & Imaizumi, K. 2011. Effect of endurance training supplemented with green tea extract on substrate metabolism during exercise in humans. *Scand J Med Sci Sports*, 21, 598–605.

Ikeuchi, M., Koyama, T., Takahashi, J. & Yazawa, K. 2006. Effects of astaxanthin supplementation on exercise-induced fatigue in mice. *Biol Pharm Bull*, 29, 2106–10.

Ito, N., Ruegg, U. T., Kudo, A., Miyagoe-Suzuki, Y. & Takeda, S. 2013. Activation of calcium signaling through Trpv1 by nnos and peroxynitrite as a key trigger of skeletal muscle hypertrophy. *Nat Med*, 19, 101–6.

Jackson, T. S., Xu, A., Vita, J. A. & Keaney Jr., J. F., 1998. Ascorbate prevents the interaction of superoxide and nitric oxide only at very high physiological concentrations. *Circ Res*, 83, 916–22.

Jakeman, P. & Maxwell, S. 1993. Effect of antioxidant vitamin supplementation on muscle function after eccentric exercise. *Eur J Appl Physiol Occup Physiol*, 67, 426–30.

Kapoor, M. P., Sugita, M., Fukuzawa, Y. & Okubo, T. 2017. Physiological effects of epigallocatechin-3-gallate (EGCG) on energy expenditure for prospective fat oxidation in humans: A systematic review and meta-analysis. *J Nutr Biochem*, 43, 1–10.

Knapik, J. J., Steelman, R. A., Hoedebecke, S. S., Austin, K. G., Farina, E. K. & Lieberman, H. R. 2016. Prevalence of dietary supplement use by athletes: Systematic review and meta-analysis. *Sports Medicine (Auckland, N.Z.)*, 46, 103–23.

Li, H. & Forstermann, U. 2013. Uncoupling of endothelial NO synthase in atherosclerosis and vascular disease. *Curr Opin Pharmacol*, 13, 161–7.

Lucchesi, P. A., Belmadani, S. & Matrougui, K. 2005. Hydrogen peroxide acts as both vasodilator and vasoconstrictor in the control of perfused mouse mesenteric resistance arteries. *J Hypertens*, 23, 571–9.

Makanae, Y., Kawada, S., Sasaki, K., Nakazato, K. & Ishii, N. 2013. Vitamin C administration attenuates overload-induced skeletal muscle hypertrophy in rats. *Acta Physiol (Oxf)*, 208, 57–65.

Mankowski, R. T., Anton, S. D., Buford, T. W. & Leeuwenburgh, C. 2015. Dietary antioxidants as modifiers of physiologic adaptations to exercise. *Med Sci Sports Exerc*, 47, 1857–68.

Margaritelis, N. V., Paschalis, V., Theodorou, A. A., Kyparos, A. & Nikolaidis, M. G. 2020. Antioxidant supplementation, redox deficiencies and exercise performance: A falsification design. *Free Radic Biol Med*, 158, 44–52.

Margaritis, I., Tessier, F., Prou, E., Marconnet, P. & Marini, J. F. 1997. Effects of endurance training on skeletal muscle oxidative capacities with and without selenium supplementation. *J Trace Elem Med Biol*, 11, 37–43.

Martin, B. J., Tan, R. B., Gillen, J. B., Percival, M. E. & Gibala, M. J. 2014. No effect of short-term green tea extract supplementation on metabolism at rest or during exercise in the fed state. *Int J Sport Nutr Exerc Metab*, 24, 656–64.

Martinez-Ferran, M., Sanchis-Gomar, F., Lavie, C. J., Lippi, G. & Pareja-Galeano, H. 2020. Do antioxidant vitamins prevent exercise-induced muscle damage? A systematic review. *Antioxidants (Basel, Switzerland)*, 9, 372.

Mason, S. A., Morrison, D., Mcconell, G. K. & Wadley, G. D. 2016. Muscle redox signalling pathways in exercise. Role of antioxidants. *Free Radic Biol Med*, 98, 29–45.

Mason, S. A., Trewin, A. J., Parker, L. & Wadley, G. D. 2020. Antioxidant supplements and endurance exercise: Current evidence and mechanistic insights. *Redox Biology*, 35, 101471.

Merry, T. L. & Ristow, M. 2016. Nuclear factor erythroid-derived 2-like 2 (NFE2L2, Nrf2) mediates exercise-induced mitochondrial biogenesis and the anti-oxidant response in mice. *J Physiol*, 594, 5195–207.

Morrison, D., Hughes, J., Della Gatta, P. A., Mason, S., Lamon, S., Russell, A. P. & Wadley, G. D. 2015. Vitamin C and E supplementation prevents some of the cellular adaptations to endurance-training in humans. *Free Radic Biol Med*, 89, 852–62.

Murase, T., Haramizu, S., Shimotoyodome, A., Nagasawa, A. & Tokimitsu, I. 2005. Green tea extract improves endurance capacity and increases muscle lipid oxidation in mice. *Am J Physiol Regul Integr Comp Physiol*, 288, R708–15.

Olesen, J., Gliemann, L., Bienso, R., Schmidt, J., Hellsten, Y. & Pilegaard, H. 2014. Exercise training, but not resveratrol, improves metabolic and inflammatory status in skeletal muscle of aged men. *J Physiol*, 592, 1873–86.

Park, S. Y., Pekas, E. J., Headid 3rd, R. J., Son, W. M., Wooden, T. K., Song, J., Layec, G., Yadav, S. K., Mishra, P. K. & Pipinos, I. I. 2020. Acute mitochondrial antioxidant intake improves endothelial function, antioxidant enzyme activity, and exercise tolerance in patients with peripheral artery disease. *Am J Physiol Heart Circ Physiol*, 319, H456–67.

Paschalis, V., Theodorou, A. A., Margaritelis, N. V., Kyparos, A. & Nikolaidis, M. G. 2018. N-acetylcysteine supplementation increases exercise performance and reduces oxidative stress only in individuals with low levels of glutathione. *Free Radic Biol Med*, 115, 288–97.

Paulsen, G., Hamarsland, H., Cumming, K. T., Johansen, R. E., Hulmi, J. J., Børsheim, E., Wiig, H., Garthe, I. & Raastad, T. 2014. Vitamin C and E supplementation alters protein signalling after a strength training session, but not muscle growth during 10 weeks of training. *J Physiol*, 592, 5391–408.

Pelletier, D. M., Lacerte, G. & Goulet, E. D. 2013. Effects of quercetin supplementation on endurance performance and maximal oxygen consumption: A meta-analysis. *Int J Sport Nutr Exerc Metab*, 23, 73–82.

Petersen, A. C., Mckenna, M. J., Medved, I., Murphy, K. T., Brown, M. J., Della Gatta, P. & Cameron-Smith, D. 2012. Infusion with the antioxidant N-acetylcysteine attenuates early adaptive responses to exercise in human skeletal muscle. *Acta Physiol (Oxf)*, 204, 382–92.

Polotow, T. G., Vardaris, C. V., Mihaliuc, A. R., Goncalves, M. S., Pereira, B., Ganini, D. & Barros, M. P. 2014. Astaxanthin supplementation delays physical exhaustion and prevents redox imbalances in plasma and soleus muscles of Wistar rats. *Nutrients*, 6, 5819–38.

Poulsen, M. M., Vestergaard, P. F., Clasen, B. F., Radko, Y., Christensen, L. P., Stodkilde-Jorgensen, H., Moller, N., Jessen, N., Pedersen, S. B. & Jorgensen, J. O. 2013. High-dose resveratrol supplementation in obese men: An investigator-initiated, randomized, placebo-controlled clinical trial of substrate metabolism, insulin sensitivity, and body composition. *Diabetes*, 62, 1186–95.

Quijano, C., Trujillo, M., Castro, L. & Trostchansky, A. 2016. Interplay between oxidant species and energy metabolism. *Redox Biol*, 8, 28–42.

Ranchordas, M. K., Rogerson, D., Soltani, H. & Costello, J. T. 2020. Antioxidants for preventing and reducing muscle soreness after exercise: A Cochrane systematic review. *Br J Sports Med* 54, 74–78.

Res, P. T., Cermak, N. M., Stinkens, R., Tollakson, T. J., Haenen, G. R., Bast, A. & Van Loon, L. J. 2013. Astaxanthin supplementation does not augment fat use or improve endurance performance. *Med Sci Sports Exerc*, 45, 1158–65.

Richardson, R. S., Donato, A. J., Uberoi, A., Wray, D. W., Lawrenson, L., Nishiyama, S. & Bailey, D. M. 2007. Exercise-induced brachial artery vasodilation: Role of free radicals. *Am J Physiol Heart Circ Physiol*, 292, H1516–22.

Ristow, M., Zarse, K., Oberbach, A., Kloting, N., Birringer, M., Kiehntopf, M., Stumvoll, M., Kahn, C. R. & Bluher, M. 2009. Antioxidants prevent health-promoting effects of physical exercise in humans. *Proc Natl Acad Sci U S A*, 106, 8665–70.

Rock, C. L., Newman, V. A., Neuhouser, M. L., Major, J. & Barnett, M. J. 2004. Antioxidant supplement use in cancer survivors and the general population. *J Nutr*, 134, 3194S.

Rossman, M. J., Trinity, J. D., Garten, R. S., Ives, S. J., Conklin, J. D., Barrett-O'keefe, Z., Witman, M. A., Bledsoe, A. D., Morgan, D. E., Runnels, S., Reese, V. R., Zhao, J., Amann, M., Wray, D. W. & Richardson, R. S. 2015. Oral antioxidants improve leg blood flow during exercise in patients with chronic obstructive pulmonary disease. *Am J Physiol Heart Circ Physiol*, 309, H977–85.

Saeedy, M., Bijeh, N. & Shoorideh, Z. 2016. Effect of six weeks of high intensity interval training and zinc supplement on serum creatine kinase and uric acid levels in futsal players. *Int J Appl Exerc Physiol*, 5, 19–27.

Scalzo, R. L., Bauer, T. A., Harrall, K., Moreau, K., Ozemek, C., Herlache, L., Mcmillin, S., Huebschmann, A. G., Dorosz, J., Reusch, J. E. B. & Regensteiner, J. G. 2018. Acute vitamin C improves cardiac function, not exercise capacity, in adults with type 2 diabetes. *Diabetol Metab Syndr*, 10, 7.

Schneider, M. P., Delles, C., Schmidt, B. M., Oehmer, S., Schwarz, T. K., Schmieder, R. E. & John, S. 2005. Superoxide scavenging effects of N-acetylcysteine and vitamin C in subjects with essential hypertension. *Am J Hypertens*, 18, 1111–7.

Schwarz, N. A., Blahnik, Z. J., Prahadeeswaran, S., Mckinley-Barnard, S. K., Holden, S. L. & Waldhelm, A. 2018. (-)-Epicatechin supplementation inhibits aerobic adaptations to cycling exercise in humans. *Front Nutr*, 5, 132.

Scribbans, T. D., Ma, J. K., Edgett, B. A., Vorobej, K. A., Mitchell, A. S., Zelt, J. G., Simpson, C. A., Quadrilatero, J. & Gurd, B. J. 2014. Resveratrol supplementation does not augment performance adaptations or fibre-type-specific responses to high-intensity interval training in humans. *Appl Physiol Nutr Metab*, 39, 1305–13.

Sena, C. M., Leandro, A., Azul, L., Seica, R. & Perry, G. 2018. Vascular oxidative stress: Impact and therapeutic approaches. *Front Physiol*, 9, 1668.

Shill, D. D., Southern, W. M., Willingham, T. B., Lansford, K. A., Mccully, K. K. & Jenkins, N. T. 2016. Mitochondria-specific antioxidant supplementation does not influence endurance exercise training-induced adaptations in circulating angiogenic cells, skeletal muscle oxidative capacity or maximal oxygen uptake. *J Physiol*, 594, 7005–14.

Steinberg, D., Parthasarathy, S., Carew, T. E., Khoo, J. C. & Witztum, J. L. 1989. Beyond cholesterol. Modifications of low-density lipoprotein that increase its atherogenicity. *N Engl J Med*, 320, 915–24.

Stepp, D. W., Ou, J., Ackerman, A. W., Welak, S., Klick, D. & Pritchard Jr., K. A. 2002. Native LDL and minimally oxidized LDL differentially regulate superoxide anion in vascular endothelium in situ. *Am J Physiol Heart Circ Physiol*, 283, H750–9.

Taddei, S., Virdis, A., Ghiadoni, L., Magagna, A. & Salvetti, A. 1998. Vitamin C improves endothelium-dependent vasodilation by restoring nitric oxide activity in essential hypertension. *Circulation*, 97, 2222–9.

Taub, P. R., Ramirez-Sanchez, I., Patel, M., Higginbotham, E., Moreno-Ulloa, A., Roman-Pintos, L. M., Phillips, P., Perkins, G., Ceballos, G. & Villarreal, F. 2016. Beneficial effects of dark chocolate on exercise capacity in sedentary subjects: Underlying mechanisms. A double blind, randomized, placebo controlled trial. *Food Funct*, 7, 3686–93.

Trewin, A. J., Petersen, A. C., Billaut, F., Mcquade, L. R., Mcinerney, B. V. & Stepto, N. K. 2013. N-acetylcysteine alters substrate metabolism during high-intensity cycle exercise in well-trained humans. *Appl Physiol Nutr Metab*, 38, 1217–27.

Trinity, J. D., Broxterman, R. M. & Richardson, R. S. 2016. Regulation of exercise blood flow: Role of free radicals. *Free Radic Biol Med*, 98, 90–102.

Wray, D. W., Nishiyama, S. K., Harris, R. A., Zhao, J., Mcdaniel, J., Fjeldstad, A. S., Witman, M. A., Ives, S. J., Barrett-O'keefe, Z. & Richardson, R. S. 2012. Acute reversal of endothelial dysfunction in the elderly after antioxidant consumption. *Hypertension*, 59, 818–24.

Wray, D. W., Uberoi, A., Lawrenson, L., Bailey, D. M. & Richardson, R. S. 2009. Oral antioxidants and cardiovascular health in the exercise-trained and untrained elderly: A radically different outcome. *Clin Sci (Lond)*, 116, 433–41.

Wu, M.-Y., Li, C.-J., Hou, M.-F. & Chu, P.-Y. 2017. New insights into the role of inflammation in the pathogenesis of atherosclerosis. *Int J Mol Sci*, 18, 2034.

Yfanti, C., Akerstrom, T., Nielsen, S., Nielsen, A. R., Mounier, R., Mortensen, O. H., Lykkesfeldt, J., Rose, A. J., Fischer, C. P. & Pedersen, B. K. 2010. Antioxidant supplementation does not alter endurance training adaptation. *Med Sci Sports Exerc*, 42, 1388–95.

Zhou, Y., Baker, J. S., Chen, X., Wang, Y., Chen, H., Davison, G. W. & Yan, X. 2019. High-dose astaxanthin supplementation suppresses antioxidant enzyme activity during moderate-intensity swimming training in mice. *Nutrients*, 11, 1244.

CHAPTER TWELVE

Nitric Oxide Biochemistry and Exercise Performance in Humans

INFLUENCE OF NITRATE SUPPLEMENTATION

Stephen J. Bailey and Andrew M. Jones

CONTENTS

Introduction / 137
Emergence of the Nitrate-Nitrite-Nitric Oxide Pathway / 138
Influence of Dietary Nitrate Supplementation on Continuous Endurance Exercise Performance / 141
Influence of Dietary Nitrate Supplementation on High-Intensity Exercise Performance / 142
Mechanisms for the Ergogenic Effect of Dietary Nitrate Supplementation / 143
Conclusion / 145
References / 145

INTRODUCTION

The free radical nitric oxide (NO) is a gaseous molecule that was first identified as an endothelium-derived vasodilator (Ignarro et al., 1987), a discovery which led to the Nobel Prize in Physiology or Medicine being awarded to Drs. Ignarro, Furchgott and Murad in 1998. It has since been established that NO is synthesised in numerous tissues and can impact a myriad of physiological processes in addition to vasodilation (Bredt, 1999). With regards to its impact on skeletal muscle physiology, NO production is increased by muscle contraction (Hirschfield et al., 2000) and regulates perfusion, excitation-contraction coupling, contractility and fatigability, and various metabolic processes such as glucose uptake and mitochondrial respiration (Moon et al., 2017a; Stamler and Meissner, 2001; Suhr et al., 2013). These wide-ranging NO-mediated physiological effects have implications for health and disease morbidity and exercise performance.

The most recognised pathway for NO synthesis is via the NO synthase (NOS) enzymes with constitutive endothelial (eNOS) and neuronal (nNOS) isoforms, as well as an inducible (iNOS) isoform, having been identified (Bredt, 1999). It is now well documented that eNOS and nNOS are ubiquitously expressed in numerous tissues (Bredt, 1999), including skeletal muscle (Frandsen et al., 2000; Kobzik et al., 1994; Wylie et al., 2019), with nNOS activity higher in fast-twitch (type II) skeletal muscle (Kobzik et al., 1994). Synthesis of NO via NOS is catalysed via the complex, five-electron oxidation of the semi-essential amino acid, L-arginine, with nicotinamide adenine dinucleotide phosphate, flavin adenine dinucleotide, flavin mononucleotide, tetrahydrobiopterin, haem and calmodulin as additional essential co-factors, to yield NO and L-citrulline (Bredt, 1999).

DOI: 10.1201/9781003051619-12

As a result, there has been great interest in the potential for dietary L-arginine supplementation to enhance NOS-derived NO and associated physiological processes including exercise performance. However, the existing evidence indicates that the efficacy of L-arginine supplementation to increase NO biomarkers and improve exercise performance in healthy adults is limited (Álvares et al., 2011). Although nitrate (NO_3^-) and nitrite (NO_2^-) were conventionally considered as inert products of NO oxidation (Moncada and Higgs, 1993), it is now recognised that these anions can be recycled back to bioactive NO and other reactive nitrogen intermediates (RNIs) under appropriate physiological conditions (Kapil et al., 2020). Consequently, dietary NO_3^- supplementation has become a popular, convenient and cost-effective approach to drive NO synthesis and signalling. This chapter will provide an overview of dietary NO_3^- metabolism and its impact on biochemical processes involving NO in the human body, the potential for exercise performance enhancement after dietary NO_3^- supplementation and the putative mechanisms that might underpin these ergogenic effects.

EMERGENCE OF THE NITRATE-NITRITE-NITRIC OXIDE PATHWAY

Since the seminal observations that NO_2^- can be reduced back to NO (Benjamin et al., 1994), there has been great interest in elevating local and systemic NO_2^- to augment NO synthesis and signalling. Given that orally administered NO_3^- is a precursor for NO_2^- synthesis in vivo (Kapil et al., 2020), and that NO_3^- is abundant in beetroot and many green leafy vegetables (Hord et al., 2009), dietary NO_3^- is now widely utilised as a nutritional intervention to drive NO synthesis. While NO_3^- is almost entirely absorbed through enterocytes in the small intestine and appears in the systemic circulation after the consumption of an oral NO_3^- bolus, the majority (~60%) is extracted by the kidneys and excreted in urine during first-pass metabolism (Wagner et al., 1984). However, ~25% of an orally administered NO_3^- bolus is absorbed from the bloodstream into the salivary glands (Spiegelhalder et al., 1976), by the NO_3^-/H^+ cotransporter, sialin (Qin et al., 2012), and concentrated at least 10-fold (Lundberg and Govoni, 2004). Subsequently, NO_3^--rich saliva is secreted into the oral cavity wherein certain taxa of the oral microflora catalyse the two-electron reduction of NO_3^- to NO_2^- (Burleigh et al., 2018; Hyde et al., 2014; Vanhatalo et al., 2018) (Figure 12.1).

During second-pass metabolism, NO_2^--rich saliva is swallowed and protonated into nitrous acid (HNO_2) after arrival in the stomach, with HNO_2 subsequently decomposed into a variety of RNIs including dinitrogen trioxide, nitrogen dioxide, NO and the nitrosonium ion (Kapil et al., 2020; Lundberg et al., 2004). Some of the RNIs derived from the acidification of NO_2^- can nitrosate thiols to yield S-nitrosothiols (RSNO) (Kapil et al., 2020; Lundberg et al., 2004) (Figure 12.1). Recent evidence suggests that formation of RSNO in the stomach and their subsequent appearance in the circulation is a key determinant of improved systemic physiological processes, such as blood pressure reduction, after ingesting a NO_3^- bolus (Oliveira-Paula and Tanus-Santos, 2019; Pinheiro et al., 2015), which could be linked to RSNO transferring their NO group to distal downstream targets through S-transnitrosylation (Oliveira-Paula and Tanus-Santos, 2019). In addition to increasing plasma [RSNO] (Carlström et al., 2010; Pinheiro et al., 2015), NO_3^- supplementation dose-dependently increases plasma $[NO_2^-]$, with peak plasma $[NO_2^-]$ attained 2–4 hours post NO_3^- ingestion (Wylie et al., 2013a), and NO_2^- has been shown to potentiate the vasodilatory effects of RSNO (Liu et al., 2018). The elevation in plasma $[NO_2^-]$ after NO_3^- supplementation can also serve as a circulating substrate pool for NO since NO_2^- can undergo a one-electron reduction to NO (van Faassen et al., 2009). This reaction is catalysed by numerous NO_2^- reductases, including deoxyhaemoglobin, deoxymyoglobin, eNOS, xanthine oxidoreductase (XOR), sulphite oxidase, aldehyde oxidase, cytochrome P450, mitochondrial amidoxime reducing component and mitochondrial electron transfer complexes (Bender and Schwarz, 2018; van Faassen et al., 2009), and is augmented in hypoxic and acidic conditions (Castello et al., 2006; Li et al., 2008; Modin et al., 2001) (Figure 12.2). There is also evidence that NO_2^- can react with non-haeme iron to form dinitrosyliron complexes (DNICs) to exert some of it physiological effects (Dungel et al., 2015; Thomas et al., 2018).

It is well documented that exercise lowers pH and the partial pressure of O_2 (PO_2) within skeletal muscle (Bailey et al., 2010; Richardson et al., 1995). Plasma $[NO_2^-]$ declines during exhaustive exercise (Kelly et al., 2014; Cocksedge et al., 2020),

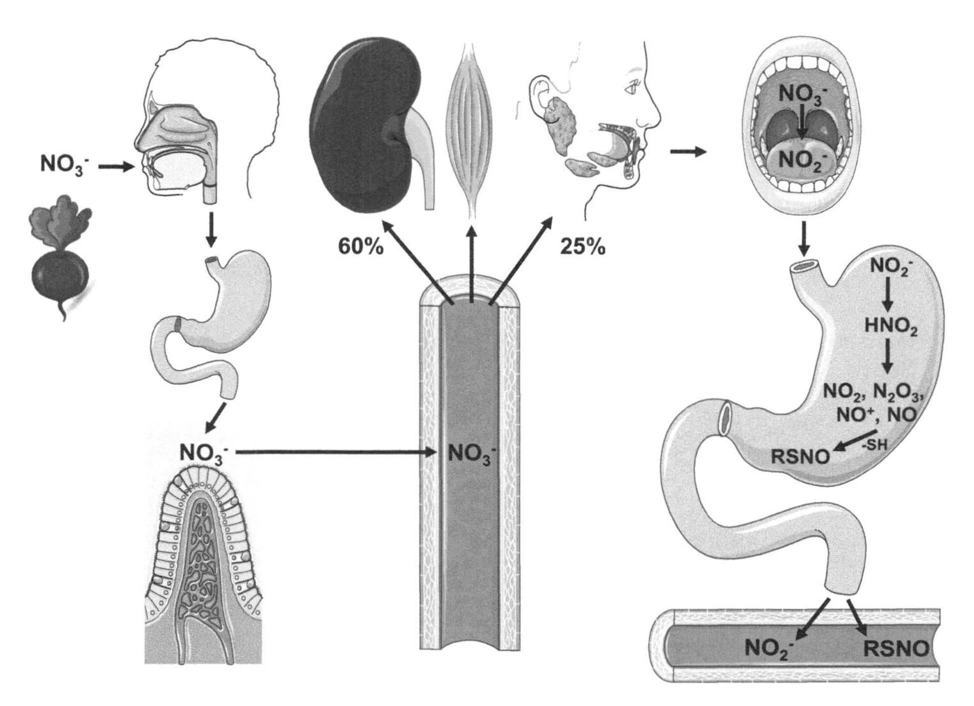

Figure 12.1 Schematic of the nitrate (NO_3^-)– nitrite (NO_2^-) – nitric oxide (NO) pathway. After ingestion of dietary NO_3^-, with most dietary NO_3^- intake in the form of NO_3^--rich vegetables and most research studies administering NO_3^--rich beetroot juice, NO_3^- is almost entirely absorbed by enterocytes in the small intestine and passes into the systemic circulation. Whilst the majority (~60%) of an oral NO_3^- bolus is extracted by the kidney and excreted in the urine, and skeletal muscle is increasingly recognised as an important NO_3^- storage site, ~25% is absorbed by, and concentrated in, the salivary glands. This NO_3^--rich saliva is secreted into the oral cavity wherein certain taxa of the oral microflora catalyse the reduction of NO_3^- to NO_2^-. After swallowing and arrival in the stomach, NO_2^- is protonated into nitrous acid (HNO_2). Subsequently, HNO_2 is decomposed into a variety of reactive nitrogen intermediates including dinitrogen trioxide (N_2O_3), nitrogen dioxide (NO_2), NO and the nitrosonium ion (NO^+), some of which can nitrosate thiols (-SH) to yield S-nitrosothiols (RSNO). It is also clear that some of the ingested NO_2^- appear in the bloodstream alongside RSNO, with NO_2^- and RSNO concentrations increased after NO_3^- ingestion. This figure was created using images from Servier Medical Art.

with a higher pre-exercise plasma [NO_2^-] (Wylie et al., 2013a) and a greater decline in plasma [NO_2^-] during exercise (Thompson et al., 2015) being correlated with improved exercise performance after NO_3^- supplementation. This suggests that NO_2^- reduction to NO is likely to be greater during exercise and that exercise performance can be enhanced by increasing the substrate for O_2-independent NO generation via dietary NO_3^- supplementation. There is evidence that reduction of NO_2^- to NO and other RNIs in the stomach is aided by polyphenols and ascorbate (Gago et al., 2007; Oliveira-Paula and Tanus-Santos, 2019; Rocha et al., 2014), and that plasma [NO_2^-] can be augmented and physiological responses improved to a greater extent after ingesting foods rich in NO_3^- and polyphenols compared to an equivalent dose of NO_3^- as a NO_3^- salt (Flueck et al., 2016; Jonvik et al., 2016). Co-ingestion of NO_3^- with polyphenols could also limit the formation of potentially carcinogenic nitrosamines when NO_2^--rich saliva arrives in the stomach (Bryan et al., 2012). Consequently, ingesting foods rich in NO_3^- and polyphenols, such as beetroot juice (Wootton-Beard and Ryan, 2011), and particularly concentrated beetroot juice (McDonagh et al., 2018), is the current recommended best practice to enhance NO synthesis and the ergogenic potential of NO_3^- supplementation. Whilst beetroot juice is rich in polyphenols and antioxidants, in addition to NO_3^- (Wootton-Beard and Ryan, 2011), dietary supplementation with NO_3^--depleted beetroot juice does not improve physiological responses and exercise performance compared to baseline conditions without dietary supplementation (Lansley et al., 2011). Therefore, while NO_3^- can interact

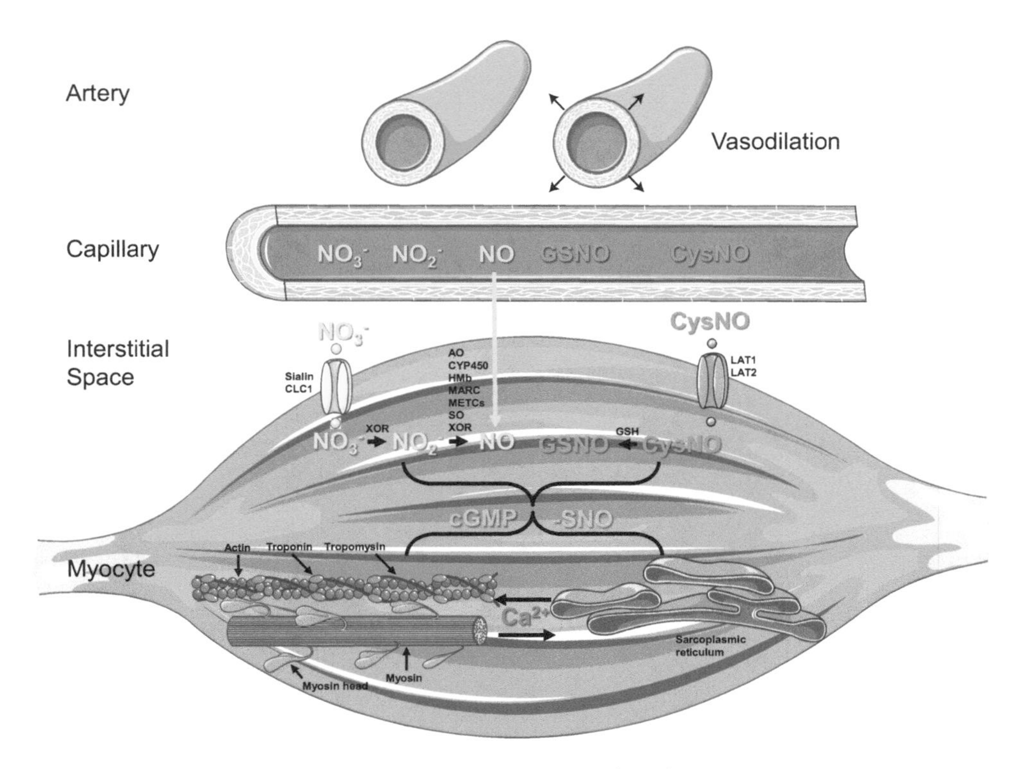

Figure 12.2 An illustration of the putative mechanisms by which nitrate (NO_3^-) supplementation impacts skeletal muscle physiology and exercise performance. Following NO_3^- supplementation, increased concentrations of NO_3^-, nitrite, (NO_2^-) nitric oxide (NO) and S-nitrosothiols (RSNO), including S-nitrosoglutathione (GSNO) and S-nitrosocysteine (CysNO), appear in the bloodstream. An increase in circulating concentrations of these reactive nitrogen intermediates may lead to vasodilation and an increase in skeletal muscle blood flow during exercise. Whilst NO gas can diffuse into the myocytes from capillaries, NO_3^- is transported into myocytes through the transporters, sialin and chloride channel 1 (CLC1). Intramyocyte NO_3^- is reduced to NO_2^- by xanthine oxidoreductase (XOR) with NO_2^- reduced to NO via numerous NO_2^- reductases including, aldehyde oxidase (AO), cytochrome P450 (CYP450), deoxygenated myoglobin (MHb), mitochondrial amidoxime-reducing component (MARC), mitochondrial electron transfer complexes (METCs), sulphite oxidase (SO) and XOR. Tissue uptake of CysNO, but not GSNO, occurs through the L-Type Amino Acid Transporters, LAT1 and LAT2. Cytosolic CysNO can subsequently transfer its NO moiety to reduced glutathione (GSH) via S-transnitrosylation to form GSNO. Subsequently, NO_2^-, NO, GSNO and CysNO can elicit physiological responses through a combination of increasing cyclic guanosine monophosphate (cGMP) and associated signalling, and S-nitrosylating protein thiols with an -SNO moiety. The existing evidence suggests that dietary NO_3^- supplementation may improve skeletal muscle physiology and exercise performance via effects of cGMP-mediated signalling and S-nitrosylation on sarcoplasmic reticulum calcium (Ca^{2+}) release and reuptake (Ca^{2+}-ATPase) channels, and myofibrillar proteins (actin, myosin, troponin and tropomysin). This figure was created using images from Servier Medical Art.

synergistically with other components present in beetroot juice to increase the synthesis of NO and other RNIs (Gago et al., 2007; Oliveira-Paula and Tanus-Santos, 2019; Rocha et al., 2014), NO_3^- is the principal active ingredient in beetroot juice that elicits physiological and ergogenic effects.

Emerging data indicate that skeletal muscle can serve as an important store of NO_3^- and NO_2^- for subsequent NO synthesis, as evidenced by higher $[NO_3^-]$ and $[NO_2^-]$ in skeletal muscle than blood (Gilliard et al., 2018; Wylie et al., 2019) and greater reduction of NO_2^- to NO in striated muscle than blood (Li et al., 2008). The NO_3^- transporter, sialin, has been identified in skeletal muscle (Wylie et al., 2019; Srihirun et al., 2020), which, together with chloride channel 1 (Srihirun et al., 2020), facilitate the concentration of NO_3^- within skeletal muscle (Figure 12.2). Skeletal muscle $[NO_3^-]$ (Gilliard et al., 2018; Wylie et al., 2019) and $[NO_2^-]$ (Gilliard et al., 2018) are increased following NO_3^- supplementation with duration-dependent increases at least up to 7 days of supplementation (Gilliard et al., 2018). In addition to its role as a

NO_2^- reductase (Zhang et al., 1998), XOR can function as a NO_3^- reductase to increase NO_2^- synthesis (Jansson et al., 2008) and is present in skeletal muscle (Wylie et al., 2019; Srihirun et al., 2020). It has been reported that the increase in skeletal muscle $[NO_2^-]$ after NO_3^- administration is enhanced by exercise and as muscle pH is lowered, with both NO_3^- reduction to NO_2^- and NO_2^- reduction to NO abolished after XOR inhibition (Piknova et al., 2016). It is, therefore, possible that increased XOR activity during exercise (Bailey et al., 2004) could contribute to enhanced NO_3^- and NO_2^- reduction in such settings. The increase in skeletal muscle $[NO_3^-]$ after NO_3^- supplementation in human skeletal muscle is lowered following the completion of exhaustive cycling exercise, suggesting that this elevated muscle NO_3^- pool is utilised as a substrate for sequential reduction to NO_2^- and then NO (Wylie et al., 2019). There is also a positive arterial-venous difference in plasma $[NO_2^-]$ across contracting skeletal muscles (Cosby et al., 2003). Since NO_2^- reduction to NO is augmented in hypoxia and acidosis (Castello et al., 2006; Li et al., 2008; Modin et al., 2001), and given that such conditions develop with the muscle microvasculature during exercise (Richardson et al., 1995), elevating circulating plasma $[NO_2^-]$ is likely to increase NO synthesis in the muscle microvasculature during exercise. It is also possible that circulating NO_2^- can be transported across plasma membranes. Indeed, NO_2^- is accumulated in heart tissue, possibly via HNO_2 (Samouilov et al., 2007), and erythrocytes, via erythrocyte anion exchange protein (Shingles et al., 1997); however, it is presently unclear whether NO_2^- can be directly transported into skeletal muscle. Based on the existing evidence, NO_3^- and NO_2^- can be increased systemically and within skeletal muscle following dietary NO_3^- supplementation with the potential to enhance NO synthesis, particularly during the hypoxic and acidic conditions that develop during exercise.

In addition to increasing circulating [RSNO] (Carlström et al., 2010; Pinheiro et al., 2015), NO_3^- supplementation can increase tissue RSNO (Carlström et al., 2010). However, while S-nitrosoglutathione (GSNO) is important for driving intracellular NO signalling and contractility in skeletal muscle (Moon et al., 2017b), and S-nitrosoalbumin is the predominant plasma RSNO, tissue uptake of these RSNO is limited (Li and Whorton, 2007; Matsumoto and Gow, 2011). Instead, L-cysteine is required as a chaperone thiol to facilitate blood to tissue RSNO flux and to propagate NO signalling (Li and Whorton, 2007). After L-cysteine undergoes S-transnitrosylation from other RSNO, S-nitrosocysteine (CysNO) is transported into tissues via the L-Type Amino Acid Transporters, LAT1 and LAT2 (Li and Whorton, 2007). Here, CysNO can S-transnitrosylate reduced glutathione, to form GSNO, and other protein thiols to provoke physiological responses (Li and Whorton, 2007; Liu et al., 2016). Since human skeletal muscle, particularly type II skeletal muscle, expresses LAT1 (Hodson et al., 2017), some of the NO-mediated physiological effects evoked by NO_3^- supplementation may be linked to greater muscle uptake of RSNO, in addition to NO_3^- (Figure 12.2).

INFLUENCE OF DIETARY NITRATE SUPPLEMENTATION ON CONTINUOUS ENDURANCE EXERCISE PERFORMANCE

In the first study to indicate that dietary NO_3^- supplementation could improve physiological responses during exercise, Larsen and colleagues (2007) reported that daily supplementation with 0.1 mmol $NaNO_3$ per kg body mass for 3 days could lower the O_2 cost of submaximal (45–85% O_{2peak}) cycling exercise in moderately trained (O_{2peak} 55 ml/kg/min) individuals (Larsen et al., 2007). This observation was corroborated following 3–6 days of supplementation with 500 mL NO_3^--rich beetroot juice, which provided a daily NO_3^- dose of 5.6 mmol (Bailey et al., 2009). In addition, beetroot juice supplementation was shown to improve the tolerable duration of constant-work-rate, exhaustive cycling exercise by 16% (Bailey et al., 2009). It was subsequently reported that cycling economy could be improved following acute ingestion of 280 mL (16.8 mmol NO_3^-), but not 70 mL (4.2 mmol NO_3^-) or 140 mL (8.4 mmol NO_3^-) of concentrated beetroot juice (BR), with exercise tolerance improved by a similar magnitude following acute ingestion of 140 and 280 mL BR, but not 70 mL BR (Wylie et al., 2013a). There is also evidence that exercise economy and tolerance can be improved following NO_3^- supplementation in a variety of different exercise modalities and settings (Pawlak-Chaouch et al., 2016). In addition to improving time-to-exhaustion during constant-work-rate exhaustive exercise (McMahon et al., 2017), acute and short-term NO_3^- supplementation can improve endurance exercise performance, as assessed with a time trial performance test, with short-term supplementation more likely to be ergogenic in such settings (Cermak et al., 2012a,b). Therefore, whilst not unequivocal (McMahon et al., 2017; Pawlak-Chaouch et al., 2016; Senefeld et al., 2020), the existing empirical evidence suggests that NO_3^- supplementation can improve exercise economy and endurance performance in moderately trained (VO_{2peak} <60 ml/kg/min) participants. Short-term supplementation of at least 6 days with a daily NO_3^- dose of 140 mL BR (8–12 mmol) and the final dose consumed approximately 3 h prior to competition (to coincide with peak plasma $[NO_2^-]$) is the current recommended best practice to maximise the ergogenic potential of BR supplementation. However, a recent study indicated that a 70 mL top-up BR dose at 1 hour during 2 hour of moderate-intensity exercise, below the lactate threshold, promoted a greater improvement in exercise economy and lowered muscle glycogen utilisation compared to BR consumed only prior to exercise (Tan et al., 2018). As such, it is possible that BR supplementation during exercise could confer additional ergogenic effects during prolonged endurance exercise.

Consistent with augmented reduction of NO_2^- to NO in conditions of acidosis and hypoxia (Castello et al., 2006; Li et al., 2008; Modin et al., 2001), NO_3^- supplementation has been reported to improve exercise economy and endurance performance in normobaric hypoxia (Muggeridge et al., 2014; Kelly et al., 2014; Cocksedge et al., 2020). Moreover, studies comparing the relative efficacy of NO_3^- supplementation to improve exercise economy and endurance performance in normoxia and hypoxia have revealed superior effects in the latter (Kelly et al., 2014; Cocksedge et al., 2020). Conversely, and notwithstanding some examples of a potential ergogenic effect in a limited number of studies (Hoon et al., 2014; Peeling et al., 2015; Rokkedal-Lausch et al., 2019; Shannon et al., 2017), the efficacy of NO_3^- supplementation to improve these variables is attenuated in well trained (VO_{2peak} >60 ml/kg/min) endurance participants (Porcelli et al., 2015; Senefeld et al., 2020). The blunted ergogenic effect of NO_3^- supplementation in well-trained endurance participants has been suggested to result from attenuated muscle deoxygenation and acidosis when exercising at a given work rate, and increased reliance on NOS-derived NO, compared to their lesser trained counterparts (Jones, 2014). There is also evidence that NO_3^- supplementation can improve exercise capacity in some (Coggan et al., 2018; Kerley et al., 2019; Kenjale et al., 2011; Ramick et al., 2021; Zamani et al., 2015), but not all (Shepherd et al., 2015a,b; Woessner et al., 2020), patient populations. Accordingly, the efficacy of NO_3^- supplementation to improve exercise economy and endurance performance appears to be linked to participant health and fitness status, and environmental conditions.

INFLUENCE OF DIETARY NITRATE SUPPLEMENTATION ON HIGH-INTENSITY EXERCISE PERFORMANCE

Initial studies conducted using murine skeletal muscle revealed that the effects of NO_3^- supplementation were fibre-type specific. Indeed, short-term (5–7 days) NO_3^- supplementation was shown to increase evoked force production in type II, but not slow-twitch (type I), muscle ex vivo (Hernández et al., 2012), and to increase skeletal muscle blood flow with this additional blood flow preferentially directed towards type II muscle fibres (Ferguson et al., 2013). These effects are likely to be mediated by a proportionally greater NO synthesis within, and in the microvasculature that surrounds, type II myocytes after NO_3^- supplementation given that PO_2 and pH are lowered more in contracting type II than type I muscle (Harkema et al., 1997; McDonough et al., 2005) and that hypoxia and acidosis facilitate NO_2^- reduction to NO (Castello et al., 2006; Li et al., 2008; Modin et al., 2001). Uptake of RSNO from the elevated circulating RSNO pool after NO_3^- supplementation (Carlström et al., 2010; Pinheiro et al., 2015) might also be greater in type II muscle (Hodson et al., 2017; Li and Whorton, 2007). Since recruitment of type II muscle fibres increases (Krustrup et al., 2009) and intramuscular pH and PO_2 decline (Bailey et al., 2010; Richardson et al., 1995) as exercise intensity increases, the ergogenic potential of NO_3^- supplementation could, therefore, be greater during high-intensity (>VO_{2peak}) exercise.

The first study to evaluate the effects of short-term NO_3^- supplementation on high-intensity exercise performance was conducted by Wylie et al. (2013b). In this study, short-term NO_3^- supplementation was reported to improve performance in the Yo-Yo Intermittent Recovery-1 test, which comprises repeated 2×20 m shuttle runs of progressively increasing velocity interspersed by a short active recovery and continued until exhaustion, in moderately trained team sports participants (Wylie et al., 2013b). This finding has since been replicated in a similar cohort (Thompson et al., 2016) as well as well-trained team sports participants (Nyakayiru et al., 2017). In a subsequent study, Wylie et al. (2016) assessed the efficacy of short-term NO_3^- supplementation to improve performance in 24×6 seconds, 7×30 seconds and 6×60 seconds sprints and only observed improved performance during the 24×6 seconds test. Improved performance during short-duration (< 10 seconds) repeated sprints has also been reported after acute (Rimer et al., 2016) and short-term NO_3^- supplementation (Porcelli et al., 2016; Thompson et al., 2015) with effects less likely following acute NO_3^- supplementation (e.g., Reynolds et al., 2020). There is also evidence for improved performance during a single short duration (≤ 30 seconds) sprint after short-term NO_3^- supplementation in moderately trained participants (e.g., Thompson et al., 2016), including a lower time to reach peak power output (Jonvik et al., 2018). Consistent with the effects on continuous endurance exercise, single and repeated sprint performance is not enhanced following NO_3^- supplementation in well-trained endurance athletes (e.g., Pawlak-Chaouch et al., 2019). However, NO_3^- supplementation has been reported to augment the physiological and performance adaptations to repeated sprint training in moderately trained participants (Thompson et al., 2017, 2018). Therefore, short-term NO_3^- supplementation appears to be ergogenic during single and repeated sprints in moderately trained and well-trained team sport athletes, but not well-trained endurance athletes.

In a non-fatigued state, NO_3^- supplementation does not improve force during a single maximum voluntary isometric contraction (Jonvik et al., 2020) but can improve contractile force during isokinetic contractions conducted at high angular velocities (Coggan and Peterson, 2018, but see Jonvik et al., 2020). During fatiguing contractions, NO_3^- supplementation improves skeletal muscle contractile function (Tillin et al., 2018) and increases repetitions to failure during resistance exercise such as bench press and back squat (San Juan et al., 2020). Type II muscle is substantially recruited during resistance exercise repetitions to failure (Morton et al., 2019), consistent with the notion that NO_3^- supplementation may be ergogenic in exercise settings that mandate significant type II muscle recruitment. Therefore, NO_3^- supplementation appears to have the potential to improve performance in sports requiring single sprints, high-velocity contractions and repeated sprints or high-intensity contractions.

MECHANISMS FOR THE ERGOGENIC EFFECT OF DIETARY NITRATE SUPPLEMENTATION

It has been reported that short-term NO_3^- supplementation improved exercise economy and tolerance during constant-work-rate exercise concomitant with a lower phosphocreatine and total adenosine 5-triphosphate turnover rate, reflecting improved skeletal muscle contractile efficiency (Bailey et al., 2010). A lower PCr cost of force production was subsequently observed in hypoxia (Vanhatalo et al., 2011), and during repeated maximal muscle contractions (Fulford et al., 2013), after NO_3^- supplementation. In addition, NO_3^- supplementation has been reported to lower the accumulation of fatigue-related metabolites, as evidenced by blunted muscle adenosine diphosphate and inorganic phosphate accumulation, and an attenuated decline in muscle pH (Bailey et al., 2010). Larsen and colleagues (2011) initially demonstrated a lower O_2 cost of ADP resynthesis (increased P/O ratio) in mitochondria harvested from human skeletal muscle after NO_3^- supplementation, reflecting improved mitochondrial respiratory efficiency. However, subsequent studies have failed to corroborate improved mitochondrial respiratory efficiency after NO_3^- supplementation (e.g., Whitfield et al., 2016). Therefore, the existing empirical evidence suggests that NO_3^- improves the efficiency of skeletal muscle contraction and lowers intramuscular metabolic perturbation during exercise rather than improving the efficiency of mitochondrial respiration.

Since NO_3^- supplementation can lower the high-energy phosphate cost of force production, and given that skeletal muscle calcium (Ca^{2+}) homeostasis is a costly energetic process (Barclay, 2015), improved exercise performance after NO_3^- supplementation could be linked to improved skeletal muscle Ca^{2+} handling. Classical studies indicated that treating ex vivo skeletal muscle with NO_3^- could improve sarcoplasmic reticulum (SR) Ca^{2+}

release (Fruen et al., 1994) and twitch force during evoked contractions (Hill and Macpherson, 1954). More recently, Hernández et al. (2012) reported that short-term NO_3^- supplementation could increase the content of the Ca^{2+} handling proteins, calsequestrin 1 (CASQ1) and the dihydropyridine receptor (DHPR), and increase evoked tetanic force production in type II, but not type I, mouse skeletal muscle. Evoked tetanic force production and cytosolic free [Ca^{2+}] were also increased after NO_3^- supplementation in single flexor digitorum brevis (FDB) muscle fibres (Hernández et al., 2012). Most recently, acute NO_2^- incubation was reported to delay fatigue development in mouse FBS fibres at a physiological PO_2 (2% O_2) concomitant with improved sarcoplasmic reticulum Ca^{2+} pumping and a better maintenance of myofilament Ca^{2+} sensitivity (Bailey et al., 2019). Whilst these studies conducted on murine skeletal muscle suggest improved skeletal muscle Ca^{2+} handling may contribute to the ergogenic of effect of NO_3^- supplementation, it has been reported that NO_3^- supplementation can improve skeletal muscle contractile function without altering CASQ1 and DHPR content in a mixed skeletal muscle fibre homogenate in humans (Whitfield et al., 2017). Instead, lower plasma [potassium] and improved performance has been reported during high-intensity intermittent exercise after NO_3^- supplementation (Wylie et al., 2013b), likely a result of NO action on the sodium-potassium pump leading to improved maintenance of muscle excitability and delayed fatigue development (Juel, 2016).

Consistent with a regulatory role for NO in exercise hyperaemia (Hellsten et al., 2012), leg blood flow during exercise is negatively correlated with a reduction in plasma $[NO_2^-]$ (Dufour et al., 2010), and NO_2^- infusion has been reported to increase forearm blood flow during exercise (Cosby et al., 2003). In addition, short-term (5 days) NO_3^- supplementation increased hind limb blood flow in the exercising rat (Ferguson et al., 2013). However, the effects of NO_3^- supplementation on forearm (Bentley et al., 2017; Craig et al., 2018; Richards et al., 2018) and leg (de Vries and DeLorey, 2019; Hughes et al., 2020) blood flow during exercise in humans is inconsistent. Therefore, further research is required to resolve the physiological mechanisms for the ergogenic effects of NO_3^- supplementation.

The molecular mechanisms that underlie the beneficial physiological and ergogenic effects of NO_3^- supplementation are unresolved but are likely to include cyclic guanosine monophosphate (cGMP)-mediated signalling and post-translational modification of protein thiols (Coggan and Peterson, 2018) (Figure 12.2). Increased [cGMP] is observed in plasma following NO_3^- supplementation (Kapil et al., 2010) and in muscle following NO_2^- administration (Srihirun et al., 2020). This increase in [cGMP] can evoke physiological responses in various tissues through activation of cGMP-dependent protein kinase (protein kinase G) and subsequent phosphorylation of numerous targets, cGMP-gated cation channels and phosphodiesterases (Francis et al., 2010). It also appears that cGMP-mediated signalling is important for improved skeletal muscle contractility following increased NO exposure (Maréchal and Gailly, 1999). In addition, the increase in plasma $[NO_2^-]$ and potential for O_2-independent generation of NO and other RNIs, as well as the increase in plasma [RSNO] and subsequent transnitrosylation, would be expected to provoke post-translational modifications of protein thiol groups (Foster et al., 2003). There is also evidence that NO_2^- administration can S-nitrosylate thiols in erythrocytes and various tissues, independently of its reduction to NO, with systemic NO_2^- administration dose-dependently increasing tissue [RSNO] (Bryan et al., 2005). It has been reported that treatment with NO_2^-, RSNO and various NO donors can modulate aspects of skeletal muscle excitation-contraction coupling ex vivo, including cross-bridge cycling, SR Ca^{2+} release, myofilament Ca^{2+} sensitivity and SR Ca^{2+} pumping, with these effects linked to various post-translational modifications of protein thiol groups including S-nitrosylation, S-glutathionylation and disulphide bonds (Andrade et al., 1998; Bailey et al., 2019; Dutka et al., 2011, 2017; Evangelista et al., 2010; Heunks et al., 2001; Maréchal and Gailly, 1999). However, and in contrast to positive effects of NO_3^- supplementation (Hernández et al., 2012), most of the empirical evidence to date indicates that administration of NO donors impairs ex vivo skeletal muscle contractile function (Andrade et al., 1998; Bailey et al., 2019; Dutka et al., 2011, 2017; Evangelista et al., 2010; Heunks et al., 2001), particularly in type II muscle (Spencer and Posterino, 2009).

Translating findings from studies assessing the effect of NO donors on skeletal muscle contractile function ex vivo to human skeletal muscle in vivo after NO_3^- supplementation is complicated by the application of a supra-physiological concentration of NO donors and/or experimental PO_2 in most previous ex vivo experiments, as well as potential interstudy differences in buffer composition and trace metals. Indeed, it appears that the effects of NO on skeletal muscle contractile function are dose-dependent (Maréchal and Gailly, 1999) and more likely to be beneficial at a physiological compared to a supraphysiological PO_2 (Eu et al., 2003). Moreover, studies reporting negative effects of NO donors on skeletal muscle contractile function have evaluated non-fatiguing contractions. In a recent study, Bailey et al. (2019) indicated that acute incubation with NO_2^- could delay fatigue development in mouse single myocytes ex vivo at a physiological PO_2, but not a supra-physiological PO_2, in association with better maintenance of myofilament Ca^{2+} sensitivity and enhanced SR Ca^{2+} pumping during the later stages of the test. Conversely, NO_2^- treatment impaired skeletal muscle force production during non-fatiguing contractions by lowering SR Ca^{2+} release (Bailey et al., 2019). These observations are consistent with experimental data from human muscle in vivo indicating that NO_3^- supplementation is more likely to improve skeletal muscle contractile function in fatigued compared to non-fatigued conditions (Tillin et al., 2018). Therefore, while NO_3^- supplementation can improve various aspects of skeletal muscle contractile function and exercise performance in some human population groups, further research is required to resolve the fundamental molecular mechanisms responsible for the effects.

CONCLUSION

The importance of NO as a multifaceted physiological signalling molecule is well documented, and it is now recognised that NO can by synthesised from the reduction of NO_2^- in addition to that derived via NOS and accessed from RSNOs and metal containing proteins such as DNICs. Dietary NO_3^- supplementation can increase circulating plasma and local muscle $[NO_2^-]$ for subsequent reduction to NO by numerous NO_2^- reductases. Reduction of NO_2^- to NO is enhanced in hypoxia and acidosis, conditions which are manifest during high-intensity exercise. Accordingly, NO_3^- supplementation has been reported to improve performance across a range of different exercise and muscle function tests in healthy adults and some patient populations, but such effects are less common in highly trained endurance athletes. To maximise the ergogenic potential of NO_3^- supplementation, short-term BR supplementation of at least 6 days, with a daily NO_3^- dose of 140 mL (8–12 mmol) and the final dose consumed approximately 3 hours prior to competition, is the current recommended best practice. The putative physiological mechanisms responsible for the ergogenic effect of NO_3^- supplementation include improved skeletal muscle O_2 delivery and contractile efficiency, and lower utilisation of finite metabolic substrates and accumulation of metabolites implicated in muscle fatigue development. The molecular bases for these effects are likely to include cGMP-mediated signalling and post-translational modification of protein thiols. However, further research is required to elucidate the molecular and physiological mechanisms for the ergogenic effects of NO_3^- supplementation. Therefore, the existing empirical evidence suggests that NO_3^- supplementation represents a practical and cost-effective dietary intervention to increase NO synthesis and improve exercise performance in healthy adults and some patient groups.

REFERENCES

Álvares, T.S., Meirelles, C.M., Bhambhani, Y.N. et al. (2011) L-Arginine as a potential ergogenic aid in healthy subjects. *Sports Medicine*. 41, 233–248.

Andrade, F.H., Reid, M.B., Allen, D.G. et al. (1998) Effect of nitric oxide on single skeletal muscle fibres from the mouse. *Journal of Physiology*. 509, 577–586.

Bailey, D.M., Young, I.S., McEneny, J. et al. (2004) Regulation of free radical outflow from an isolated muscle bed in exercising humans. *American Journal of Physiology. Heart and Circulatory Physiology*. 287, 1689–1699.

Bailey, S.J., Fulford, J., Vanhatalo, A. et al. (2010) Dietary nitrate supplementation enhances muscle contractile efficiency during knee-extensor exercise in humans. *Journal of Applied Physiology*. 109, 135–148.

Bailey, S.J., Gandra, P.G., Jones, A.M. et al. (2019) Incubation with sodium nitrite attenuates fatigue development in intact single mouse fibres at physiological PO_2. *Journal of Physiology*. 597, 5429–5443.

Bailey, S.J., Winyard, P., Vanhatalo, A. et al. (2009) Dietary nitrate supplementation reduces the O2 cost of low-intensity exercise and enhances tolerance to high-intensity exercise in humans. *Journal of Applied Physiology*. 107, 1144–1155.

Barclay, C.J. (2015) Energetics of contraction. *Comprehensive Physiology*. 5, 961–995.

Bender, D., Schwarz, G. (2018) Nitrite-dependent nitric oxide synthesis by molybdenum enzymes. *FEBS Letters*. 592, 2126–2139.

Benjamin, N., O'Driscoll, F., Dougall, H. et al. (1994) Stomach NO synthesis. *Nature*. 368: 502.

Bentley, R.F., Walsh, J.J., Drouin, P.J. et al. (2017) Dietary nitrate restores compensatory vasodilation and exercise capacity in response to a compromise in oxygen delivery in the noncompensator phenotype. *Journal of Applied Physiology*. 123, 594–605.

Bredt, D.S. (1999) Endogenous nitric oxide synthesis: Biological functions and pathophysiology. *Free Radical Research*. 31, 577–596.

Bryan, N.S., Alexander, D.D., Coughlin, J.R. et al. (2012). Ingested nitrate and nitrite and stomach cancer risk: An updated review. *Food and Chemical Toxicology*. 50, 3646–3665.

Bryan, N.S., Fernandez, B.O., Bauer, S.M. et al. (2005) Nitrite is a signaling molecule and regulator of gene expression in mammalian tissues. *Nature Chemical Biology*. 1, 290–297.

Burleigh, M.C., Liddle, L., Monaghan, C. et al. (2018) Salivary nitrite production is elevated in individuals with a higher abundance of oral nitrate-reducing bacteria. *Free Radical Biology and Medicine*. 120, 80–88.

Carlström, M., Larsen, F.J., Nyström, T. et al. (2010) Dietary inorganic nitrate reverses features of metabolic syndrome in endothelial nitric oxide synthase-deficient mice. *Proceedings of the National Academy of Sciences of the United States of America*. 107, 17716–17720.

Castello, P.R., David, P.S., McClure, T. (2006) Mitochondrial cytochrome oxidase produces nitric oxide under hypoxic conditions: Implications for oxygen sensing and hypoxic signaling in eukaryotes. *Cell Metabolism*. 3, 277–287.

Cermak, N.M., Gibala, M.J., van Loon, L.J. (2012a) Nitrate supplementation's improvement of 10-km time-trial performance in trained cyclists. *International Journal of Sport Nutrition and Exercise Metabolism*. 22, 64–71.

Cermak, N.M., Res, P., Stinkens, R. et al. (2012b) No improvement in endurance performance after a single dose of beetroot juice. *International Journal of Sport Nutrition and Exercise Metabolism*. 22, 470–478.

Cocksedge, S.P., Breese, B.C., Morgan, P.T. et al. (2020) Influence of muscle oxygenation and nitrate-rich beetroot juice supplementation on O_2 uptake kinetics and exercise tolerance. *Nitric Oxide*. 99, 25–33.

Coggan, A.R., Broadstreet, S.R., Mahmood, K. et al. (2018) Dietary nitrate increases VO(2)peak and performance but does not alter ventilation or efficiency in patients with heart failure with reduced ejection fraction. *Journal of Cardiac Failure*. 24, 65–73.

Coggan, A.R., Peterson, L.R. (2018) Dietary nitrate enhances the contractile properties of human skeletal muscle. *Exercise and Sport Science Reviews*. 46, 254–261.

Cosby, K., Partovi, K.S., Crawford, J.H. et al. (2003) Nitrite reduction to nitric oxide by deoxyhemoglobin vasodilates the human circulation. *Nature Medicine*. 9, 1498–1505.

Craig, J.C., Broxterman, R.M., Smith, J.R. et al. (2018) Effect of dietary nitrate supplementation on conduit artery blood flow, muscle oxygenation, and metabolic rate during handgrip exercise. *Journal of Applied Physiology*. 125, 254–262.

de Vries, C.J., DeLorey, D.S. (2019) Effect of acute dietary nitrate supplementation on sympathetic vasoconstriction at rest and during exercise. *Journal of Applied Physiology*. 127, 81–88.

Dufour, S.P., Patel, R.P., Brandon, A. et al. (2010) Erythrocyte-dependent regulation of human skeletal muscle blood flow: Role of varied oxyhemoglobin and exercise on nitrite, S-nitrosohemoglobin, and ATP. *American Journal of Physiology. Heart and Circulatory Physiology*. 299, 1936–1946.

Dungel, P., Perlinger, M., Weidinger, A. et al. (2015) The cytoprotective effect of nitrite is based on the formation of dinitrosyl iron complexes. *Free Radical Biology and Medicine*. 89, 300–310.

Dutka, T.L., Mollica, J.P., Lamboley, C.R. et al. (2017) S-nitrosylation and S-glutathionylation of Cys134 on troponin I have opposing competitive actions on Ca(2+) sensitivity in rat fast-twitch muscle fibers. *American Journal of Physiology. Cell Physiology*. 312, 316–327.

Dutka, T.L., Mollica, J.P., Posterino, G.S. et al. (2011) Modulation of contractile apparatus Ca^{2+} sensitivity and disruption of excitation-contraction coupling by S-nitrosoglutathione in rat muscle fibres. *Journal of Physiology*. 589, 2181–2196.

Eu, J.P., Hare, J.M., Hess, D.T. et al. (2003) Concerted regulation of skeletal muscle contractility by oxygen tension and endogenous nitric oxide. *Proceedings of the National Academy of Sciences of the United States of America*. 100, 15229–15234.

Evangelista, A.M., Rao, V.S., Filo, A.R. et al. (2010) Direct regulation of striated muscle myosins by nitric oxide and endogenous nitrosothiols. *PLoS One*. 5, e11209.

Ferguson, S.K., Hirai, D.M., Copp, S.W., et al. (2013) Impact of dietary nitrate supplementation via beetroot juice on exercising muscle vascular control in rats. *Journal of Physiology*. 591, 547–557.

Flueck, J.L., Bogdanova, A., Mettler, S. et al. (2016) Is beetroot juice more effective than sodium nitrate? The effects of equimolar nitrate dosages of nitrate-rich beetroot juice and sodium nitrate on oxygen consumption during exercise. *Applied Physiology, Nutrition, and Metabolism*. 41, 421–429.

Foster, M.W., McMahon, T.J., Stamler, J.S. (2003) S-nitrosylation in health and disease. *Trends in Molecular Medicine*. 9, 160–168.

Francis, S.H., Busch, J.L., Corbin, J.D. et al. (2010) cGMP-dependent protein kinases and cGMP phosphodiesterases in nitric oxide and cGMP action. *Pharmacological Reviews*. 62, 525–563.

Frandsen, U., Höffner, L., Betak, A. et al. (2000) Endurance training does not alter the level of neuronal nitric oxide synthase in human skeletal muscle. *Journal of Applied Physiology*. 89, 1033–1038.

Fruen, B.R., Mickelson, J.R., Roghair, T.J. et al. (1994) Anions that potentiate excitation-contraction coupling may mimic effect of phosphate on Ca^{2+} release channel. *American Journal of Physiology Cell Physiology*. 266, 1729–1735.

Fulford, J. Winyard, P.G., Vanhatalo, A. et al. (2013) Influence of dietary nitrate supplementation on human skeletal muscle metabolism and force production during maximum voluntary contractions. *Pflugers Archiv: European Journal of Physiology*. 465, 517–528.

Gago, B., Lundberg, J.O., Barbosa, R.M. et al. (2007) Red wine-dependent reduction of nitrite to nitric oxide in the stomach. *Free Radical Biology and Medicine*. 43, 1233–1242.

Gilliard, C.N., Lam, J.K., Cassel, K.S. et al. (2018) Effect of dietary nitrate levels on nitrate fluxes in rat skeletal muscle and liver. *Nitric Oxide*. 75, 1–7.

Harkema, S.J., Adams, G.R., Meyer, R.A. (1997) Acidosis has no effect on the ATP cost of contraction in cat fast- and slow-twitch skeletal muscles. *American Journal of Physiology*. 272, 485–490.

Hellsten, Y., Nyberg, M., Jensen, L.G. et al. (2012) Vasodilator interactions in skeletal muscle blood flow regulation. *Journal of Physiology*. 590, 6297–6305.

Hernández, A., Schiffer, T.A., Ivarsson, N. et al. (2012) Dietary nitrate increases tetanic $[Ca^{2+}]i$ and contractile force in mouse fast-twitch muscle. *Journal of Physiology*. 590, 3575–3583.

Heunks, L.M., Cody, M.J., Geiger, P.C. et al. (2001) Nitric oxide impairs Ca^{2+} activation and slows cross-bridge cycling kinetics in skeletal muscle. *Journal of Applied Physiology*. 91, 2233–9223.

Hill, A.V., Macpherson, L. (1954) The effect of nitrate, iodide and bromide on the duration of the active state in skeletal muscle. *Proceeding of the Royal Society of London. Series B, Biological Sciences*. 143, 81–102.

Hirschfield, W., Moody, M.R., O'Brien, W.E. et al. (2000). Nitric oxide release and contractile properties of skeletal muscles from mice deficient in type III NOS. *American Journal of Physiology. Regulatory, Integrative and Comparative Physiology*. 278, 95–100.

Hodson, N., Brown, T., Joanisse, S. et al. (2017) Characterisation of L-Type amino acid transporter 1 (LAT1) expression in human skeletal muscle by immunofluorescent microscopy. *Nutrients*. 10, 23.

Hoon, M.W., Jones, A.M., Johnson, N.A. et al. (2014) The effect of variable doses of inorganic nitrate-rich beetroot juice on simulated 2,000-m rowing performance in trained athletes. *International Journal of Sports Physiology and Performance*. 9, 615–620.

Hord, N.G., Tang, Y., Bryan N.S. (2009) Food sources of nitrates and nitrites: The physiologic context for potential health benefits. *American Journal of Clinical Nutrition*. 90, 1–10.

Hughes, W.E., Kruse, N.T., Ueda, K. et al. (2020) Dietary nitrate does not acutely enhance skeletal muscle blood flow and vasodilation in the lower limbs of older adults during single-limb exercise. *European Journal of Applied Physiology*. 120, 1357–1369.

Hyde, E.R., Andrade, F., Vaksman, Z. et al. (2014) Metagenomic analysis of nitrate-reducing bacteria in the oral cavity: implications for nitric oxide homeostasis. *PLoS One*. 9, e88645.

Ignarro, L.J., Buga, G.M., Wood, K.S. et al. (1987) Endothelium-derived relaxing factor produced and released from artery and vein is nitric oxide. *Proceedings of the National Academy of Sciences of the United States of America*. 84, 9265–9269.

Jansson, E.A., Huang, L., Malkey, R. et al. (2008) A mammalian functional nitrate reductase that regulates nitrite and nitric oxide homeostasis. *Nature Chemical Biology*. 4, 411–417.

Jones, A.M. (2014) Dietary nitrate supplementation and exercise performance. *Sports Medicine*. 44, 35–45.

Jonvik, K.L., Hoogervorst, D., Peelen, H.B. et al. (2020) The impact of beetroot juice supplementation on muscular endurance, maximal strength and countermovement jump performance. *European Journal of Sport Science*. 21, 1–8.

Jonvik, K.L., Nyakayiru, J., Pinckaers, P.J. et al. (2016) Nitrate-rich vegetables increase plasma nitrate and nitrite concentrations and lower blood pressure in healthy adults. *The Journal of Nutrition*. 146, 986–993.

Jonvik, K.L., Nyakayiru, J., Van Dijk, J.W. et al. (2018) Repeated-sprint performance and plasma responses following beetroot juice supplementation do not differ between recreational, competitive and elite sprint athletes. *European Journal of Sport Science*. 18, 524–533.

Juel, C. (2016) Nitric oxide and Na, K-ATPase activity in rat skeletal muscle. *Acta Physiologica (Oxford, England)*. 216, 447–453.

Kapil, V., Khambata, R.S., Jones, D.A. et al. (2020) The noncanonical pathway for In Vivo nitric oxide generation: The nitrate-nitrite-nitric oxide pathway. *Pharmacological Reviews*. 72, 692–766.

Kapil, V., Milsom, A.B., Okorie, M. et al. (2010) Inorganic nitrate supplementation lowers blood pressure in humans: Role for nitrite-derived NO. *Hypertension*. 56, 274–281.

Kelly, J., Vanhatalo, A., Bailey, S.J. et al. (2014) Dietary nitrate supplementation: Effects on plasma nitrite and pulmonary O_2 uptake dynamics during exercise in hypoxia and normoxia. *American Journal of Physiology. Regulatory, Integrative and Comparative Physiology*. 307, 920–930.

Kenjale, A.A., Ham, K.L., Stabler, T. et al. (2011) Dietary nitrate supplementation enhances exercise performance in peripheral arterial disease. *Journal of Applied Physiology*. 110, 1582–1591.

Kerley, C.P., James, P.E., McGowan, A. et al. (2019) Dietary nitrate improved exercise capacity in COPD but not blood pressure or pulmonary function: A 2 week, double-blind randomised, placebo-controlled crossover trial. *International Journal of Food Science and Nutrition*. 70, 222–231.

Kobzik, L., Reid, M.B., Bredt, D.S. et al. (1994) Nitric oxide in skeletal muscle. *Nature*. 372, 546–548.

Krustrup, P., Söderlund, K., Relu, M.U. et al. (2009) Heterogeneous recruitment of quadriceps muscle portions and fibre types during moderate intensity knee-extensor exercise: Effect of thigh occlusion. *Scandinavian Journal of Medicine and Science in Sports*. 19, 576–584.

Lansley, K.E., Winyard, P.G., Fulford, J. et al. (2011) Dietary nitrate supplementation reduces the O_2 cost of walking and running: A placebo-controlled study. *Journal of Applied Physiology*. 110, 591–600.

Larsen, F.J., Schiffer, T.A., Borniquel, S. et al. (2011) Dietary inorganic nitrate improves mitochondrial efficiency in humans. *Cell Metabolism*. 13, 149–159.

Larsen, F.J., Weitzberg, E., Lundberg, J.O. et al. (2007) Effects of dietary nitrate on oxygen cost during exercise. *Acta Physiologica (Oxford, England)*. 191, 59–66.

Li, H., Cui, H., Kundu, T.K. (2008) Nitric oxide production from nitrite occurs primarily in tissues not in the blood: Critical role of xanthine oxidase and aldehyde oxidase. *The Journal of Biological Chemistry*. 283, 17855–17863.

Li, S., Whorton, A.R. (2007) Functional characterization of two S-nitroso-L-cysteine transporters, which mediate movement of NO equivalents into vascular cells. *American Journal of Physiology. Cell Physiology*. 292, 1263–1271.

Liu, T., Schroeder, H.J., Wilson, S.M. et al. (2016) Local and systemic vasodilatory effects of low molecular weight S-nitrosothiols. *Free Radical Biology and Medicine*. 91, 215–223.

Liu, T., Zhang, M., Terry, M.H. et al. (2018) Nitrite potentiates the vasodilatory signaling of S-nitrosothiols. *Nitric Oxide*. 75, 60–69.

Lundberg, J.O. Govoni, M. (2004) Inorganic nitrate is a possible source for systemic generation of nitric oxide. *Free Radical Biology and Medicine*. 37, 395–400.

Lundberg, J.O., Weitzberg, E., Lundberg, J.M. et al. (1994) Intragastric nitric oxide production in humans: measurements in expelled air. *Gut*. 35, 1543–1546.

Maréchal, G., Gailly, P. (1999) Effects of nitric oxide on the contraction of skeletal muscle. *Cellular and Molecular Life Sciences*. 55, 1088–1102.

Matsumoto, A., Gow, A.J. (2011) Membrane transfer of S-nitrosothiols. *Nitric Oxide*. 25, 102–107.

McDonagh, S.T.J., Wylie, L.J., Webster, J.M.A. et al. (2018) Influence of dietary nitrate food forms on nitrate metabolism and blood pressure in healthy normotensive adults. *Nitric Oxide*. 72, 66–74.

McDonough, P., Behnke, B.J., Padilla, D.J. et al. (2005) Control of microvascular oxygen pressures in rat muscles comprised of different fibre types. *Journal of Physiology*. 563, 903–913.

McMahon, N.F., Leveritt, M.D., Pavey, T.G. (2017) The effect of dietary nitrate supplementation on endurance exercise performance in healthy adults: A systematic review and meta-analysis. *Sports Medicine*. 47, 735–756.

Modin, A., Björne, H., Herulf, M. (2001) Nitrite-derived nitric oxide: A possible mediator of 'acidic-metabolic' vasodilation. *Acta Physiologica Scandinavica*. 171, 9–16.

Moncada, S., Higgs, A. (1993) The L-arginine-nitric oxide pathway. *The New England Journal of Medicine*. 329, 2002–2012.

Moon, Y., Balke, J.E., Madorma, D. et al. (2017a) Nitric Oxide regulates skeletal muscle fatigue, fiber type, microtubule organization, and mitochondrial ATP synthesis efficiency through cGMP-dependent mechanisms. *Antioxidants and Redox Signaling*. 26, 966–985.

Moon, Y., Cao, Y. Zhu, J. et al. (2017b) GSNOR Deficiency enhances in situ skeletal muscle strength, fatigue resistance, and RyR1 S-Nitrosylation without impacting mitochondrial content and activity. *Antioxidants and Redox Signaling*. 26, 165–181.

Morton, R.W., Sonne, M.W., Farias Zuniga, A. et al. (2019) Muscle fibre activation is unaffected by load and repetition duration when resistance exercise is performed to task failure. *Journal of Physiology*. 597, 4601–4613.

Muggeridge, D.J., Howe, C.C., Spendiff, O. et al. (2014) A single dose of beetroot juice enhances cycling performance in simulated altitude. *Medicine and Science Sports and Exercise*. 46, 143–150.

Nyakayiru, J., Jonvik, K.L., Trommelen, J. et al. (2017) Beetroot juice supplementation improves high-intensity intermittent type exercise performance in trained soccer players. *Nutrients*. 9, 314.

Oliveira-Paula, G.H., Tanus-Santos, J.E. (2019) Nitrite-stimulated gastric formation of S-nitrosothiols as an antihypertensive therapeutic strategy. *Current Drug Targets*. 20, 431–443.

Pawlak-Chaouch, M., Boissière, J., Gamelin, F.X. et al. (2016) Effect of dietary nitrate supplementation on metabolic rate during rest and exercise in human: A systematic review and a meta-analysis. *Nitric Oxide*. 53, 65–76.

Pawlak-Chaouch, M., Boissière, J., Munyaneza, D. et al. (2019) Beetroot juice does not enhance supramaximal intermittent exercise performance in elite endurance athletes. *Journal of the American College of Nutrition*. 38, 729–738.

Peeling, P., Cox, G.R., Bullock, N. et al. (2015) Beetroot juice improves on-water 500 M time-trial performance, and laboratory-based paddling economy in national and international-level kayak athletes. *International Journal of Sport Nutrition and Exercise Metabolism*. 25, 278–84.

Piknova, B., Park, J.W., Lam, K.K.J. et al. (2016) Nitrate as a source of nitrite and nitric oxide during exercise hyperemia in rat skeletal muscle. *Nitric Oxide*. 55–56, 54–61.

Pinheiro, L.C., Amaral, J.H., Ferreira, G.C. et al. (2015) Gastric S-nitrosothiol formation drives the antihypertensive effects of oral sodium nitrite and nitrate in a rat model of renovascular hypertension. *Free Radical Biology and Medicine*. 87, 252–262.

Porcelli, S., Pugliese, L., Rejc, E. et al. (2016) Effects of a short-term high-nitrate diet on exercise performance. *Nutrients*. 8, 534.

Porcelli, S., Ramaglia, M., Bellistri, G. et al. (2015) Aerobic fitness affects the exercise performance responses to nitrate supplementation. *Medicine and Science Sports and Exercise*. 47, 1643–1651.

Qin, L., Liu, X., Sun, Q. et al. (2012) Sialin (SLC17A5) functions as a nitrate transporter in the plasma membrane. *Proceedings of the National Academy of Sciences of the United States of America*. 109, 13434–13439.

Ramick, M.G., Kirkman, D.L., Stock, J.M. et al. (2021) The effect of dietary nitrate on exercise capacity in chronic kidney disease: A randomized controlled pilot study. *Nitric Oxide*. 106,17–23.

Reynolds, C.M.E., Evans, M., Halpenny, C. et al. (2020) Acute ingestion of beetroot juice does not improve short-duration repeated sprint running performance in male team sport athletes. *Journal of Sports Sciences*. 38, 2063–2070.

Richards, J.C., Racine, M.L., Hearon Jr., C.M. et al. (2018) Acute ingestion of dietary nitrate increases muscle blood flow via local vasodilation during handgrip exercise in young adults. *Physiological Reports*. 6, e13572.

Richardson, R.S., Noyszewski, E.A., Kendrick, K.F., et al. (1995) Myoglobin O_2 desaturation during exercise. Evidence of limited O_2 transport. *The Journal of Clinical Investigation*. 96, 1916–1926.

Rimer, E.G., Peterson, L.R., Coggan, A.R. et al. (2016) Increase in maximal cycling power with acute dietary nitrate supplementation. *International Journal of Sports Physiology and Performance*. 11, 715–720.

Rocha, B.S., Nunes, C., Pereira, C. et al. (2014) A shortcut to wide-ranging biological actions of dietary polyphenols: Modulation of the nitrate-nitrite-nitric oxide pathway in the gut. *Food and Function*. 5, 1646–1652.

Rokkedal-Lausch, T., Franch, J., Poulsen, M.K. et al. (2019) Chronic high-dose beetroot juice supplementation improves time trial performance of well-trained cyclists in normoxia and hypoxia. *Nitric Oxide*. 85, 44–52.

Samouilov, A., Woldman, Y.Y., Zweier, J.L. et al. (2007) Magnetic resonance study of the transmembrane nitrite diffusion. *Nitric Oxide*. 16, 362–370.

San Juan, A.F., Dominguez, R., Lago-Rodríguez, Á. et al. (2020) Effects of dietary nitrate supplementation on weightlifting exercise performance in healthy adults: A systematic review. *Nutrients*. 12, 2227.

Senefeld, J.W., Wiggins, C.C., Regimbal, R.J. et al. (2020) Ergogenic effect of nitrate supplementation: A systematic review and meta-analysis. *Medicine and Science Sports and Exercise*. 52, 2250–2261.

Shannon, O.M., Barlow, M.J., Duckworth, L. et al. (2017) Dietary nitrate supplementation enhances short but not longer duration running time-trial performance. *European Journal of Applied Physiology*. 117, 775–785.

Shepherd, A.I., Gilchrist, M., Winyard, P.G. et al. (2015a) Effects of dietary nitrate supplementation on the oxygen cost of exercise and walking performance in individuals with type 2 diabetes: A randomized, double-blind, placebo-controlled crossover trial. *Free Radical Biology and Medicine*. 86, 200–208.

Shepherd, A.I., Wilkerson, D.P., Dobson, L. et al. (2015b) The effect of dietary nitrate supplementation on the oxygen cost of cycling, walking performance and resting blood pressure in individuals with chronic obstructive pulmonary disease: A double blind placebo controlled, randomised control trial. *Nitric Oxide*. 48, 31–7.

Shingles, R., Roh, M.H., McCarty, R.E. (1997) Direct measurement of nitrite transport across erythrocyte membrane vesicles using the fluorescent probe, 6-methoxy-N-(3-sulfopropyl) quinolinium. *J Bioenergetics and Biomembranes*. 29, 611–616.

Spencer, T., Posterino, G.S. (2009) Sequential effects of GSNO and H_2O_2 on the Ca^{2+} sensitivity of the contractile apparatus of fast- and slow-twitch skeletal muscle fibers from the rat. *American Journal of Physiology. Cell Physiology*. 296, 1015–1023.

Spiegelhalder, B., Eisenbrand, G., Preussmann, R. (1976) Influence of dietary nitrate on nitrite content of human saliva: Possible relevance to in vivo formation of N-nitroso compounds. *Food and Cosmetics Toxicology*. 14, 545–548.

Srihirun, S., Park, J.W., Teng, R. (2020) Nitrate uptake and metabolism in human skeletal muscle cell cultures. *Nitric Oxide*. 94, 1–8.

Stamler, J.S., Meissner, G. (2001) Physiology of nitric oxide in skeletal muscle. *Physiological Reviews*. 81, 209–237.

Suhr, F., Gehlert, S., Grau, M. et al. (2013) Skeletal muscle function during exercise-fine-tuning of diverse subsystems by nitric oxide. *International Journal of Molecular Sciences*. 14, 7109–7139.

Tan, R., Wylie, L.J., Thompson, C. et al. (2018) Beetroot juice ingestion during prolonged moderate-intensity exercise attenuates progressive rise in O(2) uptake. *Journal of Applied Physiology*. 124, 1254–1263.

Thomas, D.D., Corey, C., Hickok, J. et al. (2018) Differential mitochondrial dinitrosyliron complex formation by nitrite and nitric oxide. *Redox Biology*. 15, 277–283.

Thompson, C., Vanhatalo, A. Jell, H. et al. (2016) Dietary nitrate supplementation improves sprint and high-intensity intermittent running performance. *Nitric Oxide*. 61, 55–61.

Thompson, C., Vanhatalo, A., Kadach, S. et al. (2018) Discrete physiological effects of beetroot juice and potassium nitrate supplementation following 4-wk sprint interval training. *Journal of Applied Physiology*. 124, 1519–1528.

Thompson, C., Wylie, L.J., Blackwell, J.R. et al. (2017) Influence of dietary nitrate supplementation on physiological and muscle metabolic adaptations to sprint interval training. *Journal of Applied Physiology*. 122, 642–652.

Thompson, C., Wylie, L.J., Fulford, J. et al. (2015) Dietary nitrate improves sprint performance and cognitive function during prolonged intermittent exercise. *European Journal of Applied Physiology*. 115, 1825–1834.

Tillin, N.A., Moudy, S., Nourse, K.M. et al. (2018) Nitrate supplement benefits contractile forces in fatigued but not unfatigued muscle. *Medicine and Science in Sports and Exercise*. 50, 2122–2131.

van Faassen, E.E., Bahrami, S., Feelisch, M. (2009) Nitrite as regulator of hypoxic signaling in mammalian physiology. *Medical Research Reviews*. 29, 683–741.

Vanhatalo, A., Blackwell, J.R., L'Heureux, J.E. et al. (2018) Nitrate-responsive oral microbiome modulates nitric oxide homeostasis and blood pressure in humans. *Free Radical Biology and Medicine*. 124, 21–30.

Vanhatalo, A., Fulford, J., Bailey, S.J. et al. (2011) Dietary nitrate reduces muscle metabolic perturbation and improves exercise tolerance in hypoxia. *Journal of Physiology*. 589, 5517–5528.

Wagner, D.A., Young, V.R., Tannenbaum, S.R. et al. (1984) Mammalian nitrate biochemistry: metabolism and endogenous synthesis. *IARC Scientific Publications*. 57, 247–253.

Whitfield, J., Gamu, D., Heigenhauser, G.J.F. et al. (2017) Beetroot juice increases human muscle force without changing Ca^{2+}-handling proteins. *Medicine and Science in Sports and Exercise*. 49, 2016–2024.

Whitfield, J., Ludzki, A., Heigenhauser, G.J. et al. (2016) Beetroot juice supplementation reduces whole body oxygen consumption but does not improve indices of mitochondrial efficiency in human skeletal muscle. *Journal of Physiology*. 594, 421–435.

Woessner, M.N., Neil, C., Saner, N.J. et al. (2020) Effect of inorganic nitrate on exercise capacity, mitochondria respiration, and vascular function in heart failure with reduced ejection fraction. *Journal of Applied Physiology*. 128, 1355–1364.

Wootton-Beard, P.C., Ryan, L. (2011) A beetroot juice shot is a significant and convenient source of bioaccessible antioxidants. *Journal of Functional Foods*. 3, 329–34.

Wylie, L.J., Bailey, S.J., Kelly, J. et al. (2016) Influence of beetroot juice supplementation on intermittent exercise performance. *European Journal of Applied Physiology*. 116, 415–425.

Wylie, L.J., Kelly, J., Bailey, S.J. et al. (2013a) Beetroot juice and exercise: pharmacodynamic and dose-response relationships. *Journal of Applied Physiology*. 115, 325–336.

Wylie, L.J., Mohr, M., Krustrup, P. et al. (2013b) Dietary nitrate supplementation improves team sport-specific intense intermittent exercise performance. *European Journal of Applied Physiology*. 113, 1673–1684.

Wylie, L.J., Park, J.W., Vanhatalo, A. et al. (2019) Human skeletal muscle nitrate store: Influence of dietary nitrate supplementation and exercise. *Journal of Physiology*. 597, 5565–5576.

Zamani, P., Rawat, D., Shiva-Kumar, P. et al. (2015) Effect of inorganic nitrate on exercise capacity in heart failure with preserved ejection fraction. *Circulation*. 131, 371–380.

Zhang, Z., Naughton, D., Winyard, P.G. et al. (1998) Generation of nitric oxide by a nitrite reductase activity of xanthine oxidase: A potential pathway for nitric oxide formation in the absence of nitric oxide synthase activity. *Biochemical and Biophysical Research Communications*. 249, 767–772.

CHAPTER THIRTEEN

(Poly)phenols in Exercise Performance and Recovery

MORE THAN AN ANTIOXIDANT?

Tom Clifford and Glyn Howatson

CONTENTS

Introduction / 153
Mechanisms of Action / 153
(Poly)phenols and Exercise Performance / 156
(Poly)phenols and Exercise Recovery / 158
Practical Application and Summary / 160
References / 160

INTRODUCTION

The term phytochemical refers to compounds found in plants (Frank et al., 2019). Phytochemicals are distinguished by their chemical structures and can be sub-divided into four higher-order classes: phenols and polyphenols, terpenoids, alkaloids and sulphur compounds (Crozier et al., 2007). Phenols and polyphenols have received the most attention in the literature (collectively termed (poly)phenols) (Frank et al., 2019) and are abundant in fruits and vegetables, with foods such as tea, coffee, wine, grapes and cocoa being exceptionally rich sources (see Table 13.1). Although widely known as antioxidants, it is now recognized that many dietary (poly)phenols are pleiotropic compounds with a wide range of biological activities that include antioxidant, anti-inflammatory, anti-atherogenic, anti-carcinogenic, anti-microbial, vasodilatory and chemoprotective (Pandey and Rizvi, 2009). Given their diverse functionality, it is unsurprising that interest in (poly)phenols continues to expand. In this chapter, we discuss the putative mechanisms underpinning the performance-enhancing and recovery effects of (poly)phenols, followed by a critical précis of the underlying evidence.

MECHANISMS OF ACTION

Depending on the environment, (poly)phenols can have both pro-oxidant and antioxidant actions in vitro (Halliwell, 2008); however, it is their antioxidant effects that are often used to support their ergogenic effects. In this sense, the major mechanism by which they are proposed to modulate exercise is by reducing muscle fatigue or muscle damage caused by reactive oxygen and nitrogen species (RONS) (Reid, 2016; Margaritelis et al., 2020). Numerous clinical trials have found that (poly)phenols, including anthocyanins (Bell et al., 2014b), flavanols (Wiswedel et al., 2004) and curcumin (Kawanishi et al., 2013), lower biomarkers of lipid or protein oxidation, which lends support to this theory. However, whether

DOI: 10.1201/9781003051619-13

TABLE 13.1

(Poly)phenol subclasses, examples of dietary sources and use in studies.

(Poly)phenol subclass	Major dietary sources	Examples and use in studies
Flavonoids		
Flavones	Carrots, peppers, cabbage	Luteolin (Gelabert-Rebato et al., 2019)
Flavonones	Lemons, limes, grapefruit	Hesperidin (Estruel-Amades et al., 2019)
Flavonols	Apples, onions, broccoli	Quercetin (Nieman et al., 2007a; Nieman et al., 2010)
Flavanols	Tea, grapes, cocoa	Catechins (Kerksick, Kreider and Willoughby, 2010; Jówko et al., 2011, 2012)
Isoflavones	Soybeans, legumes, nuts	Genistein (Choquette et al., 2013; Seeley, Jacobs and Signorile, 2020)
Anthocyanins	Cherries, blackberries, blueberries	Montmorency tart cherries (Howatson et al., 2010; Bell et al., 2014a, 2016; Keane et al., 2018) Blackcurrants (Cook et al., 2015; Theodorus Willems et al., 2015; Willems et al., 2016)
Procyanidins/tannins	Wine, tea, coffee	Pomegranate (Trombold et al., 2010, 2011)
	Non-flavonoids	
Phenolic acids	Coffee, nuts, seeds	Chlorogenic acid (Nieman et al., 2018)
Stilbenes	Wine, grapes, peanuts	Resveratrol (Voduc et al., 2014; Alway et al., 2017)
Lignans	Linseed, wholegrains	Herpetrione (Jin et al., 2016)
Chalcones	Tomatoes, beer	Ashitaba (Kweon et al., 2019)
Curcumin	Turmeric spice	Curcumin (Drobnic et al., 2014; Tanabe et al., 2015)

(poly)phenols act as antioxidants after absorption through the gastrointestinal (GI) tract is a matter of debate. Many (poly)phenols are metabolized to sulphurated, methylated or glucuronidated forms following ingestion which may disable their antioxidant function (see Figure 13.1) (Scalbert et al., 2002; Halliwell, 2008). It is conceivable that (poly)phenols are direct antioxidants in the stomach or small intestine prior to phase 1 metabolism, but how this would translate to enhanced performance or recovery is unclear. It is possible that a more favourable redox environment in the gut milieu could aid the absorption of nutrients that enhance muscle recovery, such as amino acids. This is one of the mechanisms by which probiotics are purported to attenuate markers of muscle damage following exercise (Jäger et al., 2016, 2018). However, this remains speculative and to our knowledge has not been examined with (poly)phenols. Generally, many (poly)phenols have poor bioavailability meaning the dose that reaches the target cell is likely too low for direct free radical scavenging (Forman, Davies and Ursini, 2014). As such, the contemporary view is that most, if not all, (poly)phenols exert their antioxidant effects, outside the gut at least, indirectly, by inducing redox enzymes (Forman, Davies and Ursini, 2014). Indeed, there is now compelling evidence that (poly)phenols induce superoxide dismutase, catalase, glutathione peroxidase and other redox enzymes (e.g., thioredoxin) by activating the transcription factor nuclear factor-erythroid-2-related factor 2 (Nrf2) (Na and Surh, 2008). For instance, (poly) phenols that have, amongst other properties, electrophilic Michael acceptors (e.g., curcumin, sulforaphane) or that are modified to acquire this property (e.g., quercetin, epigallocatechin gallate) can activate Nrf2 by modifying a cysteine residue on kelch-like ECH-associated protein 1 (Keap-1), which, in resting conditions, sequesters Nrf2 in the cytosol (Christensen and Christensen, 2013). Alternatively, some (poly)phenols activate Keap-1-Nrf2 signalling via phosphorylation of upstream kinases or are able to via multiple mechanisms (Na and Surh, 2008). Disabling Keap-1 enables Nrf2 to translocate to the nucleus and transactivate a battery of enzymes that detoxify potentially damaging RONS (e.g., hypochlorous acid) and defend against challenges to the cellular redox environment (Huang, Nguyen and Pickett, 2000). This suggests that, somewhat paradoxically, the

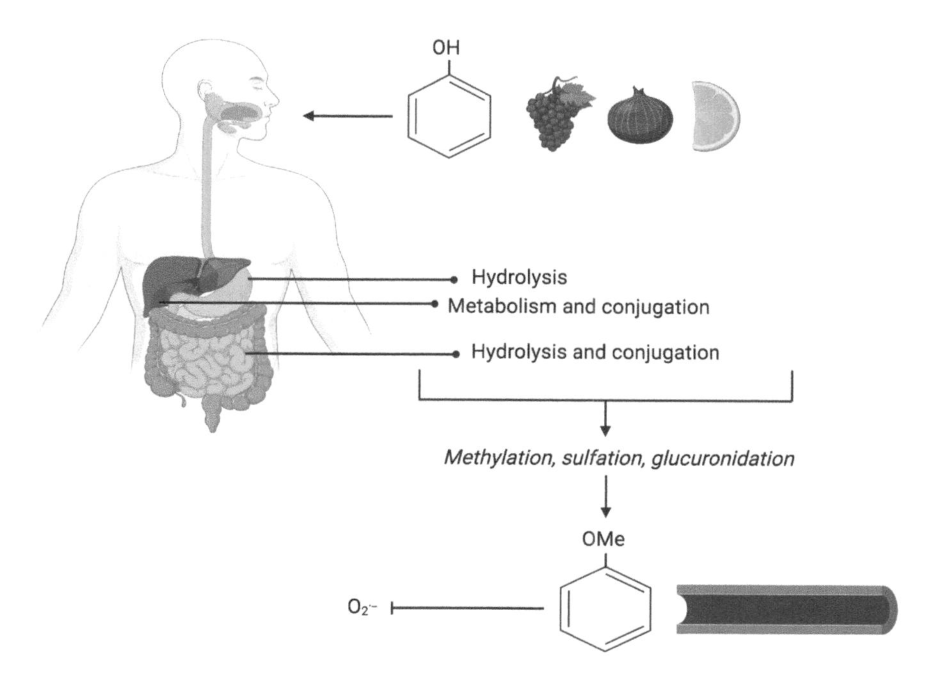

Figure 13.1 Illustrative example of (poly)phenol metabolism. During digestion, phenols and polyphenols are released from the food matrix and subjected to several biotransformation reactions, including oxidation, reduction, hydrolysis and conjugation. Common conjugation reactions are methylation (as depicted in the figure; OH is a hydroxyl group; OMe representing methylation; $O_2^{\bullet-}$ is superoxide), sulphation and glucuronidation, and these can occur in the small intestine, liver and circulation. This extensive metabolism means the (poly)phenols available to cells may no longer possess antioxidant activity or at least their potency has been greatly reduced. It is important to note that the biological activity of the conjugated metabolites that appear in the circulation is not well understood (Scalbert et al., 2002).

antioxidant effects of (poly)phenols are more likely to be pro-oxidant effects. It is important to stress that while these effects have been well-described in vitro and in rodents, there is currently limited evidence that (poly)phenols activate Nrf2 or its target genes in humans (Clifford et al., 2021). Cell culture studies should also be interpreted cautiously; many (poly)phenols oxidize to H_2O_2 in cell media, making it difficult to discern whether the observed cellular changes are due to the H_2O_2 generated or the (poly)phenols (Halliwell, 2008).

Many (poly)phenols have been shown to attenuate inflammation. Part of their anti-inflammatory effects is mediated by interfering with key pro-inflammatory signalling cascades (Kobayashi et al., 2016). The mechanisms for such effects are still being unravelled but could involve activation of Nrf2 (Kobayashi et al., 2016; Zhang and Tsao, 2016). Indeed, by activating Nrf2 many (poly)phenols have been shown to inhibit the NLR family pyrin domain containing 3 (NLPR3) inflammasome that governs pro-inflammatory signalling following stress (Zhang and Tsao, 2016). Disrupting this pathway decreases nuclear factor kappa-light-chain-enhancer of activated B cells (NF-κB) expression and subsequent pro-inflammatory signalling, which might limit damage to skeletal muscle tissue and accelerate regeneration following exercise (Pizza et al., 2005). Nonetheless, these mechanisms are yet to be validated in humans, and some have questioned whether dampening the inflammatory response could delay rather than accelerate muscle regeneration (Lundberg and Howatson, 2018).

It has been suggested that (poly)phenols and their downstream metabolites might also enhance exercise performance by augmenting vascular function. The theory is that increasing endothelial function could improve muscle perfusion and the oxidative energy contribution during exercise (Keane et al., 2018; Bowtell and Kelly, 2019). Most phytochemicals are purported to enhance vasodilation by increasing nitric oxide (NO) bioactivity, in part by suppressing superoxide producing nicotinamide adenine dinucleotide phosphate (NADPH) oxidase and subsequently peroxynitrate ($ONOO^-$) formation (Ghosh and Scheepens, 2009). $ONOO^-$ is a highly toxic radical that can oxidize macromolecules and initiate inflammatory responses; thus, reducing the formation $ONOO^-$ is a viable mechanism by which some (poly)phenols could enhance performance or expedite recovery. These biological effects are supported by numerous clinical trials that reported increased endothelial function and/or surrogate markers of NO following (poly)phenol intake (Oyama et al., 2010; Li et al., 2015). However, evidence that (poly)phenols improve local perfusion or NO production during and after exercise is limited and inconsistent (Keane et al., 2018; Morgan, Barton and Bowtell, 2019). Notwithstanding, NO-dependent vasodilation remains a prominent explanation for how many (poly)phenols might enhance exercise performance (Bowtell and Kelly, 2019).

Another mechanism by which (poly)phenols might enhance exercise performance is by augmenting mitochondrial biogenesis. Rodent and in vitro studies have found that (poly)phenols activate two important molecular pathways for mitochondrial biogenesis, sirtuin-1 (SIRT 1) and peroxisome proliferator-activated receptor-γ coactivator 1-α (PGC1α) (Sandoval-Acuña, Ferreira and Speisky, 2014). In a rodent study, the polyphenol quercetin increased cytochrome C enzyme concentration, a marker of mitochondrial biogenesis, and this was paralleled by improvements in exercise performance (Davis et al., 2009). Follow-up studies with more valid markers of mitochondrial biogenesis such as citrate synthase activity and cardiolipin content (Larsen et al., 2012) are still needed. Similar ergogenic effects have been reported with green tea catechins (Murase et al., 2006). However, aside from one study with quercetin (Nieman et al., 2010), there is currently little evidence that (poly) phenols enhance mitochondrial biogenesis in humans.

It has been also proposed that certain (poly) phenols, especially green tea catechins, can increase fat oxidation by augmenting the expression of genes involved in fat metabolism, or by decreasing catechol-O-methyltransferase (COMT), an enzyme that inhibits norepinephrine and β-oxidation (Lin and Lin-Shiau, 2006). Presumably, an increased reliance on fat oxidation would spare finite muscle glycogen stores which are depleted after 90–120 minutes of high-intensity endurance exercise (Murase et al., 2006). There is some evidence to suggest that green tea catechins modify fat oxidation and energy expenditure at rest (Amiot, Riva and Vinet, 2016), but as of yet there is little evidence of these effects occuring during exercise, at least in humans (Hodgson, Randell and Jeukendrup, 2013; Randell et al., 2013). More recently, anthocyanins from blackcurrant were shown to enhance fat oxidation during exercise; however, the underpinning mechanisms were not explored (Cook et al., 2015, 2017).

In summary, there are several mechanisms by which (poly)phenols could modulate exercise performance and recovery (see Figure 13.2 for overview). However, as will become apparent, most studies focus on the outcome and often neglect to interrogate the underlying mechanisms, meaning our understanding of how these compounds modulate athletic performance and recovery is still very much in its infancy.

(POLY)PHENOLS AND EXERCISE PERFORMANCE

In most studies, (poly)phenols are administered as juices, whole foods and powders that contain a mixture of compounds. It is beyond the scope of this chapter to discuss them all in detail, rather, to briefly review the evidence for the most frequently studied interventions and provide an overview of the potential for (poly)phenols more generically to exert a positive effect on exercise performance and recovery. For more detailed reviews, the reader is directed elsewhere (Bowtell and Kelly, 2019; Mason et al., 2020).

Several studies have examined whether supplements derived from the ellagitannin and anthocyanin-rich pomegranate fruit can enhance exercise performance. In one of the first studies (Trexler et al., 2014), it was found that 1,000 mg of pomegranate extract enhanced brachial artery blood flow and extended time to exhaustion during a running task 30 minutes post-ingestion. However, Trinity et al. (Trinity et al., 2014) could not replicate these findings; they found no effect of 500 mL of pomegranate juice (equivalent to 1,800 mg of polyphenols; 95% ellagitannins) on cycling time trial performance after 7 days of supplementation in 12 well-trained cyclists. Recently, Crum et al. (Crum et al., 2017) reported no effects

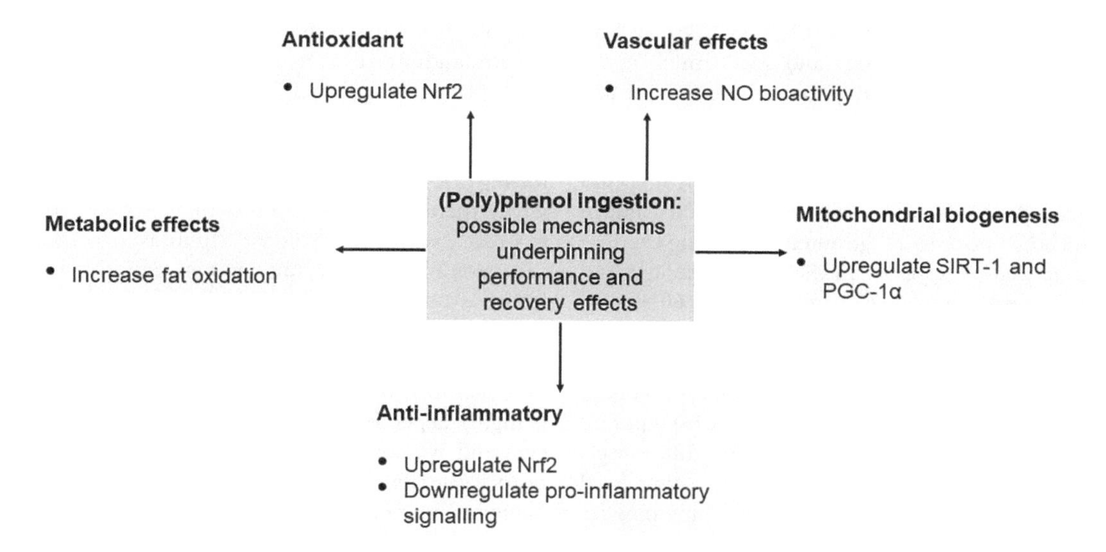

Figure 13.2 Mechanisms by which (poly)phenols may modulate exercise performance and exercise recovery. NO, nitric oxide; Nrf2, nuclear factor-erythroid-2-related factor 2; PGC1α, peroxisome proliferator-activated receptor-γ coactivator 1-α; SIRT 1, sirtuin-1.

of pomegranate extract (1,000 mg) on time trial performance in eight male cyclists. Another recent study found no ergogenic effects with pomegranate extract intake (1,000 mg) on repeated sprint and resistance training performance (Roelofs et al., 2017). However, this study did report increases in blood flow and vessel diameter 30 minutes following pomegranate ingestion, corroborating the findings from Trexler et al. (2014). In aggregate, whilst there is some evidence that pomegranate intake might modulate vascular function, there is little evidence that these effects translate to performance improvements.

Blackcurrant is another phytochemical-rich fruit that has been investigated for its ergogenic effects. Blackcurrant contains various anthocyanins and phenolic acids and is thought to mediate its ergogenic effects by modulating blood flow (Cook and Willems, 2019), increasing fat oxidation (Hiles et al., 2020), and attenuating oxidative stress (herein used to indicate oxidative damage to macromolecules) (Hurst et al., 2020). In controlled trials, blackcurrant supplementation has consistently been shown to enhance exercise performance. For example, Cook et al. (2015) found that a 7-day intake of anthocyanin-rich black currant extracts (300 mg) improved 16.1 Km cycling time trial performance by ~2.4% compared to a placebo. Using the same intervention, Perkins et al. (2015) reported a staggering 11% increase in distance covered by recreationally active males during 16 × 19 second sprints on a treadmill. A 7–10-day intake of anthocyanin-rich blackcurrant supplements has also been shown to enhance 4 km cycling performance (Murphy, Cook and Willems, 2017), intermittent running performance (Willems et al., 2016; Godwin, Cook and Willems, 2017) and sport climbing performance (Potter et al., 2020). Accordingly, ample evidence suggests blackcurrant intake enhances performance in a variety of exercise tasks and populations. However, the underlying mechanisms were rarely considered and remain ambiguous.

Like blackcurrants, tart cherries contain various flavonoid polyphenols but are especially rich in anthocyanins, a group of water-soluble pigments responsible for their red colour (Bell et al., 2014b). Numerous studies show that tart cherries, often provided as a juice concentrate, can modulate vascular function (Keane et al., 2016), lower oxidative stress (Bell et al., 2014a, 2015) and attenuate inflammation (Bell et al., 2016; Jackman et al., 2018). Nonetheless, evidence that tart cherry juice enhances exercise performance is equivocal. In a recent study (Davis and Bellar, 2019), 6 days of cherry intake (500 mg/day freeze-dried powder) did not alter muscle oxygenation or extend time to exhaustion during a high-intensity cycling task. Similarly, Clifford et al. (2013) found no effects of 3 days of tart cherry (200 mg/day in capsules) on 20 km cycling time trial performance, and McCormick et al. (McCormick et al., 2016) found no effects of tart Montmorency cherry juice (90 mL/day for 6 days) on a battery of performance tests in water polo players. By contrast, Keane et al. (2018) found that 60 mL of Montmorency cherry juice intake 1.5 hours prior to a high-intensity cycling exercise enhanced peak power in the first 20 seconds and total work completed during the final 60 second sprint. However, time to exhaustion and muscle oxygenation measured by near-infrared spectroscopy (NIRS) were unaffected. Recently, Morgan et al. (Morgan, Barton and Bowtell, 2019) reported improvements in performance and NIRS-muscle oxygenation during a 16 km cycling time trial after ingesting Montmorency tart cherry powder capsules for 6 days prior. Interestingly, despite the overall inconsistent findings, when the studies were pooled in a recent meta-analysis the results indicated a small positive effect of tart cherry juice on endurance performance (Gao and Chilibeck, 2020). However, only 2/10 of the included studies reported statistically significant changes so this finding should be interpreted carefully.

Collectively, studies examining the effects of (poly)phenol-rich supplements on exercise performance have produced mixed findings. Except for anthocyanin-rich blackcurrants, there is weak and inconsistent evidence that (poly)phenols are given alone (see Mason et al., 2020) or as whole-food combinations enhance exercise performance. Overall, more well-designed human intervention studies that also measure the underpinning mechanisms are needed before (poly) phenols can be confidently recommended as an ergogenic strategy for athletic performance.

(POLY)PHENOLS AND EXERCISE RECOVERY

Research indicates that inflammation and oxidative stress occur following strenuous exercise (Paulsen et al., 2012). These processes are necessary for the structural and functional adaptation of skeletal muscle and other tissue, and hence critical for the resolution of muscle function (Owens et al., 2019). Somewhat paradoxically, nutrition strategies that target the post-exercise inflammatory and oxidative stress response to accelerate recovery could interfere with these processes and thereby inhibit molecular adaptations (Owens et al., 2019). Thus, if the primary purpose is to maximise the training stimulus, athletes and practitioners might need to consider a "periodised" approach to supplementation to support training and competition demands that can maintain the balance for adaptation and optimising recovery. However, there are scenarios when there might be a need to intervene; for example, when the exercise stimulus becomes excessive and impairs function and performance for an extended period, or if the need to accelerate recovery is greater than the need for an adaptive response (e.g., tournament scenario). Consequently, the aim of a recovery intervention is highly dependent upon the purpose of the session and whether it is more important to adapt or to accelerate recovery in preparation for subsequent training or competition.

To examine the effects of (poly)phenols on exercise recovery, many studies have utilised a model that evokes significant muscle damage. This is usually characterised by exercise that has a substantial eccentric muscle contraction (i.e., the muscle lengthening under tension) component and/or is of a long duration or high intensity. However, there are examples where the exercise challenge is metabolically taxing (cycling) but has little or no eccentric component (Bell et al., 2014a, 2015). Other examples include recovery from marathon running (Howatson et al., 2010), intermittent sprint sports (Bell et al., 2016) and resistance exercise (Connolly, McHugh and Padilla-Zakour, 2006; Trombold et al., 2010; Bowtell et al., 2011). It is unlikely that (poly)phenol interventions will reduce the magnitude of mechanical damage on the skeletal muscle myofilaments, but given the putative antioxidative and anti-inflammatory properties of many (poly) phenols (via the above-described mechanisms), it makes the expectation tenable that exercise-induced increases in RONS and/or inflammation could be attenuated and help return function at an accelerated rate (Howatson and van Someren, 2008; Bell et al., 2014b). We provide some cursory examples to illustrate the potential benefits of (poly)phenols in exercise recovery.

The first examples are quercetin, which is a flavonoid polyphenol present in berries, grapes, tomatoes and teas that has been shown to have reasonable bioavailability (Egert et al., 2008), and catechins, which are present in tea and cocoa. Both polyphenols have been the subject of several investigations for their potential benefits on aspects of exercise recovery. In a study with quercetin, participants performed 3 days of high-intensity cycling for 3 hours per day to evoke an inflammatory response (McAnulty et al., 2008). Despite 6 weeks of supplementation, quercetin performed no better than the placebo. In another study, quercetin attenuated pro-inflammatory cytokine mRNA expression following cycling exercise (Nieman et al., 2007a); however, others have failed to find any benefits of quercetin on the recovery of muscle function, markers of inflammation, oxidative stress or muscle soreness (Nieman et al., 2007b; O'Fallon et al., 2012). Catechins were shown to reduce post-exercise muscle soreness, but all other indices of muscle function and damage remained unaltered compared with a control (Kerksick, Kreider and Willoughby, 2010). Others reported that green tea catechins reduced lipid oxidation (malondialdehyde) following high-intensity exercise; however, changes in soreness and muscle function were not reported (Jówko et al., 2015). Overall, catechins and quercetin warrant greater research to elucidate their potential, but at present their application in managing exercise recovery is unsupported.

Pomegranate has also been examined for its effects on exercise recovery. Two studies in recreational exercisers examined the application of a pomegranate extract following muscle-damaging exercise (Trombold et al., 2010, 2011) and showed improved return of function, but there were no measures of oxidative stress and inflammation to determine the underlying mechanisms. These data were later supported by Ammar et al. (2016) in resistance trained volunteers, which illustrated its potential application to well-trained cohorts. Although these data appear promising, they also provide no mechanistic insights.

Conversely, tart Montmorency cherries have been consistently shown to facilitate recovery following numerous types of exercise. Numerous studies have shown accelerated recovery following heavy resistance training (eccentric contractions) in untrained (Connolly, McHugh and Padilla-Zakour, 2006) and well-trained (Bowtell et al., 2011) volunteers; following long-distance running (Howatson et al., 2010) after intermittent sprint (Bell et al., 2016) and cycling exercise (Bell et al., 2014a). Although not all studies have shown a positive effect (Abbott et al., 2019; Lamb et al., 2019), on balance, the evidence is supportive. The positive effects have almost always been attributable to the reduction in oxidative stress and inflammation; however, few studies have measured these parameters. Some attempt has been made to understand the influence on inflammation via systemic measures of c-reactive protein and pro-inflammatory cytokines, but the results are inconsistent. Importantly to the theme of this book, oxidative stress has been measured in very few studies, probably because of the assay complexity and methodological limitations. For example, one study (Howatson et al., 2010) showed a reduction lipid peroxidation (thiobarbituric acid; TBARS) following a marathon run, and another (Bowtell et al., 2011) showed a reduction in protein oxidation (protein carbonyls) following resistance training. At face value, these data seem to provide some insight into the possible mechanisms underpinning the recovery of muscle function. However, as previously highlighted in this book, these methodologies are fraught with limitations and are heavily influenced by uric acid concentrations, for example. In general, the evidence-base has shown some positive effects with cherry supplementation; nonetheless, there is still limited understanding of how these (poly)phenols exert their effects.

An important, but nonetheless often-overlooked limitation in many recovery studies is the use of a crossover design, where a single limb is used in the first arm of the trial and the contralateral limb used in the second arm of the trial. This design, particularly when muscle damage is present, has conclusively been shown to confer a contralateral repeated bout effect (Howatson and van Someren, 2007; Starbuck, Eston and Kraemer, 2012) and hence protect from muscle damage/stress in the contralateral limb; a fact that is often ignored or overlooked. What is also interesting to note is that studies with cherries and other (poly)phenols use a loading phase, whereby participants consume the intervention for several days or weeks before the exercise bout (Howatson et al., 2010; Nieman et al., 2010). The idea being that dosing with (poly)phenols will induce molecular responses or be stored in tissues. Somewhat contradictory to the latter contention is the fact that most (poly)phenols are absorbed and peak within the first few hours of intake, and then appear to be rapidly excreted (Scalbert and Williamson, 2000). Indeed, high systemic concentrations of (poly)phenols can only be maintained by regular consumption (Scalbert and Williamson, 2000). This does not rule out the possibility that tissue does not store (poly)phenols, but this has not been sufficiently examined in humans. Alternatively, several days of intake could improve redox status at baseline, as shown with the antioxidants N-acetylcysteine (NAC) and vitamin C (Paschalis et al., 2016, 2018). This, in turn, could lessen post-exercise oxidative stress and any oxidant-mediated damage to skeletal muscle. However, a caveat with this theory is that antioxidants might only improve redox status, as measured by systemic glutathione levels, when baseline levels are low (Paschalis et al., 2016, 2018). In this scenario, a pre-load would only be anticipated to enhance recovery in those with low glutathione levels, and to our knowledge, this is yet to be examined. Notwithstanding, the studies that used several days of loading seem to show the most benefit for exercise recovery, and thus the mechanisms behind this phenomenon warrant further attention.

As described in previous sections, how (poly) phenols exert their effects remains unclear. From a recovery perspective, the prevailing theory is that (poly)phenols attenuate oxidative stress and/or inflammation, both of which are associated with the two cardinal symptoms of exercise-induced muscle damage, muscle soreness and loss of muscle function (Howatson and Van Someren, 2008; Bowtell and Kelly, 2019; Mason et al., 2020). However, these mechanistic effects are not necessarily well-supported; indeed, such effects have largely been inferred from in vitro and animal studies (Zerba, Komorowski and Faulkner, 1990; Pizza et al., 2005), or from systemic changes in human studies (Paulsen et al., 2012). Furthermore, many of the markers used to determine oxidative stress are not robust enough to adequately reflect the magnitude of macro-molecule damage caused by the exercise bout (Cobley et al., 2017). Interestingly, while many studies suggest that (poly)phenols attenuate oxidative stress by scavenging free radicals, it seems more likely that (poly)phenols attenuate oxidative stress (and possibly inflammation) by activating endogenous antioxidants via Nrf2 signalling (Forman, Davies and Ursini, 2014). Thus, from a mechanistic perspective, future studies should probably begin to explore whether Nrf2 activation can explain the potential benefits of (poly)phenols on recovery from muscle-damaging exercise.

PRACTICAL APPLICATION AND SUMMARY

Interventions involving the use pharmacological analogues are widespread and are often consumed in doses that are well in excess of the recommended daily allowance that could result in unwanted side-effects (Lundberg and Howatson, 2018). Importantly for athletic populations, this behaviour increases the risk of consuming a contaminated product. The use of (poly)phenol-rich foods is growing in interest and represents a realistic alternative for numerous areas of sport and exercise nutrition, not least to enhance performance and manage recovery. The wider application is that if free radical production and inflammation can be managed using these foods, it has much wider implications for clinical populations, especially those with chronic inflammatory disorders. Other (poly)phenol-rich foods (e.g., chokeberry, curcumin, acai, concord grapes and blackcurrants) that have not been explored thoroughly could also provide a means to manage exercise recovery. There is no evidence that (poly)phenols have a detrimental effect on performance or recovery; at worst, these foods provide vital nutrients – at best, performance and recovery could be augmented. Pragmatically, a diet rich in (poly)phenols (fruit and vegetables) may be the best strategy to support both performance and recovery.

An important focus of future work will be unravelling the mechanisms by which (poly)phenols might enhance performance and accelerate recovery. There is compelling evidence in animals and in vitro models that various (poly)phenols activate Nrf2 and endogenous antioxidants, and this represents a plausible mechanism to explain, at least in part, their ergogenic effects. Studies are now needed in humans to confirm these findings. In addition, more robust measures of oxidative stress and inflammation are needed. Regarding the former, most of the frequently used methods are inadequate (e.g., TBARS), and this has undoubtedly held back our understanding of how (poly) phenols might influence oxidative stress. A better understanding of the underlying mechanisms will be essential for the development of more targeted and effective (poly)phenol interventions.

REFERENCES

Abbott, W. et al. (2019) 'Tart cherry juice: No effect on muscle function loss or muscle soreness in professional soccer players after a match', *International Journal of Sports Physiology and Performance*, 15, pp. 1–6. doi: 10.1123/ijspp.2019-0221. United States: Human Kinetics.

Alway, S. E. et al. (2017) 'Resveratrol enhances exercise-induced cellular and functional adaptations of skeletal muscle in older men and women', *The Journals of Gerontology. Series A, Biological Sciences and Medical Sciences*, 72(12), pp. 1595–1606. doi: 10.1093/gerona/glx089. Division of Exercise Physiology, Department of Human Performance and Applied Exercise Science, West Virginia University School of Medicine, Morgantown; West Virginia Clinical and Translational Science Institute, Morgantown; Center for Neuroscience, Morg: published on behalf of the Gerontological Society of America by Oxford University Press.

Amiot, M. J., Riva, C. and Vinet, A. (2016) 'Effects of dietary polyphenols on metabolic syndrome features in humans: A systematic review', *Obesity Reviews*. doi: 10.1111/obr.12409.

Ammar, A. et al. (2016) 'Pomegranate supplementation accelerates recovery of muscle damage and soreness and inflammatory markers after a weightlifting training session', *PLoS ONE*. doi: 10.1371/journal.pone.0160305.

Bell, P. G. et al. (2014a) 'Montmorency cherries reduce the oxidative stress and inflammatory responses to repeated days high-intensity stochastic cycling', *Nutrients*. doi: 10.3390/nu6020829.

Bell, P. G. et al. (2014b) 'The role of cherries in exercise and health', *Scandinavian Journal of Medicine and Science in Sports*. doi: 10.1111/sms.12085.

Bell, P. G. et al. (2015) 'Recovery facilitation with montmorency cherries following high-intensity, metabolically challenging exercise', *Applied Physiology, Nutrition and Metabolism*. doi: 10.1139/apnm-2014-0244.

Bell, P. G. et al. (2016) 'The effects of montmorency tart cherry concentrate supplementation on recovery following prolonged, intermittent exercise', *Nutrients*. doi: 10.3390/nu8070441.

Bowtell, J. and Kelly, V. (2019) 'Fruit-derived polyphenol supplementation for athlete recovery and performance', *Sports Medicine*. doi: 10.1007/s40279-018-0998-x.

Bowtell, J. L. et al. (2011) 'Montmorency cherry juice reduces muscle damage caused by intensive strength exercise', *Medicine & Science in Sports & Exercise*, 43(8), pp. 1544–1551.

Choquette, S. et al. (2013) 'Soy isoflavones and exercise to improve physical capacity in postmenopausal women', *Climacteric*. doi: 10.3109/13697137.2011.643515.

Christensen, L. P. and Christensen, K. B. (2013) 'The role of direct and indirect polyphenolic antioxidants in protection against oxidative stress', In *Polyphenols in Human Health and Disease*. doi: 10.1016/B978-0-12–398456-2.00023-2.

Clifford, T. et al. (2021) 'The effect of dietary phytochemicals on nuclear factor erythroid 2-related factor 2 (Nrf2) activation: A systematic review of human intervention trials', *Molecular Biology Reports*. doi: 10.1007/s11033-020-06041-x.

Clifford, T., Mitchell, N. and Scott, A. (2013) 'The Influence of different sources of polyphenols on sub-maximal cycling and time trial performance. *J Athl Enhancement*, 2(6). doi: 10.4172/2324-9080.1000130.

Cobley, J. N. et al. (2017) 'Exercise redox biochemistry: Conceptual, methodological and technical recommendations', *Redox Biology*. doi: 10.1016/j.redox.2017.03.022.

Connolly, D. A. J., McHugh, M. P. and Padilla-Zakour, O. I. (2006) 'Efficacy of a tart cherry juice blend in preventing the symptoms of muscle damage', *British Journal of Sports Medicine*, 40(8), pp. 679–683.

Cook, M. D. et al. (2015) 'New Zealand blackcurrant extract improves cycling performance and fat oxidation in cyclists', *European Journal of Applied Physiology*. doi: 10.1007/s00421-015-3215-8.

Cook, M. D. et al. (2017) 'Dose effects of New Zealand blackcurrant on substrate oxidation and physiological responses during prolonged cycling', *European Journal of Applied Physiology*. doi: 10.1007/s00421-017-3607-z.

Cook, M. D. and Willems, M. E.T. (2019) 'Dietary anthocyanins: A review of the exercise performance effects and related physiological responses', *International Journal of Sport Nutrition and Exercise Metabolism*. doi: 10.1123/ijsnem.2018-0088.

Crozier, A. et al. (2007) 'Secondary metabolites in fruits, vegetables, beverages and other plant-based dietary components', In *Plant Secondary Metabolites: Occurrence, Structure and Role in the Human Diet*. doi: 10.1002/9780470988558.ch7.

Crum, E. M. et al. (2017) 'The effect of acute pomegranate extract supplementation on oxygen uptake in highly-trained cyclists during high-intensity exercise in a high altitude environment', *Journal of the International Society of Sports Nutrition*. doi: 10.1186/s12970-017-0172-0.

Davis, G. R. and Bellar, D. (2019) 'Montmorency cherry supplement does not affect aerobic exercise performance in healthy men', *International Journal for Vitamin and Nutrition Research*. doi: 10.1024/0300-9831/a000575.

Davis, J. M. et al. (2009) 'Quercetin increases brain and muscle mitochondrial biogenesis and exercise tolerance', *American Journal of Physiology - Regulatory Integrative and Comparative Physiology*. doi: 10.1152/ajpregu.90925.2008.

Drobnic, F. et al. (2014) 'Reduction of delayed onset muscle soreness by a novel curcumin delivery system (Meriva®): A randomised, placebo-controlled trial', *Journal of the International Society of Sports Nutrition*. doi: 10.1186/1550-2783-11-31.

Egert, S. et al. (2008) 'Daily quercetin supplementation dose-dependently increases plasma quercetin concentrations in healthy humans'. *Journal of Nutrition*. doi: 10.1093/jn/138.9.1615.

Estruel-Amades, S. et al. (2019) 'Protective effect of hesperidin on the oxidative stress induced by an exhausting exercise in intensively trained rats', *Nutrients*. doi: 10.3390/nu11040783.

Forman, H. J., Davies, K. J. A. and Ursini, F. (2014) 'How do nutritional antioxidants really work: Nucleophilic tone and para-hormesis versus free radical scavenging in vivo', *Free Radical Biology and Medicine*. doi: 10.1016/j.freeradbiomed.2013.05.045.

Frank, J. et al. (2019) 'Terms and nomenclature used for plant-derived components in nutrition and related research: Efforts toward harmonization', *Nutrition Reviews*. doi: 10.1093/nutrit/nuz081.

Gao, R. and Chilibeck, P. D. (2020) 'Effect of tart cherry concentrate on endurance exercise performance: A meta-analysis', *Journal of the American College of Nutrition*. doi: 10.1080/07315724.2020.1713246.

Gelabert-Rebato, M. et al. (2019) 'Enhancement of exercise performance by 48 hours, and 15-day supplementation with mangiferin and luteolin in men', *Nutrients*. doi: 10.3390/nu11020344.

Ghosh, D. and Scheepens, A. (2009) 'Vascular action of polyphenols', *Molecular Nutrition and Food Research*. doi: 10.1002/mnfr.200800182.

Godwin, C., Cook, M. and Willems, M. (2017) 'Effect of New Zealand blackcurrant extract on performance during the running based anaerobic sprint test in trained youth and recreationally active male football players', *Sports*. doi: 10.3390/sports5030069.

Halliwell, B. (2008) 'Are polyphenols antioxidants or pro-oxidants? What do we learn from cell culture and in vivo studies?', *Archives of Biochemistry and Biophysics*. doi: 10.1016/j.abb.2008.01.028.

Hiles, A. M. et al. (2020) 'Dietary supplementation with New Zealand blackcurrant extract enhances fat oxidation during submaximal exercise in the heat', *Journal of Science and Medicine in Sport*. doi: 10.1016/j.jsams.2020.02.017.

Hodgson, A. B., Randell, R. K. and Jeukendrup, A. E. (2013) 'The effect of green tea extract on fat oxidation at rest and during exercise: Evidence of efficacy and proposed mechanisms', *Advances in Nutrition*. doi:10.3945/an.112.003269.

Howatson, G. et al. (2010) 'Influence of tart cherry juice on indices of recovery following marathon running Howatson et al. Cherry juice supplementation and Marathon running', *Scandinavian Journal of Medicine & Science in Sports*, 20(6), pp. 843–852.

Howatson, G. and van Someren, K. A. (2007) 'Evidence of a contralateral repeated bout effect after maximal eccentric contractions', *European Journal of Applied Physiology*. doi: 10.1007/s00421-007-0489-5.

Howatson, G. and Van Someren, K. A. (2008) 'The prevention and treatment of exercise-induced muscle damage', *Sports Medicine*. doi: 10.2165/00007256-200838060-00004.

Huang, H. C., Nguyen, T. and Pickett, C. B. (2000) 'Regulation of the antioxidant response element by protein kinase C-mediated phosphorylation of NF-E2-related factor 2', *Proceedings of the National Academy of Sciences of the United States of America*, 97(23), pp. 12475–12480. United States ([Erratum in: Proc Natl Acad Sci U S A 2001 Jan 2;98(1):379]).

Hurst, R. D. et al. (2020) 'Daily consumption of an anthocyanin-rich extract made from new zealand blackcurrants for 5 weeks supports exercise recovery through the management of oxidative stress and inflammation: A randomized placebo controlled pilot study', *Frontiers in Nutrition*. doi: 10.3389/fnut.2020.00016.

Jackman, S. R. et al. (2018) 'Tart cherry concentrate does not enhance muscle protein synthesis response to exercise and protein in healthy older men', *Experimental Gerontology*. doi: 10.1016/j.exger.2018.06.007.

Jäger, R. et al. (2016) 'Probiotic Bacillus coagulans GBI-30, 6086 reduces exercise-induced muscle damage and increases recovery', *PeerJ*. doi: 10.7717/peerj.2276.

Jäger, R. et al. (2018) 'Probiotic Bacillus coagulans GBI-30, 6086 improves protein absorption and utilization', *Probiotics and Antimicrobial Proteins*. doi: 10.1007/s12602-017-9354-y.

Jin, S.-Y. et al. (2016) 'Lignans-rich extract from Herpetospermum caudigerum alleviate physical fatigue in mice', *Chinese Journal of Integrative Medicine*. doi: 10.1007/s11655-016-2254-2.

Jówko, E. et al. (2011) 'Green tea extract supplementation gives protection against exercise-induced oxidative damage in healthy men', *Nutrition research (New York, N.Y.)*, 31(11), pp. 813–821. doi: 10.1016/j.nutres.2011.09.020. Department of Biochemistry, Faculty of Physical Education and Sport in Biala Podlaska, Jozef Pilsudski University of Physical Education in Warsaw, Akademicka 2, 21–500 Biala Podlaska, Poland. ewa.jowko@awf-bp.edu.pl: Elsevier Science.

Jówko, E. et al. (2012) 'Effect of a single dose of green tea polyphenols on the blood markers of exercise-induced oxidative stress in soccer players', *International Journal of Sport Nutrition & Exercise Metabolism*, 22(6), pp. 486–496.

Jówko, E. et al. (2015) 'The effect of green tea extract supplementation on exercise-induced oxidative stress parameters in male sprinters', *European Journal of Nutrition*. doi: 10.1007/s00394-014-0757-1.

Kawanishi, N. et al. (2013) 'Curcumin attenuates oxidative stress following downhill running-induced muscle damage', *Biochemical and Biophysical Research Communications*. doi: 10.1016/j.bbrc.2013.10.119.

Keane, K. M. et al. (2016) 'Effects of Montmorency tart cherry (Prunus Cerasus L.) consumption on vascular function in men with early hypertension', *American Journal of Clinical Nutrition*. doi: 10.3945/ajcn.115.123869.

Keane, K. M. et al. (2018) 'Effects of montmorency tart cherry (L. Prunus Cerasus) consumption on nitric oxide biomarkers and exercise performance', *Scandinavian Journal of Medicine and Science in Sports*. doi: 10.1111/sms.13088.

Kerksick, C. M., Kreider, R. B. and Willoughby, D. S. (2010) 'Intramuscular adaptations to eccentric exercise and antioxidant supplementation', *Amino Acids*. doi: 10.1007/s00726-009-0432-7.

Kobayashi, E. H. et al. (2016) 'Nrf2 suppresses macrophage inflammatory response by blocking proinflammatory cytokine transcription', *Nature Communications*. doi: 10.1038/ncomms11624.

Kweon, M. et al. (2019) 'A chalcone from ashitaba (Angelica keiskei) stimulates myoblast differentiation and inhibits dexamethasone-induced muscle atrophy', *Nutrients*. doi: 10.3390/nu11102419.

Lamb, K. L. et al. (2019) 'No effect of tart cherry juice or pomegranate juice on recovery from exercise-induced muscle damage in non-resistance trained men', *Nutrients*, 11(7). doi: 10.3390/nu11071593. Food and Nutrition Group, Sheffield Hallam University, Sheffield S1 1WB, UK: MDPI Publishing.

Larsen, S. et al. (2012) 'Biomarkers of mitochondrial content in skeletal muscle of healthy young human subjects', *Journal of Physiology*. doi: 10.1113/jphysiol.2012.230185.

Li, S. H. et al. (2015) 'Effect of grape polyphenols on blood pressure: A meta-analysis of randomized controlled trials', *PLoS ONE*. doi: 10.1371/journal.pone.0137665.

Lin, J. K. and Lin-Shiau, S. Y. (2006) 'Mechanisms of hypolipidemic and anti-obesity effects of tea and tea polyphenols', *Molecular Nutrition and Food Research*. doi: 10.1002/mnfr.200500138.

Lundberg, T. R. and Howatson, G. (2018) 'Analgesic and anti-inflammatory drugs in sports: Implications for exercise performance and training adaptations', *Scandinavian Journal of Medicine and Science in Sports*. doi: 10.1111/sms.13275.

Margaritelis, N. V. et al. (2020) 'Redox basis of exercise physiology', *Redox Biology*. doi: 10.1016/j.redox.2020.101499.

Mason, S. A. et al. (2020) 'Antioxidant supplements and endurance exercise: Current evidence and mechanistic insights', *Redox Biology*. doi: 10.1016/j.redox.2020.101471.

McAnulty, L. S. et al. (2008) 'Chronic quercetin ingestion and exercise-induced oxidative damage and inflammation', *Applied Physiology, Nutrition & Metabolism*, 33(2), pp. 254–262.

McCormick, R. et al. (2016) 'Effect of tart cherry juice on recovery and next day performance in well-trained Water Polo players', *Journal of the International Society of Sports Nutrition*. doi: 10.1186/s12970-016-0151-x.

Morgan, P. T., Barton, M. J. and Bowtell, J. L. (2019) 'Montmorency cherry supplementation improves 15-km cycling time-trial performance', *European Journal of Applied Physiology*. doi: 10.1007/s00421-018-04058-6.

Murase, T. et al. (2006) 'Green tea extract improves running endurance in mice by stimulating lipid utilization during exercise', *American Journal of Physiology - Regulatory Integrative and Comparative Physiology*. doi: 10.1152/ajpregu.00752.2005.

Murphy, C., Cook, M. and Willems, M. (2017) 'Effect of New Zealand blackcurrant extract on repeated cycling time trial performance', *Sports*. doi: 10.3390/sports5020025.

Na, H.-K. and Surh, Y.-J. (2008) 'Modulation of Nrf2-mediated antioxidant and detoxifying enzyme induction by the green tea polyphenol EGCG', *Food and Chemical Toxicology*. doi:10.1016/j.fct.2007.10.006.

Nieman, D. C. et al. (2007a) 'Quercetin's influence on exercise-induced changes in plasma cytokines and muscle and leukocyte cytokine mRNA', *Journal of Applied Physiology*, 103(5), pp. 1728–1735. doi: 10.1152/japplphysiol.00707.2007. Department of Health, Leisure, and Exercise Science, PO Box 32071, Appalachian State Univ., Boone, NC 28608, USA. niemandc@appstate.edu: American Physiological Society.

Nieman, D.C. et al. (2007b) 'Quercetin ingestion does not alter cytokine changes in athletes competing in the Western States Endurance Run.', *Journal of Interferon & Cytokine Research: The Official Journal of the International Society for Interferon and Cytokine Research*, 27(12), pp. 1003–1011. doi: 10.1089/jir.2007.0050. Department of Biology, Health, Leisure, and Exercise Science, Fischer Hamilton/Nycom Biochemistry Laboratory, Appalachian State University, Boone, North Carolina. 28608, USA. niemandc@appstate.edu: Mary Ann Liebert.

Nieman, D. C. et al. (2010) 'Quercetin's influence on exercise performance and muscle mitochondrial biogenesis', *Medicine and Science in Sports and Exercise*. doi: 10.1249/MSS.0b013e3181b18fa3.

Nieman, D. C. et al. (2018) 'Influence of 2-weeks ingestion of high chlorogenic acid coffee on mood state, performance, and postexercise inflammation and oxidative stress: A randomized, placebo-controlled trial', *International Journal of Sport Nutrition and Exercise Metabolism*. doi: 10.1123/ijsnem.2017-0198.

O'Fallon, K. S. et al. (2012) 'Effects of quercetin supplementation on markers of muscle damage and inflammation after eccentric exercise', *International Journal of Sport Nutrition and Exercise Metabolism*, 22(6), pp. 430–437. Dept of Kinesiology, University of Massachusetts, Amherst, MA, USA: Human Kinetics Publishers.

Owens, D. J. et al. (2019) 'Exercise-induced muscle damage: What is it, what causes it and what are the nutritional solutions?', *European Journal of Sport Science*. doi: 10.1080/17461391.2018.1505957.

Oyama, J. I. et al. (2010) 'Green tea catechins improve human forearm vascular function and have potent anti-inflammatory and anti-apoptotic effects in smokers', *Internal Medicine*. doi: 10.2169/internalmedicine.49.4048.

Pandey, K. B. and Rizvi, S. I. (2009) 'Plant polyphenols as dietary antioxidants in human health and disease', *Oxidative Medicine and Cellular Longevity*. doi: 10.4161/oxim.2.5.9498.

Paschalis, V. et al. (2016) 'Low vitamin C values are linked with decreased physical performance and increased oxidative stress: Reversal by vitamin C supplementation', *European Journal of Nutrition*. doi: 10.1007/s00394-014-0821-x.

Paschalis, V. et al. (2018) 'N-acetylcysteine supplementation increases exercise performance and reduces oxidative stress only in individuals with low levels of glutathione', *Free Radical Biology and Medicine*. doi: 10.1016/j.freeradbiomed.2017.12.007.

Paulsen, G. et al. (2012) 'Leucocytes, cytokines and satellite cells: What role do they play in muscle damage and regeneration following eccentric exercise?', *Exercise Immunology Review*, 18, pp. 42–97.

Perkins, I. C. et al. (2015) 'New Zealand blackcurrant extract improves high-intensity intermittent running', *International Journal of Sport Nutrition and Exercise Metabolism*. doi: 10.1123/ijsnem.2015-0020.

Pizza, F. X. et al. (2005) 'Neutrophils contribute to muscle injury and impair its resolution after lengthening contractions in mice', *Journal of Physiology*. doi: 10.1113/jphysiol.2004.073965.

Potter, J. A. et al. (2020) 'Effects of New Zealand blackcurrant extract on sport climbing performance', *European Journal of Applied Physiology*. doi: 10.1007/s00421-019-04226-2.

Randell, R. K. et al. (2013) 'No effect of 1 or 7 d of green tea extract ingestion on fat oxidation during exercise', *Medicine and Science in Sports and Exercise*. doi: 10.1249/MSS.0b013e31827dd9d4.

Reid, M. B. (2016) 'Redox interventions to increase exercise performance', *Journal of Physiology*. doi: 10.1113/JP270653.

Roelofs, E. J. et al. (2017) 'Effects of pomegranate extract on blood flow and vessel diameter after high-intensity exercise in young, healthy adults', *European Journal of Sport Science*. doi: 10.1080/17461391.2016.1230892.

Sandoval-Acuña, C., Ferreira, J. and Speisky, H. (2014) 'Polyphenols and mitochondria: An update on their increasingly emerging ROS-scavenging independent actions', *Archives of Biochemistry and Biophysics*. doi: 10.1016/j.abb.2014.05.017.

Scalbert, A. et al. (2002) 'Absorption and metabolism of polyphenols in the gut and impact on health', *Biomedicine and Pharmacotherapy*. doi: 10.1016/S0753-3322(02)00205-6.

Scalbert, A. and Williamson, G. (2000) 'Dietary intake and bioavailability of polyphenols', *The Journal of Nutrition*. doi: 10.1093/jn/130.8.2073s.

Seeley, A. D., Jacobs, K. A. and Signorile, J. F. (2020) 'Acute soy supplementation improves 20-km time trial performance, power, and speed', *Medicine & Science in Sports & Exercise*, 52(1), pp. 170–177.

Starbuck, C., Eston, R. G. and Kraemer, W. J. (2012) 'Exercise-induced muscle damage and the repeated bout effect: Evidence for cross transfer', *European Journal of Applied Physiology*. doi: 10.1007/s00421-011-2053-6.

Tanabe, Y. et al. (2015) 'Attenuation of indirect markers of eccentric exercise-induced muscle damage by curcumin', *European Journal of Applied Physiology*. doi: 10.1007/s00421-015-3170-4.

Theodorus Willems, M. E. et al. (2015) 'Beneficial physiological effects with blackcurrant intake in endurance athletes', *International Journal of Sport Nutrition & Exercise Metabolism*, 25(4), pp. 367–374.

Trexler, E.T. et al. (2014) 'Effects of pomegranate extract on blood flow and running time to exhaustion', *Applied Physiology, Nutrition and Metabolism*. doi: 10.1139/apnm-2014-0137.

Trinity, J. D. et al. (2014) 'Impact of polyphenol antioxidants on cycling performance and cardiovascular function', *Nutrients*. doi: 10.3390/nu6031273.

Trombold, J. R. et al. (2010) 'Ellagitannin consumption improves strength recovery 2–3 d after eccentric exercise', *Medicine and Science in Sports and Exercise*. doi: 10.1249/MSS.0b013e3181b64edd.

Trombold, J. R. et al. (2011) 'The effect of pomegranate juice supplementation on strength and soreness after eccentric exercise', *Journal of Strength and Conditioning Research*. doi: 10.1519/JSC.0b013e318220d992.

Voduc, N. et al. (2014) 'Effect of resveratrol on exercise capacity: A randomized placebo-controlled crossover pilot study', *Applied Physiology, Nutrition & Metabolism*, 39(10), pp. 1183–1187.

Willems, M. et al. (2016) 'Beneficial effects of New Zealand blackcurrant extract on maximal sprint speed during the loughborough intermittent shuttle test', *Sports*. doi: 10.3390/sports4030042.

Wiswedel, I. et al. (2004) 'Flavanol-rich cocoa drink lowers plasma F2-isoprostane concentrations in humans', *Free Radical Biology and Medicine*. doi: 10.1016/j.freeradbiomed.2004.05.013.

Zerba, E., Komorowski, T. E. and Faulkner, J. A. (1990) 'Free radical injury to skeletal muscles of young, adult, and old mice', *American Journal of Physiology - Cell Physiology*. doi: 10.1152/ajpcell.1990.258.3.c429.

Zhang, H. and Tsao, R. (2016) 'Dietary polyphenols, oxidative stress and antioxidant and anti-inflammatory effects', *Current Opinion in Food Science*. doi: 10.1016/j.cofs.2016.02.002.

CHAPTER FOURTEEN

Exercise

A STRATEGY TO TARGET OXIDATIVE STRESS IN CANCER

Amélie Rébillard, Cindy Richard and Suzanne Dufresne

CONTENTS

Introduction / 167
The Benefits of Physical Activity in Cancer Survival / 167
Oxidative Stress as a Key Mechanism? / 168
Oxidative Stress and Cancer / 168
Exercise Modulation of Oxidative Stress in Cancer Patients / 169
Exercise, Tumor Growth, and Oxidative Stress: Possible Impact on Treatments? / 170
Physical Activity, Oxidative Stress and Cancer-Induced Muscle Wasting / 172
Limitations / 174
Concluding Remarks / 174
Acknowledgments / 174
Funding / 174
References / 174

INTRODUCTION

Exercise is increasingly recognized as a powerful tool for limiting the risk of developing cancer and for providing psychological, physiological, and survival benefits in cancer patients (Campbell et al., 2019; Patel et al., 2019). Exercise corresponds to a subcategory of physical activity (PA) that is structured and planned (Caspersen et al., 1985). The effects of training are highly dependent on the type, intensity, duration, and frequency of exercises performed. It challenges whole-body homeostasis and gives rise to widespread adaptations in cells, tissues, and organ systems (Hawley et al., 2014). Notably, while a single bout of exercise can generate an acute rise in reactive oxygen species (ROS) levels, regular exercise leads to enhanced antioxidant defenses through physiological adaptations (Ji, 1999). Given the major role of oxidative stress (OS) in cancer, we hypothesize that exercise can generate many benefits in cancer patients, in particular by targeting OS. In this chapter, we will first describe the benefits of PA on cancer survival, and second, we will evaluate the potential role played by OS in these effects, focusing both on tumor growth, anticancer treatment response, and muscle wasting.

THE BENEFITS OF PHYSICAL ACTIVITY IN CANCER SURVIVAL

Exercise is now recognized as an effective strategy in the rehabilitation of cancer patients, resulting in improvements in psychological and physiological

DOI: 10.1201/9781003051619-14

function. In 2010, the American College of Sports Medicine (ACSM) recommended to engage in a minimum of 150 minute/week of aerobic activity as well as to perform at least two resistance training sessions per week and daily stretching exercises (Schmitz et al., 2010). More recently, the ACSM updated these guidelines based on a growing number of clinical trials and provided more refined recommendations aiming to improve distinct cancer-related health outcomes (Campbell et al., 2019). In regard to patients survival, however, most studies published to date are observational studies which evaluated broad PA levels (rather specific exercise trainings) which include occupational and household activities, active transportation, and recreational or leisure-time activities (comprising exercise). PA has to be shown to help cancer survivors cope with cancer treatments, can improve their long-term health, and in some cases may reduce the risk of cancer-specific mortality (Courneya and Friedenreich, 2011; Friedenreich et al., 2020). Over the past decade, the impact of PA levels on cancer survival has raised considerable interest. More than 130 studies have investigated this association in several cancer types, mainly breast, colorectal, or prostate cancer (Christensen et al., 2018; Patel et al., 2019; Friedenreich et al., 2020). Recently, systematic reviews and meta-analyses have distinguished the effects of PA performed either before or after diagnosis on the risk of cancer mortality (Patel et al., 2019; Friedenreich et al., 2020), which provide valuable additional insights. It was notably shown that the practice of PA prior to diagnosis reduces mortality from six different types of cancer, i.e., breast, colorectal, hematological, liver, lung, and stomach cancer. PA after diagnosis also provides benefits as it is associated with a reduction in mortality from breast, colorectal, and prostate cancer, emphasizing that it is never too late to engage in PA. With regard to overall mortality, similar effects were observed: high levels of PA in pre-diagnosis are associated with reduced overall mortality for breast, colorectal, hematological, and prostate cancer while after diagnosis, this practice would reduce overall mortality in patients with breast, colorectal, gynecological, pediatric, glioma, hematological, kidney, lung, prostate, and stomach cancer. The benefits of PA on overall survival may be due, in part, to improved physical and psychological fitness, thus helping to reduce risk factors for other chronic diseases. There is some evidence that regular PA of approximately 10–15 metabolic equivalent of task (MET) hours per week (corresponding to approximately 30–60 minute per day of moderate to high-intensity PA) should be targeted to maximize survival benefits in breast cancer (Friedenreich et al., 2020), but the optimal dose of PA and timing of this intervention remain to be refined. All these results are very encouraging, but the underlying mechanisms need to be identified, in order to improve the management care of cancer patients. In addition to clinical studies, exercise was reported to slow down tumor growth in a wide variety of murine models of cancer (Ashcraft et al., 2016) and, more recently, to improve the antitumor effect of cancer treatments (Ashcraft et al., 2019; Dufresne et al., 2020; Morrell et al., 2019; Wennerberg et al., 2020). These preclinical models are also an opportunity to unravel the molecular mechanisms involved in the anticancer effects of exercise. Among them, improved immune function, normalization of tumor vascularization, metabolism modulation, and OS are promising leads that continue to be explored in preclinical and clinical studies (Pedersen et al., 2015; Hojman et al., 2018).

OXIDATIVE STRESS AS A KEY MECHANISM?

Oxidative Stress and Cancer

OS is defined as a "disruption of the redox balance toward an increase in prooxidant over the capacity of antioxidants, leading to a perturbation of redox signaling and control and/or molecular damage" (Sies and Jones, 2007). As described by various authors (Fiaschi and Chiarugi, 2012; Luo et al., 2009; Zhou et al., 2020), OS contributes to many of the cancer hallmarks (Hanahan and Weinberg, 2011). An increase in the levels of lipid peroxidation adducts (4-hydroxy-2-nonenal (4-HNE), malondialdehyde (MDA)), protein carbonyls, or oxidative DNA lesions (mainly 8-oxo-7,8-dihydro-2′-deoxyguanosine (8-oxodG)) is detected in tumors (Ohtake et al., 2018; Qing et al., 2019), skeletal muscle (Puig-Vilanova et al., 2015; Ramamoorthy et al., 2009), and fluids in cancer patients (Kosova et al., 2014; Lagadu et al., 2010; Sheridan et al., 2009; Singh et al., 2016). These damages are correlated with disease progression and poorer prognosis (Arakaki et al., 2016; Araki et al., 2018; Boakye et al., 2020; Dirican et al., 2016; Dziaman et al., 2014; Sato et al., 2010). A reduced systemic antioxidant capacity is also observed in different types of cancers. Particularly, thiol levels (total and/or native) are decreased in patients with colorectal, prostate, lung cancer (Dirican et al., 2016; Gào et al., 2020; Karatas et al., 2019), or oral squamous carcinoma, and such changes are associated with poorer overall survival (Patel et al., 2007). Depending on the species involved, their concentrations, and localization, ROS can play a dual role. In low-to-moderate concentrations, ROS can act as second messengers and regulate various signaling pathways involved in proliferation, migration, differentiation, and angiogenesis (Circu and Aw, 2010; England and Cotter, 2005; Pouysségur and Mechta-Grigoriou, 2006; Schieber and Chandel, 2014). Conversely, high concentrations of ROS cause significant cellular damage to lipids, proteins, and DNA, leading to apoptosis. Of note, these concentrations are dependent on the species involved and the context. Cancer cells as well as tumor microenvironment (TME) components have the ability to maintain ROS levels that promote their proliferation and survival while avoiding death (Fiaschi and Chiarugi, 2012; Hayes et al., 2020; Reuter et al., 2010; Sosa et al., 2013).

Hence, the challenge is to precisely determine the ROS level in cancer cells to identify whether therapeutic strategy should aim to decrease ROS to limit proliferation or conversely increase ROS to induce apoptosis (Aggarwal et al., 2019; Gorrini et al., 2013; Perillo et al., 2020; Postovit et al., 2018).

Exercise Modulation of Oxidative Stress in Cancer Patients

During exercise, the organism is exposed to OS as well as thermal, mechanical, and metabolic stress. In the late 1970s, exercise, and in particular acute exercise, was described as being associated with increased oxidative damage (Dillard et al., 1978). This pioneer work mainly focused on lipid damage as well as systemic antioxidant defenses. Animal models were then used to assess this response in other tissues such as skeletal muscle or liver (Davies et al., 1982), revealing that skeletal muscle during contraction generates ROS. In the following decades, the role of ROS in fatigue and force production was better understood (Novelli et al., 1990), and redox signaling has started to be studied (Powers et al., 2010). While intense acute exercise increases ROS production (mainly superoxide $(O_2^{\bullet-})$), and oxidative damage in blood or skeletal muscle, repetition of moderate-intensity exercise may generate an adaptive response by increasing antioxidant defenses (e.g., GPx, MnSOD) (Gomez-Cabrera et al., 2008; Powers et al., 2020, 2016).

As observed in healthy population, it can be hypothesized that regular exercise could modify OS levels among cancer patients (Arena et al., 2020), possibly having an impact on the disease and its side effects (Figure 14.1).

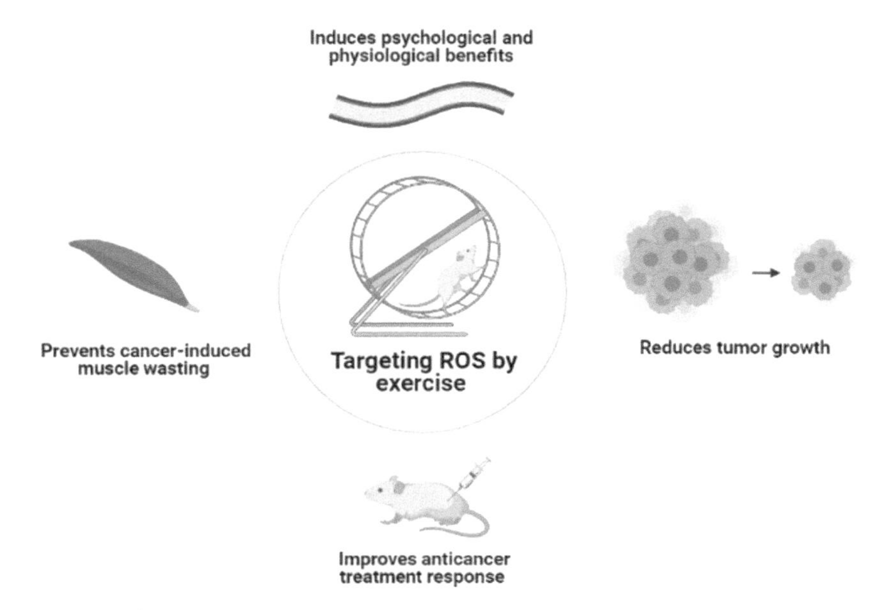

Figure 14.1 Targeting ROS in cancer: the potential effects of exercise. By regulating OS levels in the context of cancer, exercise could induce psychological and physiological benefits, reduce tumor growth, improve treatment efficacy, and prevent cancer-related muscle loss.

Indeed, reductions in OS markers (8-OHdG and protein carbonyl) as well as increases in total antioxidant capacity (TAC) measured in the plasma have been reported in response to an eight-week exercise intervention combining aerobic and resistance sessions in head and neck cancer patients (Yen et al., 2020). These results should however be interpreted with caution as TAC assessment is considered as a flawed measure (Cobley et al., 2017). Nevertheless, Allgayer et al. also suggested that cancer patients engaging in exercise could modulate their OS levels. In a randomized control trial, they demonstrated that 2 weeks of moderate-intensity exercise training reduced urinary 8-OHdG levels in colorectal cancer patients (Allgayer et al., 2008). Taken together, these studies suggest that exercise can modulate oxidative damage in plasma and urine of cancer patients. However, when patients underwent high-intensity physical training, a nonsignificant increase in 8-OHdG levels was reported (Allgayer et al., 2008), suggesting exercise intensity is an important parameter.

Interestingly, as the American College of Sports Medicine remark, the impact of exercise on cancer outcomes such as fatigue and physical functioning appears to be highly dependent on its intensity: low-intensity activities may provide only weak improvements, while moderate- to vigorous-intensity activities could provide the strongest effects (Campbell et al., 2019). It can be hypothesized that this difference is explained in part by the intensity-dependent effect of exercise on OS levels. In women with breast cancer, Tomasello et al. explored the effects of dragon boat racing, a moderate-to-high intensity activity, or walking, a low-intensity activity, realized twice per week for ≥ 7 months on systemic OS. None of the intervention counteracted high levels of OS associated with cancer but both dragon boat racing and walking increased antioxidant defenses (i.e., enzymatic activities of superoxide dismutase and glutathione peroxidase, reduced glutathione levels) (Tomasello et al., 2017). Importantly, dragon boat racing seemed to enhance the antioxidant response to a greater extent than walking (Tomasello et al., 2017), suggesting that intensity could explain some of the benefits associated with exercise.

In addition, redox modifications by exercise have been linked to cancer-related outcomes such as fatigue. Ten weeks of exercise training (a combination of aerobic and resistance training) was shown to increase antioxidant capacity and decrease protein carbonyls in the plasma of cancer patients compared to healthy subjects. These parameters were significantly correlated with reductions in affective, sensory, and cognitive fatigue, as evaluated by the Piper Fatigue Inventory Scale. Furthermore, this exercise program also induced a reduction in circulating 8-OHdG levels which were significantly correlated with increased $\dot{V}O_{2\ peak}$ and muscular strength (Repka and Hayward, 2016). Since $\dot{V}O_{2\ peak}$ is a well-established independent predictor of mortality, these data raise the question of the relationship between exercise modulation of OS and cancer mortality. To date, no clinical studies have addressed this issue. Indeed, while oxidative damage or antioxidant defenses have been studied in plasma or urine, much less is known about other tissues, such as tumor or skeletal muscle.

Exercise, Tumor Growth, and Oxidative Stress: Possible Impact on Treatments?

Recently, Louzada and colleagues proposed a hypothetical model of successive ROS waves explaining how contracting skeletal muscles could communicate with distant tissues (Fortunato and Louzada, 2020; Louzada et al., 2020). As previously described, an initial wave of ROS is produced in skeletal muscle during exercise. When a muscle contracts, it can (1) promote the release of myokines or extracellular vesicles which could be delivered to remote tissues, (2) generate extracellular lipid peroxidation products known to act as signaling mediators, (3) contribute to the production of certain metabolites (lactate, pyruvate, etc.). All these mechanisms can be triggered by the first ROS wave and lead to the generation of a second wave of ROS in remote noncontracting tissues. This creates a transient prooxidant environment which in turn modulates intracellular signaling (Fortunato & Louzada, 2020; Louzada et al., 2020). This interesting model could hypothetically be applied to the context of cancer: The release of mediators by the muscles in response to exercise could trigger redox signaling in the tumor tissue thereby impacting proliferation, survival, angiogenesis, or cell death (Aoi et al., 2013; Gannon et al., 2015; Hojman et al., 2011; Huang et al., 2020) (Figure 14.2).

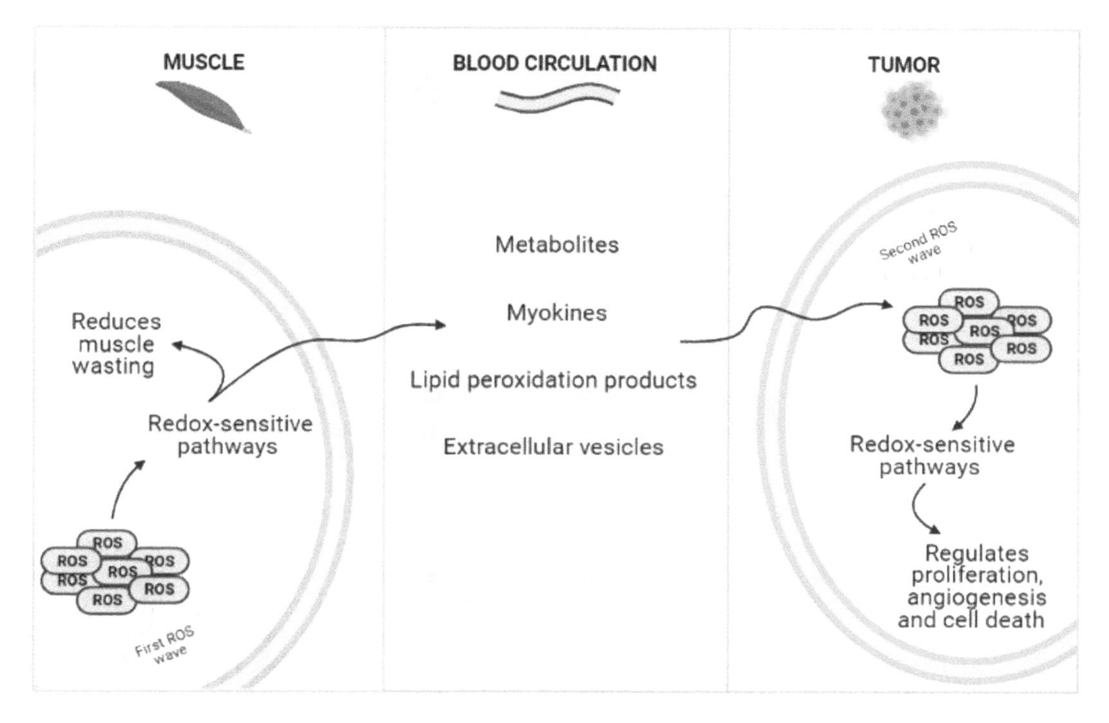

Figure 14.2 The redox-dependent crosstalk between contracting skeletal muscle and tumor. During exercise, the first wave of ROS is produced in skeletal muscle, leading to the activation of ROS-mediated signaling pathways. When repeated regularly, physiological adaptations may occur and could limit cancer-induced muscle wasting. Moreover, activation of redox-sensitive signaling pathways in response to acute exercise could participate in the release of various mediators in blood circulation. Thus, metabolites, myokines, lipid peroxidation products, and extracellular vesicles released by the skeletal muscle during contraction could hypothetically stimulate ROS production in the tumor. The production of ROS in tumor tissue would represent a second wave of ROS which could affect different signaling pathways involved in angiogenesis as well as proliferation and cell death. (Adapted from Fortunato and Louzada, 2020; Louzada et al., 2020).

In support of this theory, several studies have suggested that some of the exercise anticancer effects could be explained by the existence of a crosstalk between muscle and tumor tissues (Aoi et al., 2013; Hojman et al., 2011; Pedersen et al., 2020; Le Guennec et al., 2020). Interestingly, Le Guennec and collaborators showed that spontaneous PA counteracted tumor growth through a crosstalk between tumor, muscle, and adipose tissue in obese and ovariectomized mammary tumor-bearing mice (Le Guennec et al., 2020). Adipokines and growth factors such as VEGF-α and EGF were suggested to play a major role in this dialogue. Using multiple factorial analyses, the authors reported that higher levels of PA induced by an enriched environment decreased inflammation as well as reduced the antioxidant response in the TME (Le Guennec et al., 2020). These results suggest that voluntary PA, by targeting inflammation and antioxidant defenses in the tumor tissue, could slow down tumor growth. They may seem counterintuitive at first glance since an increase in antioxidant defenses is usually described in response to physical training. However, this adaptation has been described mainly in muscle (Gomez-Cabrera et al., 2008).

Importantly, other studies have shown that exercise is able to reduce oxidative damage. A few years ago, our team showed that aerobic exercise training in Copenhagen rats bearing subcutaneous AT-1 tumor reduced cancer cells proliferation and delayed prostate cancer growth. This antiproliferative effect of exercise training was associated with a decrease in intratumor levels of DNA oxidative damage (i.e. 8-OHdG), independently of a modulation in protein or lipid oxidative damage (Gueritat et al., 2014). Interestingly, when exercise training was associated with antioxidant (pomegranate juice consumption), intratumor oxidative damage, proliferation, and tumor growth were similar compared to untreated cancer mice, suggesting that this combination blunted the effect of isolated strategies (Gueritat et al., 2014). This study was the first to suggest a link between OS changes in tumor and the antiproliferative effect of exercise.

In line with these results, it was later shown that voluntary wheel running in a 4T1 breast tumor model improved tumor perfusion, decreased hypoxia (Betof et al., 2015), and lowered intra-tumoral 8-OHdG levels (Ashcraft et al., 2019), leading to a reduction in tumor growth. Interestingly, evidence from a large number of studies has demonstrated an interplay between ROS and hypoxia, in which hypoxia may enhance ROS production and ROS may lead to adaptation to hypoxia through HIF-1 alpha stabilization (Chandel et al., 2000; Guzy and Schumacker, 2006; Kulkarni et al., 2007; Sabharwal and Schumacker, 2014). Taken together, these findings support OS as a potential direct or indirect mechanism implicated in the survival benefits associated with exercise in cancer.

It can further be suggested that OS modulation by exercise can also impact the response to anticancer treatments (Ashcraft et al., 2019; Assi et al., 2020). In the same model as described above, Betof et al. reported that voluntary wheel running improved tumor sensitivity to chemotherapy (Betof et al., 2015). A few years later, this group also discussed unpublished data indicating that voluntary wheel running combined with radiation therapy was more effective than radiation therapy alone. This effect could be explained by hypoxia modulation since the tumor hypoxic fraction was lower in mice that exercised prior to radiation (Ashcraft et al., 2019). Interestingly, the authors also noted that exercise lowered oxidative DNA damage, but this did not negatively interact with the radiotherapy response. Indeed, the modulation of OS by exercise may represent a double-edged sword for radiation therapy. Radiation therapy kills cancer cells in part by generating a high production of ROS. The acute ROS production in response to a single bout of exercise might enhance the response to radiotherapy by increasing ROS levels, creating subsequent cell death. On the other hand, chronic (or regular) exercise could induce an increase in systemic and tissue antioxidant defenses (Radak et al., 2001), which could potentially limit the effectiveness of radiation therapy. Preclinical evidence however support the safety of exercise in combination with radiotherapy as two other preclinical studies showed that exercise training improved radiotherapy efficiency (Dufresne et al., 2020; Wennerberg et al., 2020). Dufresne and collaborators demonstrated that exercise training enhanced the response to radiotherapy in a murine model of prostate cancer through increased infiltration of immune natural killer (NK) cells to the tumor tissue and higher apoptotic death (Dufresne et al., 2020). Similarly, Wennerberg reported in a murine model of breast cancer that treadmill running improved response to a combination of anti-PD1/radiation therapy, through a decrease in intratumoral myeloid-derived suppressor cells (MDSCs) infiltration and an increase in NK and CD8 T cell activation (Wennerberg et al., 2020). All these results show that exercise training could modulate the TME and target some of the metabolic and immunological dysregulations in cancer. Importantly, these effects could be related to OS. First, the leaky and immature blood vessels in the TME can promote hypoperfusion, which participates in poor infiltration of antitumoral T cells. The resulting hypoxia preferentially favors infiltration of immunosuppressive cells such as MDSCs (Jain, 2014; Schaaf et al., 2018). Moreover, ROS have been described as a regulator of TME (Weinberg et al., 2019) and MDSC-mediated immune suppression (Ohl and Tenbrock, 2018). Interestingly, recent literature indicated that strategies aiming at reprogramming the components of the TME could sensitize tumors to anticancer treatments (Datta et al., 2019; Kaymak et al., 2021; Li et al., 2018; Mpekris et al., 2020). These suggestions are speculative but offer some general directions for future research.

Although limited, these results show a positive association between exercise, OS, and cancer outcomes (tumor growth, anticancer treatment response), which comfort the importance of exercise in cancer patient health care. Further studies are required to confirm these benefits, to clarify the underlying mechanisms and to determine whether this association can be extended to other cancers. Together, these data will help to define more specific exercise recommendations for cancer patients.

Physical Activity, Oxidative Stress and Cancer-Induced Muscle Wasting

As previously mentioned, OS is commonly observed in skeletal muscle during cancer. These damages were reported in several clinical studies conducted in patients with lung (Puig-Vilanova et al., 2015; Ramamoorthy et al., 2009), colorectal, and gastrointestinal cancer (Brzeszczyńska et al., 2016; Ramamoorthy et al., 2009) as well as in preclinical experiments using breast (Alves et al., 2020; Guarnier et al., 2010; Hentilä et al., 2018; Padilha et al., 2017), colon (Assi et al., 2016; Ballarò et al., 2019; Ham et al., 2014; Sullivan-Gunn et al., 2011), lung (Brown et al., 2017), liver (Mastrocola et al., 2008), and pancreatic (Rupert et al., 2020) cancer models. OS was described as a key regulator of muscle wasting in different experimental settings (Ábrigo et al., 2018; Aquila et al., 2020; Penna et al., 2020) and can regulate multiple mechanisms including the protein synthesis/degradation balance (Dodd et al., 2010; Gomes-Marcondes and Tisdale, 2002; Tan et al., 2015; Yang et al., 2010), autophagy processes (Aucello et al., 2009; McClung et al., 2010; Smuder et al., 2018), apoptosis (Smuder et al., 2010), and mitochondrial dysfunction (Ábrigo et al., 2018). Interestingly, these signaling pathways appear to be differently activated depending on the skeletal muscle fiber types. Indeed, oxidative muscle fibers are more resistant to these signals than glycolytic fibers (Alves et al., 2020; Marin-Corral et al., 2010; Salazar-Degracia et al., 2017), in part due to their higher levels of antioxidant defenses (Powers et al., 2011). Although the involvement of OS in muscle wasting observed in cancer patients has not been clearly proven, several preclinical observations highlight the importance of limiting ROS production in skeletal muscles to counteract muscle wasting.

Similar to what is observed in aging or chronic diseases, it can be hypothesized that regular exercise may reduce muscular OS levels (Musci et al., 2019; Sallam and Laher, 2015). For example, Ballarò et al. have demonstrated that 12 days of moderate aerobic and resistance training decreased protein carbonylation, increased catalase levels, and restored Nuclear respiratory factor 2 (Nrf2) signaling in gastrocnemius muscle, leading to the maintenance of muscle mass and function in C26 tumor-bearing mice (Ballarò et al., 2019). However, whether the combination of resistance and aerobic training provided additional benefits than either exercise modality performed alone was not investigated in these mice. Two other studies performed in breast cancer murine models suggest that both modalities may be able to regulate redox signaling in the muscle tissue. Eight weeks of resistance exercise alone reduced lipid oxidative damage (MDA and total lipid hydroperoxides levels) and improved GSH:GSSG ratio in the soleus muscle of Walker-256 tumor-bearing rats. These benefits were coupled with an improvement in systemic inflammation and contributed to preserve body weight and muscle strength (Padilha et al., 2017). As shown by the group of Alves, aerobic exercise seemed to induce similar effects. Ten days of aerobic interval training led to a reduction in protein carbonylation and a normalization of the GSH:GSSG ratio in the plantaris muscle of Walker-256 tumor-bearing rats (Alves et al., 2020). The modulation of OS by exercise was associated with attenuated muscle atrophy, improved skeletal muscle contractile function, and improved running capacity as well as prolonged lifespan (Alves et al., 2020). These findings suggest that exercise may increase survival by limiting muscle mass loss through OS modulation. Importantly, these effects were not attributed to smaller tumor volumes, as exercise did not impact tumor growth in this study (Alves et al., 2020). These results are in accordance with recent observational studies in cancer patients showing that loss of skeletal muscle mass is associated with short progression-free survival and poor response to anticancer treatments (Cortellini et al., 2019; Shiroyama et al., 2019; Tsukagoshi et al., 2020).

Overall, several preclinical studies support the ability of exercise to protect skeletal muscle from oxidative damage and limit skeletal muscle wasting. To date, no clinical studies have investigated this response in cancer patients. It is important to note that in the animal models mentioned, exercise interventions were either initiated 1 day after tumor injection (Alves et al., 2020; Ballarò et al., 2019) or 6 weeks prior to tumor cell injection and then maintained for 12 days (Padilha et al., 2017). Therefore, they explored the ability of exercise to prevent the onset of muscle damage rather than its ability to counteract existing damage. Hence, the question of the role of exercise in reducing preexisting muscle damage remains unanswered. Such investigations appear, however, essential as 50%–80% of advanced cancer patients experience bodyweight loss and muscle wasting (Ryan et al., 2016). Although exercise is considered to be safe and tolerable in patients with advanced cancer (Heywood et al., 2017), additional preclinical and clinical studies should be conducted to help refine the role of exercise in the management of cancer patients with muscle wasting.

LIMITATIONS

Most studies evaluating the modulation of OS by exercise in the context of cancer are preclinical studies, which may not translate to what is observed in humans. Furthermore, the few clinical studies evaluating OS levels in response to exercise in cancer patients have used blood and urine samples, which may not be reflective of what is observed in the tumor or muscles. Finally, further advances in methodology for assessing OS are essential as they will contribute to a better understanding of the response to exercise in cancer.

CONCLUDING REMARKS

Since ROS play a central role in cancer and cancer-related organ damage, targeting these molecules may be a relevant treatment option for cancer patients but evidence argues against antioxidant supplementation during cancer (Assi, 2017; Assi et al., 2020, 2016). Exercise appears to be an effective strategy to modulate systemic, muscular, and even tumor OS, thereby protecting skeletal muscles, slowing down tumor growth, and improving response to cancer treatments. Benefits might be dependent on a redox-dependent dialogue between the contracting muscle and remote tissues. These results are promising and must be confirmed to determine the optimal dose and type of exercise to target in order to maximize the benefits for cancer patients. Finally, these findings are mainly obtained from studies using animal models and further investigations are needed to determine if they apply to humans.

ACKNOWLEDGMENTS

Figure 14.1–14.2 were created with Biorender.com The authors thank Mohamad Assi from Université Catholique de Louvain (UCL), de Duve Institute, 1200 Brussels, Belgium.

FUNDING

A.R. is supported by University Rennes 2, C.R. is supported by ANRT and Spormed, and S.D. is supported by ENS Rennes and Fondation ARC.

REFERENCES

Ábrigo, J., Elorza, A.A., Riedel, C.A., Vilos, C., Simon, F., Cabrera, D., Estrada, L., Cabello-Verrugio, C., 2018. Role of oxidative stress as key regulator of muscle wasting during cachexia. *Oxid. Med. Cell. Longev.* 2018, 2063179. https://doi.org/10.1155/2018/2063179.

Aggarwal, V., Tuli, H.S., Varol, A., Thakral, F., Yerer, M.B., Sak, K., Varol, M., Jain, A., Khan, Md.A., Sethi, G., 2019. Role of reactive oxygen species in cancer progression: Molecular mechanisms and recent advancements. *Biomolecules* 9. https://doi.org/10.3390/biom9110735.

Allgayer, H., Owen, R.W., Nair, J., Allgayer, H., Owen, R.W., Nair, J., Spiegelhalder, B., Streit, J., Reichel, C., Bartsch, H., 2008. Short-term moderate exercise programs reduce oxidative DNA damage as determined by high-performance liquid chromatography-electrospray ionization-mass spectrometry in patients with colorectal carcinoma following primary treatment. *Scand. J. Gastroenterol.* 43, 971–978. https://doi.org/10.1080/00365520701766111.

Alves, C.R.R., das Neves, W., de Almeida, N.R., Eichelberger, E.J., Jannig, P.R., Voltarelli, V.A., Tobias, G.C., Bechara, L.R.G., de Paula Faria, D., Alves, M.J.N., Hagen, L., Sharma, A., Slupphaug, G., Moreira, J.B.N., Wisloff, U., Hirshman, M.F., Negrão, C.E., de Castro Jr., G., Chammas, R., Swoboda, K.J., Ruas, J.L., Goodyear, L.J., Brum, P.C., 2020. Exercise training reverses cancer-induced oxidative stress and decrease in muscle COPS2/TRIP15/ALIEN. *Mol. Metab.* 39, 101012. https://doi.org/10.1016/j.molmet.2020.101012.

Aoi, W., Naito, Y., Takagi, T., Tanimura, Y., Takanami, Y., Kawai, Y., Sakuma, K., Hang, L.P., Mizushima, K., Hirai, Y., Koyama, R., Wada, S., Higashi, A., Kokura, S., Ichikawa, H., Yoshikawa, T., 2013. A novel myokine, secreted protein acidic and rich in cysteine (SPARC), suppresses colon tumorigenesis via regular exercise. *Gut* 62, 882–889. https://doi.org/10.1136/gutjnl-2011-300776.

Aquila, G., Re Cecconi, A.D., Brault, J.J., Corli, O., Piccirillo, R., 2020. Nutraceuticals and exercise against muscle wasting during cancer cachexia. *Cells* 9, 2536. https://doi.org/10.3390/cells9122536.

Arakaki, H., Osada, Y., Takanashi, S., Ito, C., Aisa, Y., Nakazato, T., 2016. Oxidative stress is associated with poor prognosis in patients with follicular lymphoma. *Blood* 128, 1787–1787. https://doi.org/10.1182/blood.V128.22.1787.1787.

Araki, O., Matsumura, Y., Inoue, T., Karube, Y., Maeda, S., Kobayashi, S., Chida, M., 2018. Association of perioperative redox balance on long-term outcome in patients undergoing lung resection. *Ann. Thorac. Cardiovasc. Surg.* 24, 13–18. https://doi.org/10.5761/atcs.oa.17-00127.

Arena, S.K., Doherty, D.J., Bellford, A., Hayman, G., 2020. Effects of aerobic exercise on oxidative stress in patients diagnosed with cancer: A narrative review. *Cureus* 11. https://doi.org/10.7759/cureus.5382.

Ashcraft, K.A., Peace, R.M., Betof, A.S., Dewhirst, M.W., Jones, L.W., 2016. Efficacy and mechanisms of aerobic exercise on cancer initiation, progression, and metastasis: A critical systematic review of in vivo preclinical data. *Cancer Res.* 76, 4032–4050. https://doi.org/10.1158/0008-5472.CAN-16-0887.

Ashcraft, K.A., Warner, A.B., Jones, L.W., Dewhirst, M.W., 2019. Exercise as adjunct therapy in cancer. *Semin. Radiat. Oncol.* 29, 16–24. https://doi.org/10.1016/j.semradonc.2018.10.001.

Assi, M., 2017. The differential role of reactive oxygen species in early and late stages of cancer. *Am. J. Physiol. Regul. Integr. Comp. Physiol.* 313, R646–R653. https://doi.org/10.1152/ajpregu.00247.2017.

Assi, M., Derbré, F., Lefeuvre-Orfila, L., Rébillard, A., 2016. Antioxidant supplementation accelerates cachexia development by promoting tumor growth in C26 tumor-bearing mice. *Free Radic. Biol. Med.* 91, 204–214. https://doi.org/10.1016/j.freeradbiomed.2015.12.019.

Assi, M., Dufresne, S., Rébillard, A., 2020. Exercise shapes redox signaling in cancer. *Redox Biol.* 35. https://doi.org/10.1016/j.redox.2020.101439.

Aucello, M., Dobrowolny, G., Musarò, A., 2009. Localized accumulation of oxidative stress causes muscle atrophy through activation of an autophagic pathway. *Autophagy* 5, 527–529. https://doi.org/10.4161/auto.5.4.7962.

Ballarò, R., Penna, F., Pin, F., Gómez-Cabrera, M.C., Viña, J., Costelli, P., 2019. Moderate exercise improves experimental cancer cachexia by modulating the redox homeostasis. *Cancers* 11. https://doi.org/10.3390/cancers11030285.

Betof, A.S., Lascola, C.D., Weitzel, D., Landon, C., Scarbrough, P.M., Devi, G.R., Palmer, G., Jones, L.W., Dewhirst, M.W., 2015. Modulation of murine breast tumor vascularity, hypoxia and chemotherapeutic response by exercise. *J. Natl. Cancer Inst.* 107. https://doi.org/10.1093/jnci/djv040.

Boakye, D., Jansen, L., Schöttker, B., Jansen, E.H.J.M., Schneider, M., Halama, N., Gào, X., Chang-Claude, J., Hoffmeister, M., Brenner, H., 2020. Blood markers of oxidative stress are strongly associated with poorer prognosis in colorectal cancer patients. *Int. J. Cancer* 147, 2373–2386. https://doi.org/10.1002/ijc.33018.

Brown, J.L., Rosa-Caldwell, M.E., Lee, D.E., Blackwell, T.A., Brown, L.A., Perry, R.A., Haynie, W.S., Hardee, J.P., Carson, J.A., Wiggs, M.P., Washington, T.A., Greene, N.P., 2017. Mitochondrial degeneration precedes the development of muscle atrophy in progression of cancer cachexia in tumour-bearing mice. *J. Cachexia Sarcopenia Muscle* 8, 926–938. https://doi.org/10.1002/jcsm.12232.

Brzeszczyńska, J., Johns, N., Schilb, A., Degen, S., Degen, M., Langen, R., Schols, A., Glass, D.J., Roubenoff, R., Greig, C.A., Jacobi, C., Fearon, K.C., Ross, J.A., 2016. Loss of oxidative defense and potential blockade of satellite cell maturation in the skeletal muscle of patients with cancer but not in the healthy elderly. *Aging* 8, 1690–1702. https://doi.org/10.18632/aging.101006.

Campbell, K.L., Winters-Stone, K.M., Wiskemann, J., May, A.M., Schwartz, A.L., Courneya, K.S., Zucker, D.S., Matthews, C.E., Ligibel, J.A., Gerber, L.H., Morris, G.S., Patel, A.V., Hue, T.F., Perna, F.M., Schmitz, K.H., 2019. Exercise guidelines for cancer survivors: Consensus statement from international multidisciplinary roundtable. *Med. Sci. Sports Exerc.* 51, 2375–2390. https://doi.org/10.1249/MSS.0000000000002116.

Caspersen, C.J., Powell, K.E., Christenson, G.M., 1985. Physical activity, exercise, and physical fitness: Definitions and distinctions for health-related research. *Public Health Rep.* 100, 126–131.

Chandel, N.S., McClintock, D.S., Feliciano, C.E., Wood, T.M., Melendez, J.A., Rodriguez, A.M., Schumacker, P.T., 2000. Reactive oxygen species generated at mitochondrial complex III stabilize hypoxia-inducible factor-1alpha during hypoxia: a mechanism of O_2 sensing. *J. Biol. Chem.* 275, 25130–25138. https://doi.org/10.1074/jbc.M001914200.

Christensen, J.F., Simonsen, C., Hojman, P., 2018. Exercise training in cancer control and treatment. *Compr. Physiol.* 9, 165–205. https://doi.org/10.1002/cphy.c180016.

Circu, M.L., Aw, T.Y., 2010. Reactive oxygen species, cellular redox systems, and apoptosis. *Free Radic. Biol. Med.* 48, 749–762. https://doi.org/10.1016/j.freeradbiomed.2009.12.022.

Cobley, J.N., Close, G.L., Bailey, D.M., Davison, G.W., 2017. Exercise redox biochemistry: Conceptual, methodological and technical recommendations. *Redox Biol.* 12, 540–548. https://doi.org/10.1016/j.redox.2017.03.022.

Cortellini, A., Verna, L., Porzio, G., Bozzetti, F., Palumbo, P., Masciocchi, C., Cannita, K., Parisi, A., Brocco, D., Tinari, N., Ficorella, C., 2019. Predictive value of skeletal muscle mass for immunotherapy with nivolumab in non-small cell lung cancer patients: A "hypothesis-generator" preliminary report. *Thorac. Cancer* 10, 347–351. https://doi.org/10.1111/1759-7714.12965.

Courneya, K.S., Friedenreich, C. (Eds.), 2011. *Physical Activity and Cancer, Recent Results in Cancer Research*. Springer-Verlag, Berlin. https://doi.org/10.1007/978-3-642-04231-7.

Datta, M., Coussens, L.M., Nishikawa, H., Hodi, F.S., Jain, R.K., 2019. Reprogramming the tumor microenvironment to improve immunotherapy: Emerging strategies and combination therapies. *Am. Soc. Clin. Oncol. Educ. Book* 39, 165–174. https://doi.org/10.1200/EDBK_237987.

Davies, K.J., Quintanilha, A.T., Brooks, G.A., Packer, L., 1982. Free radicals and tissue damage produced by exercise. *Biochem. Biophys. Res. Commun.* 107, 1198–1205. https://doi.org/10.1016/s0006-291x(82)80124-1.

Dillard, C.J., Litov, R.E., Savin, W.M., Dumelin, E.E., Tappel, A.L., 1978. Effects of exercise, vitamin E, and ozone on pulmonary function and lipid peroxidation. *J. Appl. Physiol.* 45, 927–932. https://doi.org/10.1152/jappl.1978.45.6.927.

Dirican, N., Dirican, A., Sen, O., Aynali, A., Atalay, S., Bircan, H.A., Oztürk, O., Erdogan, S., Cakir, M., Akkaya, A., 2016. Thiol/disulfide homeostasis: A prognostic biomarker for patients with advanced non-small cell lung cancer? *Redox Rep. Commun. Free Radic. Res.* 21, 197–203. https://doi.org/10.1179/1351000215Y.0000000027.

Dodd, S.L., Gagnon, B.J., Senf, S.M., Hain, B.A., Judge, A.R., 2010. ROS-mediated activation of NF-kappaB and Foxo during muscle disuse. *Muscle Nerve* 41, 110–113. https://doi.org/10.1002/mus.21526.

Dufresne, S., Guéritat, J., Chiavassa, S., Noblet, C., Assi, M., Rioux-Leclercq, N., Rannou-Bekono, F., Lefeuvre-Orfila, L., Paris, F., Rébillard, A., 2020. Exercise training improves radiotherapy efficiency in a murine model of prostate cancer. *FASEB J. Off. Publ. Fed. Am. Soc. Exp. Biol.* 34, 4984–4996. https://doi.org/10.1096/fj.201901728R.

Dziaman, T., Banaszkiewicz, Z., Roszkowski, K., Gackowski, D., Wisniewska, E., Rozalski, R., Foksinski, M., Siomek, A., Speina, E., Winczura, A., Marszalek, A., Tudek, B., Olinski, R., 2014. 8-Oxo-7,8-dihydroguanine and uric acid as efficient predictors of survival in colon cancer patients. *Int. J. Cancer* 134, 376–383. https://doi.org/10.1002/ijc.28374.

England, K., Cotter, T.G., 2005. Direct oxidative modifications of signalling proteins in mammalian cells and their effects on apoptosis. *Redox Rep. Commun. Free Radic. Res.* 10, 237–245. https://doi.org/10.1179/135100005X70224.

Fiaschi, T., Chiarugi, P., 2012. Oxidative stress, tumor microenvironment, and metabolic Reprogramming: A diabolic liaison [WWW Document]. *Int. J. Cell Biol.* https://doi.org/10.1155/2012/762825.

Fortunato, R.S., Louzada, R.A., 2020. Muscle redox signaling: Engaged in sickness and in health. *Antioxid. Redox Signal.* 33, 539–541. https://doi.org/10.1089/ars.2020.8095.

Friedenreich, C.M., Stone, C.R., Cheung, W.Y., Hayes, S.C., 2020. Physical activity and mortality in cancer survivors: A systematic review and meta-analysis. *JNCI Cancer Spectr.* 4. https://doi.org/10.1093/jncics/pkz080.

Gannon, N.P., Vaughan, R.A., Garcia-Smith, R., Bisoffi, M., Trujillo, K.A., 2015. Effects of the exercise-inducible myokine irisin on malignant and non-malignant breast epithelial cell behavior in vitro. *Int. J. Cancer* 136, E197–E202. https://doi.org/10.1002/ijc.29142.

Gào, X., Wilsgaard, T., Jansen, E.H.J.M., Xuan, Y., Anusruti, A., Brenner, H., Schöttker, B., 2020. Serum total thiol levels and the risk of lung, colorectal, breast and prostate cancer: A prospective case-cohort study. *Int. J. Cancer* 146, 1261–1267.https://doi.org/10.1002/ijc.32428.

Gomes-Marcondes, M.C.C., Tisdale, M.J., 2002. Induction of protein catabolism and the ubiquitin-proteasome pathway by mild oxidative stress. *Cancer Lett.* 180, 69–74. https://doi.org/10.1016/s0304-3835(02)00006-x.

Gomez-Cabrera, M.-C., Domenech, E., Viña, J., 2008. Moderate exercise is an antioxidant: upregulation of antioxidant genes by training. *Free Radic. Biol. Med.* 44, 126–131. https://doi.org/10.1016/j.freeradbiomed.2007.02.001.

Gorrini, C., Harris, I.S., Mak, T.W., 2013. Modulation of oxidative stress as an anticancer strategy. *Nat. Rev. Drug Discov.* 12, 931–947. https://doi.org/10.1038/nrd4002.

Guarnier, F.A., Cecchini, A.L., Suzukawa, A.A., Maragno, A.L.G.C., Simão, A.N.C., Gomes, M.D., Cecchini, R., 2010. Time course of skeletal muscle loss and oxidative stress in rats with Walker 256 solid tumor. *Muscle Nerve* 42, 950–958. https://doi.org/10.1002/mus.21798.

Gueritat, J., Lefeuvre-Orfila, L., Vincent, S., Cretual, A., Ravanat, J.-L., Gratas-Delamarche, A., Rannou-Bekono, F., Rebillard, A., 2014. Exercise training combined with antioxidant supplementation prevents the antiproliferative activity of their single treatment in prostate cancer through inhibition of redox adaptation. *Free Radic. Biol. Med.* 77, 95–105. https://doi.org/10.1016/j.freeradbiomed.2014.09.009.

Guzy, R.D., Schumacker, P.T., 2006. Oxygen sensing by mitochondria at complex III: The paradox of increased reactive oxygen species during hypoxia. *Exp. Physiol.* 91, 807–819. https://doi.org/10.1113/expphysiol.2006.033506.

Ham, D.J., Murphy, K.T., Chee, A., Lynch, G.S., Koopman, R., 2014. Glycine administration attenuates skeletal muscle wasting in a mouse model of cancer cachexia. *Clin. Nutr.* 33, 448–458. https://doi.org/10.1016/j.clnu.2013.06.013.

Hanahan, D., Weinberg, R.A., 2011. Hallmarks of cancer: The next generation. *Cell* 144, 646–674. https://doi.org/10.1016/j.cell.2011.02.013.

Hawley, J.A., Hargreaves, M., Joyner, M.J., Zierath, J.R., 2014. Integrative biology of exercise. *Cell* 159, 738–749. https://doi.org/10.1016/j.cell.2014.10.029.

Hayes, J.D., Dinkova-Kostova, A.T., Tew, K.D., 2020. Oxidative stress in cancer. *Cancer Cell* 38, 167–197. https://doi.org/10.1016/j.ccell.2020.06.001.

Hentilä, J., Nissinen, T.A., Korkmaz, A., Lensu, S., Silvennoinen, M., Pasternack, A., Ritvos, O., Atalay, M., Hulmi, J.J., 2018. Activin receptor ligand blocking and cancer have distinct effects on protein and redox homeostasis in skeletal muscle and liver. *Front. Physiol.* 9, 1917. https://doi.org/10.3389/fphys.2018.01917.

Heywood, R., McCarthy, A.L., Skinner, T.L., 2017. Safety and feasibility of exercise interventions in patients with advanced cancer: A systematic review. *Support. Care Cancer* 25, 3031–3050. https://doi.org/10.1007/s00520-017-3827-0.

Hojman, P., Dethlefsen, C., Brandt, C., Hansen, J., Pedersen, L., Pedersen, B.K., 2011. Exercise-induced muscle-derived cytokines inhibit mammary cancer cell growth. *Am. J. Physiol. Endocrinol. Metab.* 301, E504–510. https://doi.org/10.1152/ajpendo.00520.2010.

Hojman, P., Gehl, J., Christensen, J.F., Pedersen, B.K., 2018. Molecular mechanisms linking exercise to cancer prevention and treatment. *Cell Metab.* 27, 10–21. https://doi.org/10.1016/j.cmet.2017.09.015.

Huang, C.-W., Chang, Y.-H., Lee, H.-H., Wu, J.-Y., Huang, J.-X., Chung, Y.-H., Hsu, S.-T., Chow, L.-P., Wei, K.-C., Huang, F.-T., 2020. Irisin, an exercise myokine, potently suppresses tumor proliferation, invasion, and growth in glioma. *FASEB J.* 34, 9678–9693. https://doi.org/10.1096/fj.202000573RR.

Jain, R.K., 2014. Antiangiogenesis strategies revisited: From starving tumors to alleviating hypoxia. *Cancer Cell* 26, 605–622. https://doi.org/10.1016/j.ccell.2014.10.006.

Ji, L.L., 1999. Antioxidants and oxidative stress in exercise. *Proc. Soc. Exp. Biol. Med.* 222, 283–292. https://doi.org/10.1046/j.1525-1373.1999.d01-145.x.

Karatas, F., Acat, M., Sahin, S., Inci, F., Karatas, G., Neselioglu, S., Haskul, I., Erel, O., 2019. The prognostic and predictive significance of serum thiols and disulfide levels in advanced non-small cell lung cancer. *Aging Male* 23, 1–10. https://doi.org/10.1080/13685538.2018.1559805.

Kaymak, I., Williams, K.S., Cantor, J.R., Jones, R.G., 2021. Immunometabolic interplay in the tumor microenvironment. *Cancer Cell* 39, 28–37. https://doi.org/10.1016/j.ccell.2020.09.004.

Kosova, F., Temeltaş, G., Arı, Z., Lekili, M., 2014. Possible relations between oxidative damage and apoptosis in benign prostate hyperplasia and prostate cancer patients. *Tumour Biol.* 35, 4295–4299. https://doi.org/10.1007/s13277-013-1560-y.

Kulkarni, A.C., Kuppusamy, P., Parinandi, N., 2007. Oxygen, the lead actor in the pathophysiologic drama: Enactment of the trinity of normoxia, hypoxia, and hyperoxia in disease and therapy. *Antioxid. Redox Signal.* 9, 1717–1730. https://doi.org/10.1089/ars.2007.1724.

Lagadu, S., Lechevrel, M., Sichel, F., Breton, J., Pottier, D., Couderc, R., Moussa, F., Prevost, V., 2010. 8-oxo-7,8-dihydro-2'-deoxyguanosine as a biomarker of oxidative damage in oesophageal cancer patients: Lack of association with antioxidant vitamins and polymorphism of hOGG1 and GST. *J. Exp. Clin. Cancer Res. CR* 29, 157. https://doi.org/10.1186/1756-9966-29-157.

Le Guennec, D., Hatte, V., Farges, M.-C., Rougé, S., Goepp, M., Caldefie-Chezet, F., Vasson, M.-P., Rossary, A., 2020. Modulation of inter-organ signalling in obese mice by spontaneous physical activity during mammary cancer development. *Sci. Rep.* 10, 8794. https://doi.org/10.1038/s41598-020-65131-9.

Li, Y., Patel, S.P., Roszik, J., Qin, Y., 2018. Hypoxia-driven immunosuppressive metabolites in the tumor microenvironment: New approaches for combinational immunotherapy. *Front. Immunol.* 9. https://doi.org/10.3389/fimmu.2018.01591.

Louzada, R.A., Bouviere, J., Matta, L.P., Werneck-de-Castro, J.P., Dupuy, C., Carvalho, D.P., Fortunato, R.S., 2020. Redox signaling in widespread health benefits of exercise. *Antioxid. Redox Signal.* 33, 745–760. https://doi.org/10.1089/ars.2019.7949.

Luo, J., Solimini, N.L., Elledge, S.J., 2009. Principles of cancer therapy: Oncogene and non-oncogene addiction. *Cell* 136, 823–837. https://doi.org/10.1016/j.cell.2009.02.024.

Marin-Corral, J., Fontes, C.C., Pascual-Guardia, S., Sanchez, F., Olivan, M., Argilés, J.M., Busquets, S., López-Soriano, F.J., Barreiro, E., 2010. Redox balance and carbonylated proteins in limb and heart muscles of cachectic rats. *Antioxid. Redox Signal.* 12, 365–380. https://doi.org/10.1089/ars.2009.2818.

Mastrocola, R., Reffo, P., Penna, F., Tomasinelli, C.E., Boccuzzi, G., Baccino, F.M., Aragno, M., Costelli, P., 2008. Muscle wasting in diabetic and in tumor-bearing rats: Role of oxidative stress. *Free Radic. Biol. Med.* 44, 584–593. https://doi.org/10.1016/j.freeradbiomed.2007.10.047.

McClung, J.M., Judge, A.R., Powers, S.K., Yan, Z., 2010. p38 MAPK links oxidative stress to autophagy-related gene expression in cachectic muscle wasting. *Am. J. Physiol. Cell Physiol.* 298, C542–549. https://doi.org/10.1152/ajpcell.00192.2009.

Morrell, M.B.G., Alvarez-Florez, C., Zhang, A., Kleinerman, E.S., Savage, H., Marmonti, E., Park, M., Shaw, A., Schadler, K.L., 2019. Vascular modulation through exercise improves chemotherapy efficacy in Ewing sarcoma. *Pediatr. Blood Cancer* 66, e27835. https://doi.org/10.1002/pbc.27835.

Mpekris, F., Voutouri, C., Baish, J.W., Duda, D.G., Munn, L.L., Stylianopoulos, T., Jain, R.K., 2020. Combining microenvironment normalization strategies to improve cancer immunotherapy. *Proc. Natl. Acad. Sci.* 117, 3728–3737. https://doi.org/10.1073/pnas.1919764117.

Musci, R.V., Hamilton, K.L., Linden, M.A., 2019. Exercise-induced mitohormesis for the maintenance of skeletal muscle and healthspan extension. *Sports Basel Switz.* 7. https://doi.org/10.3390/sports7070170.

Novelli, G.P., Bracciotti, G., Falsini, S., 1990. Spin-trappers and vitamin E prolong endurance to muscle fatigue in mice. *Free Radic. Biol. Med.* 8, 9–13. https://doi.org/10.1016/0891-5849(90)90138-9.

Ohl, K., Tenbrock, K., 2018. Reactive oxygen species as regulators of MDSC-mediated immune suppression. *Front. Immunol.* 9. https://doi.org/10.3389/fimmu.2018.02499.

Ohtake, S., Kawahara, T., Ishiguro, Y., Takeshima, T., Kuroda, S., Izumi, K., Miyamoto, H., Uemura, H., 2018. Oxidative stress marker 8-hydroxyguanosine is more highly expressed in prostate cancer than in benign prostatic hyperplasia. *Mol. Clin. Oncol.* 9, 302–304. https://doi.org/10.3892/mco.2018.1665.

Padilha, C.S., Borges, F.H., Costa Mendes da Silva, L.E., Frajacomo, F.T.T., Jordao, A.A., Duarte, J.A., Cecchini, R., Guarnier, F.A., Deminice, R., 2017. Resistance exercise attenuates skeletal muscle oxidative stress, systemic pro-inflammatory state, and cachexia in Walker-256 tumor-bearing rats. *Appl. Physiol. Nutr. Metab.* 42, 916–923. https://doi.org/10.1139/apnm-2016-0436.

Patel, A.V., Friedenreich, C.M., Moore, S.C., Hayes, S.C., Silver, J.K., Campbell, K.L., Winters-Stone, K., Gerber, L.H., George, S.M., Fulton, J.E., Denlinger, C., Morris, G.S., Hue, T., Schmitz, K.H., Matthews, C.E., 2019. American college of sports medicine roundtable report on physical activity, sedentary behavior, and cancer prevention and control. *Med. Sci. Sports Exerc.* 51, 2391–2402. https://doi.org/10.1249/MSS.0000000000002117.

Patel, B.P., Rawal, U.M., Dave, T.K., Rawal, R.M., Shukla, S.N., Shah, P.M., Patel, P.S., 2007. Lipid peroxidation, total antioxidant status, and total thiol levels predict overall survival in patients with oral squamous cell carcinoma. *Integr. Cancer Ther.* 6, 365–372. https://doi.org/10.1177/1534735407309760.

Pedersen, K.S., Gatto, F., Zerahn, B., Nielsen, J., Pedersen, B.K., Hojman, P., Gehl, J., 2020. Exercise-mediated lowering of glutamine availability suppresses tumor growth and attenuates muscle wasting. *iScience* 23. https://doi.org/10.1016/j.isci.2020.100978.

Pedersen, L., Christensen, J.F., Hojman, P., 2015. Effects of exercise on tumor physiology and metabolism. *Cancer J. Sudbury Mass* 21, 111–116. https://doi.org/10.1097/PPO.0000000000000096.

Penna, F., Ballarò, R., Costelli, P., 2020. The redox balance: A target for interventions against muscle wasting in cancer cachexia? *Antioxid. Redox Signal.* 33, 542–558. https://doi.org/10.1089/ars.2020.8041.

Perillo, B., Di Donato, M., Pezone, A., Di Zazzo, E., Giovannelli, P., Galasso, G., Castoria, G., Migliaccio, A., 2020. ROS in cancer therapy: The bright side of the moon. *Exp. Mol. Med.* 52, 192–203. https://doi.org/10.1038/s12276-020-0384-2.

Postovit, L., Widmann, C., Huang, P., Gibson, S.B., 2018. Harnessing oxidative stress as an innovative target for cancer therapy [WWW Document]. *Oxid. Med. Cell. Longev.* https://doi.org/10.1155/2018/6135739.

Pouysségur, J., Mechta-Grigoriou, F., 2006. Redox regulation of the hypoxia-inducible factor. *Biol. Chem.* 387, 1337–1346. https://doi.org/10.1515/BC.2006.167.

Powers, S.K., Deminice, R., Ozdemir, M., Yoshihara, T., Bomkamp, M.P., Hyatt, H., 2020. Exercise-induced oxidative stress: Friend or foe? *J. Sport Health Sci.* 9, 415–425. https://doi.org/10.1016/j.jshs.2020.04.001.

Powers, S.K., Duarte, J., Kavazis, A.N., Talbert, E.E., 2010. Reactive oxygen species are signalling molecules for skeletal muscle adaptation. *Exp. Physiol.* 95, 1–9. https://doi.org/10.1113/expphysiol.2009.050526.

Powers, S.K., Ji, L.L., Kavazis, A.N., Jackson, M.J., 2011. Reactive oxygen species: impact on skeletal muscle. *Compr. Physiol.* 1, 941–969. https://doi.org/10.1002/cphy.c100054.

Powers, S.K., Radak, Z., Ji, L.L., 2016. Exercise-induced oxidative stress: Past, present and future. *J. Physiol.* 594, 5081–5092. https://doi.org/10.1113/JP270646.

Puig-Vilanova, E., Rodriguez, D.A., Lloreta, J., Ausin, P., Pascual-Guardia, S., Broquetas, J., Roca, J., Gea, J., Barreiro, E., 2015. Oxidative stress, redox signaling pathways, and autophagy in cachectic muscles of male patients with advanced COPD and lung cancer. *Free Radic. Biol. Med.* 79, 91–108. https://doi.org/10.1016/j.freeradbiomed.2014.11.006.

Qing, X., Shi, D., Lv, X., Wang, B., Chen, S., Shao, Z., 2019. Prognostic significance of 8-hydroxy-2′-deoxyguanosine in solid tumors: A meta-analysis. *BMC Cancer* 19, 997. https://doi.org/10.1186/s12885-019-6189-9.

Radak, Z., Taylor, A.W., Ohno, H., Goto, S., 2001. Adaptation to exercise-induced oxidative stress: from muscle to brain. *Exerc. Immunol. Rev.* 7, 90–107.

Ramamoorthy, S., Donohue, M., Buck, M., 2009. Decreased Jun-D and myogenin expression in muscle wasting of human cachexia. *Am. J. Physiol. Endocrinol. Metab.* 297, E392–E401. https://doi.org/10.1152/ajpendo.90529.2008.

Repka, C.P., Hayward, R., 2016. Oxidative stress and fitness changes in cancer patients after exercise training. *Med. Sci. Sports Exerc.* 48, 607–614. https://doi.org/10.1249/MSS.0000000000000821.

Reuter, S., Gupta, S.C., Chaturvedi, M.M., Aggarwal, B.B., 2010. Oxidative stress, inflammation, and cancer: How are they linked? *Free Radic. Biol. Med.* 49, 1603–1616. https://doi.org/10.1016/j.freeradbiomed.2010.09.006.

Rupert, J.E., Bonetto, A., Narasimhan, A., Liu, Y., O'Connell, T.M., Koniaris, L.G., Zimmers, T.A., 2020. IL-6 trans-signaling and crosstalk among tumor, muscle and fat mediate pancreatic cancer cachexia. bioRxiv 2020.09.16.300798. https://doi.org/10.1101/2020.09.16.300798.

Ryan, A.M., Power, D.G., Daly, L., Cushen, S.J., Ní Bhuachalla, Ē., Prado, C.M., 2016. Cancer-associated malnutrition, cachexia and sarcopenia: the skeleton in the hospital closet 40 years later. *Proc. Nutr. Soc.* 75, 199–211. https://doi.org/10.1017/S002966511500419X.

Sabharwal, S.S., Schumacker, P.T., 2014. Mitochondrial ROS in cancer: Initiators, amplifiers or an Achilles' heel? *Nat. Rev. Cancer* 14, 709–721. https://doi.org/10.1038/nrc3803.

Salazar-Degracia, A., Busquets, S., Argilés, J.M., López-Soriano, F.J., Barreiro, E., 2017. Formoterol attenuates increased oxidative stress and myosin protein loss in respiratory and limb muscles of cancer cachectic rats. *PeerJ* 5, e4109. https://doi.org/10.7717/peerj.4109.

Sallam, N., Laher, I., 2015. Exercise modulates oxidative stress and inflammation in aging and cardiovascular diseases. *Oxid. Med. Cell. Longev.* 2016, e7239639. https://doi.org/10.1155/2016/7239639.

Sato, T., Takeda, H., Otake, S., Yokozawa, J., Nishise, S., Fujishima, S., Orii, T., Fukui, T., Takano, J., Sasaki, Y., Nagino, K., Iwano, D., Yaoita, T.,

Kawata, S., 2010. Increased plasma levels of 8-hydroxydeoxyguanosine are associated with development of colorectal tumors. *J. Clin. Biochem. Nutr.* 47, 59–63. https://doi.org/10.3164/jcbn.10-12.

Schaaf, M.B., Garg, A.D., Agostinis, P., 2018. Defining the role of the tumor vasculature in antitumor immunity and immunotherapy. *Cell Death Dis.* 9. https://doi.org/10.1038/s41419-017-0061-0.

Schieber, M., Chandel, N.S., 2014. ROS function in redox signaling and oxidative stress. *Curr. Biol. CB* 24, R453–R462. https://doi.org/10.1016/j.cub.2014.03.034.

Schmitz, K.H., Courneya, K.S., Matthews, C., Demark-Wahnefried, W., Galvão, D.A., Pinto, B.M., Irwin, M.L., Wolin, K.Y., Segal, R.J., Lucia, A., Schneider, C.M., Von Gruenigen, V.E., Schwartz, A.L., 2010. American college of sports medicine roundtable on exercise guidelines for cancer survivors. *Med. Sci. Sports Exerc.* 42, 1409–1426.

Sheridan, J., Wang, L.-M., Tosetto, M., Sheahan, K., Hyland, J., Fennelly, D., O'Donoghue, D., Mulcahy, H., O'Sullivan, J., 2009. Nuclear oxidative damage correlates with poor survival in colorectal cancer. *Br. J. Cancer* 100, 381–388. https://doi.org/10.1038/sj.bjc.6604821.

Shiroyama, T., Nagatomo, I., Koyama, S., Hirata, H., Nishida, S., Miyake, K., Fukushima, K., Shirai, Y., Mitsui, Y., Takata, S., Masuhiro, K., Yaga, M., Iwahori, K., Takeda, Y., Kida, H., Kumanogoh, A., 2019. Impact of sarcopenia in patients with advanced non–small cell lung cancer treated with PD-1 inhibitors: A preliminary retrospective study. *Sci. Rep.* 9, 1–7. https://doi.org/10.1038/s41598-019-39120-6.

Sies, H., Jones, D.P., 2007. Encyclopedia of stress. *Oxidative Stress* 3, 45–8.

Singh, A.K., Pandey, P., Tewari, M., Pandey, H.P., Gambhir, I.S., Shukla, H.S., 2016. Free radicals hasten head and neck cancer risk: A study of total oxidant, total antioxidant, DNA damage, and histological grade. *J. Postgrad. Med.* 62, 96–101. https://doi.org/10.4103/0022-3859.180555.

Smuder, A.J., Kavazis, A.N., Hudson, M.B., Nelson, W.B., Powers, S.K., 2010. Oxidation enhances myofibrillar protein degradation via calpain and caspase-3. *Free Radic. Biol. Med.* 49, 1152–1160. https://doi.org/10.1016/j.freeradbiomed.2010.06.025.

Smuder, A.J., Sollanek, K.J., Nelson, W.B., Min, K., Talbert, E.E., Kavazis, A.N., Hudson, M.B., Sandri, M., Szeto, H.H., Powers, S.K., 2018. Crosstalk between autophagy and oxidative stress regulates proteolysis in the diaphragm during mechanical ventilation. *Free Radic. Biol. Med.* 115, 179–190. https://doi.org/10.1016/j.freeradbiomed.2017.11.025.

Sosa, V., Moliné, T., Somoza, R., Paciucci, R., Kondoh, H., LLeonart, M.E., 2013. Oxidative stress and cancer: An overview. *Ageing Res. Rev.* 12, 376–390. https://doi.org/10.1016/j.arr.2012.10.004.

Sullivan-Gunn, M.J., Campbell-O'Sullivan, S.P., Tisdale, M.J., Lewandowski, P.A., 2011. Decreased NADPH oxidase expression and antioxidant activity in cachectic skeletal muscle. *J. Cachexia Sarcopenia Muscle* 2, 181–188. https://doi.org/10.1007/s13539-011-0037-3.

Tan, P.L., Shavlakadze, T., Grounds, M.D., Arthur, P.G., 2015. Differential thiol oxidation of the signaling proteins Akt, PTEN or PP2A determines whether Akt phosphorylation is enhanced or inhibited by oxidative stress in C2C12 myotubes derived from skeletal muscle. *Int. J. Biochem. Cell Biol.* 62, 72–79. https://doi.org/10.1016/j.biocel.2015.02.015.

Tomasello, B., Malfa, G.A., Strazzanti, A., Gangi, S., Di Giacomo, C., Basile, F., Renis, M., 2017. Effects of physical activity on systemic oxidative/DNA status in breast cancer survivors. *Oncol. Lett.* 13, 441–448. https://doi.org/10.3892/ol.2016.5449.

Tsukagoshi, M., Yokobori, T., Yajima, T., Maeno, T., Shimizu, K., Mogi, A., Araki, K., Harimoto, N., Shirabe, K., Kaira, K., 2020. Skeletal muscle mass predicts the outcome of nivolumab treatment for non-small cell lung cancer. *Medicine (Baltimore)* 99, e19059. https://doi.org/10.1097/MD.0000000000019059.

Weinberg, F., Ramnath, N., Nagrath, D., 2019. Reactive oxygen species in the tumor microenvironment: An overview. *Cancers* 11. https://doi.org/10.3390/cancers11081191.

Wennerberg, E., Lhuillier, C., Rybstein, M.D., Dannenberg, K., Rudqvist, N.-P., Koelwyn, G.J., Jones, L.W., Demaria, S., 2020. Exercise reduces immune suppression and breast cancer progression in a preclinical model. *Oncotarget* 11, 452–461. https://doi.org/10.18632/oncotarget.27464.

Yang, S.Y., Hoy, M., Fuller, B., Sales, K.M., Seifalian, A.M., Winslet, M.C., 2010. Pretreatment with insulin-like growth factor I protects skeletal muscle cells against oxidative damage via PI3K/Akt and ERK1/2 MAPK pathways. *Lab. Investig. J. Tech. Methods Pathol.* 90, 391–401. https://doi.org/10.1038/labinvest.2009.139.

Yen, C.-J., Hung, C.-H., Tsai, W.-M., Cheng, H.-C., Yang, H.-L., Lu, Y.-J., Tsai, K.-L., 2020. Effect of exercise training on exercise tolerance and level of oxidative stress for head and neck cancer patients following chemotherapy. *Front. Oncol.* 10, 1536. https://doi.org/10.3389/fonc.2020.01536.

Zhou, L., Zhang, Z., Huang, Z., Nice, E., Zou, B., Huang, C., 2020. Revisiting cancer hallmarks: Insights from the interplay between oxidative stress and non-coding RNAs. *Mol. Biomed.* 1, 4. https://doi.org/10.1186/s43556-020-00004-1.

CHAPTER FIFTEEN

Oxidative Stress and Exercise Tolerance in Cystic Fibrosis

Cassandra C. Derella, Adeola A. Sanni and Ryan A. Harris

CONTENTS

Cystic Fibrosis Overview / 183
Inflammation and Oxidative Stress in CF / 183
Prognostic Values of Exercise Testing CF / 184
Benefits of Exercise in CF / 184
Mechanistic Insight into Exercise Intolerance in CF / 185
Pulmonary Function / 185
Cardiovascular Function / 186
Skeletal Muscle Function / 188
Conclusions / 188
References / 189

CYSTIC FIBROSIS OVERVIEW

Cystic fibrosis (CF) is caused by a mutation in the cystic fibrosis transmembrane conductance regulator (CFTR) gene, which alters the regulation of chloride entering and leaving the cells (Radlovic, 2012). The CFTR gene mutation alters the production, expression, and activity of the CFTR chloride channel and although millions of people are carriers of the CFTR mutation (National Heart, 2018), only ~35,000 people in the United States (Centers for Disease Control and Prevention, 2020) and ~100,000 people worldwide currently have overt CF. Cystic fibrosis was once regarded as a lung disease due to the pulmonary manifestations and the fact that lung infections are the primary cause of morbidity and mortality. However, given that CFTR is also ubiquitously expressed throughout the body in organs such as the intestines, kidneys, pancreases, and liver, CF is now recognized as a multisystemic condition that requires a multidisciplinary approach to the presenting CF-related diabetes, pancreatic insufficiency, and GI malabsorption (Xue et al., 2016). In addition, CFTR is expressed on other cell types including endothelial cells, skeletal muscle, macrophages, lymphocytes, and mast cells (Xue et al., 2016).

Inflammation and Oxidative Stress in CF

Due to the multiple systemic consequences of CF, systemic inflammation and oxidative stress are common phenotypes in people with CF. There is convincing evidence that people with CF are in a chronic state of oxidative stress (Galli et al., 2012; Ntimbane et al., 2009). Although the exact mechanisms of oxidative stress in CF have yet to be fully elucidated, systemic inflammation can directly contribute to the redox imbalance by increasing inflammasomes and free radical production in tissues (Galli et al., 2012). In support, the frequent acute pulmonary exacerbations that people with CF experience not only increase the innate immune response but also facilitate an increase in both lung and systemic oxidative stress (McGrath et al., 1999). In addition,

DOI: 10.1201/9781003051619-15

the presence of a defective CFTR appears to directly produce and contribute to a redox imbalance in epithelial cells and in extracellular fluids (Galli et al., 2012). Indeed, as CF progresses over time and disease severity worsens, the heightened chronic inflammatory condition further exacerbates the state of oxidative stress. Moreover, people with CF experience intestinal malabsorption of endogenous fat-soluble vitamins and antioxidants, a consequence of CF that limits the defense when ingesting exogenous antioxidants (Childers et al., 2007). For this reason, individuals with CF routinely supplement with antioxidants to help increase the absorption of fats and prevent malnutrition; the latter is a prognostic indicator (Wood et al., 2003; Stallings et al., 2008). Nonetheless, an imbalance of pro- and antioxidants is still present in CF, and the state of chronic oxidative stress likely contributes to overall lower wellness, poor quality of life, and reductions in exercise tolerance.

To improve the quality of life and increase longevity in CF, modulator therapies have been introduced to help synthesize, translocate, and correct the dysfunctional CFTR protein. In addition, using exercise as an adjunct therapy has gained traction for improving aerobic capacity, reducing inflammation, promoting lung health, acting as an airway clearance technique (ACT), and contributing to increased survival among people with CF (Nixon et al., 1992; Orenstein et al., 1989). Despite the overall benefits of both acute and chronic exercise, people with CF may not be adhering to the exercise recommendations and guidance that can improve overall clinical outcomes.

PROGNOSTIC VALUES OF EXERCISE TESTING CF

In general, cardiopulmonary exercise testing (CPET) is a valuable clinical tool that is utilized to uncover pathology for dyspnea on exertion in vulnerable patient populations (i.e., coronary artery disease, chronic lung disease). Exercise testing provides an integrative physiological response allowing researchers and clinicians to evaluate how the pulmonary, cardiovascular, neurological, and skeletal muscle systems interact and work together during periods of increased physiological demand.

Exercise testing has been evaluated in people with CF for decades and maximal oxygen uptake ($\dot{V}O_2$), the primary outcome from CPET, was identified to positively correlate with functional status and overall well-being (Orenstein et al., 1989). However, it was not until 1992 that the true prognostic value of exercise testing in CF was uncovered. After conducting routine exercise tests for several years of follow-up in 100 people with CF, exercise capacity was found to predict mortality in people with CF, independent of their biological age or lung function (Nixon et al., 1992). Specifically, people with CF that exhibited a higher $\dot{V}O_2$ peak had a longer life expectancy, whereas a greater severity of disease and mortality was associated with those with a lower exercise capacity (Nixon et al., 1992). Since then, reports have demonstrated that lower exercise capacity is associated with a steeper decline in pulmonary function over time, reduced quality of life, a greater prevalence of pulmonary exacerbations, and higher rates of mortality (van de Weert-van Leeuwen et al., 2014). The high clinical utility and prognostic value of conducting maximal exercise tests in CF have led to the strong recommendation by the European Respiratory Society and the Cystic Fibrosis Foundation that regular CPET be conducted on a cycle ergometer using the Godfrey protocol to identify potential mechanisms that limit exercise capacity in CF (Hebestreit et al., 2015). Despite the fact that exercise intolerance is a hallmark of disease progression in CF, exercise testing is not routinely performed in the CF clinic, and exercise tolerance is not a primary clinical outcome that is the target of therapy in CF.

BENEFITS OF EXERCISE IN CF

A decline in exercise capacity in CF is related to reduced lung function, increased vascular dysfunction, and increased mortality. Accordingly, exercise has always been recommended for people with CF, and there is even debate as to whether or not it can be used as a substitute technique for airway clearance. In a research survey of 488 people with CF and health care professionals, 54% of the people with CF reported they had already incorporated exercise into their ACT, 24% use only exercise as their ACT, and 48% were willing to replace their ACT with exercise alone. Among the health care professionals, 93% already recommend incorporating exercise into ACT and 73% support a trial to replace ACT with exercise (Rowbotham et al., 2020). In another survey of 692 respondents with CF, 43% suggest that exercise could be used as a substitute for airway clearance, while 43% report that they have been using exercise alone as an ACT (Ward et al., 2019). Nonetheless, treadmill exercise was as effective at increasing peak expiratory flow and improving mucous clearance compared to traditional airway clearance devices (Dwyer et al., 2017), suggesting that exercise may be used as a substitute and not just an adjuvant therapy for airway clearance in CF.

Although reports suggest that oxidative stress is increased during submaximal exercise (Tucker et al., 2018), moderate exercise can also facilitate the production of endogenous antioxidants and help reduce cell and systemic oxidative stress (Vassalle et al., 2015). Interestingly, a single bout of maximal exercise on a cycle ergometer can also acutely improve forced vital capacity (FVC), forced expiratory volume in 1 second (FEV1), and lung clearance index (LCI) in CF (Tucker et al., 2017). In addition to the improvements following a single exercise session, the chronic response to exercise in CF appears to be more beneficial. Following 2-weeks of exercise therapy, which included two sessions of progressive cycling and one session of bronchial hygiene treatment per day, an improvement in pulmonary function was observed in people with exacerbated CF (Cerny, 1989). In addition, 6-months of partially supervised endurance and strength training performed 3 days/week for 30 minutes significantly improves FEV1 (Kriemler et al., 2013). An improvement in aerobic capacity, anaerobic capacity, and quality of life have been reported after 12-weeks of resistance training in children with CF (Klijn et al., 2004). In addition, an improvement in muscle strength, muscle endurance, and speed was observed following aerobic exercise and chest physiotherapy three times a week for 6 weeks in children with CF (Elbasan et al., 2012). Exercise not only slows down the rate of decline in pulmonary function in people with CF, but it can also improve physical fitness and quality of life in this patient population (Schneiderman-Walker et al., 2000). Taken together, engaging in exercise can lead to improvements in overall wellness and clinical outcomes and supports the use of exercise as adjunct therapy in the treatment of CF.

MECHANISTIC INSIGHT INTO EXERCISE INTOLERANCE IN CF

In theory, four domains can limit the physiological uptake of oxygen and affect exercise capacity: (1) oxygen (O_2) content in the air; (2) pulmonary function – the inability to transport O_2 from the air into circulation; (3) cardiovascular or circulatory function – ability to transport O_2 throughout the body and clear out carbon dioxide (CO_2); and (4) skeletal muscle function – the ability of the muscle to consume O_2 and produce energy/work (Urquhart and Saynor, 2018). Interestingly, CF can impact each of these aforementioned physiological functions (Lands and Hebestreit, 2015) which in turn can also be a source of ROS and play a role in exercise intolerance (see Figure 15.1). However, the consequences of systemic oxidative stress on exercise intolerance in CF have yet to be fully elucidated (Ntimbane et al., 2009; Wood et al., 2003). Under the assumption that O_2 content in the air is not a limiting factor, the role of each of the remaining biological systems and how oxidative stress may contribute to exercise intolerance in CF is described in the following sections.

Pulmonary Function

The lungs are structured to handle the exchange of respiratory gases and can adapt quickly to the sudden increase in O_2 demand with exercise (Weibel, 2009). With each breath, air enters the body and travels down the pulmonary tree to the alveoli. Alongside this path is a vascular network that ultimately arrives at a capillary bed filled with mixed venous blood surrounding each alveolus – this is where the exchange of respiratory gases (O_2 and CO_2) occurs (Petersson and Glenny, 2014). Increased production of reactive oxygen species (e.g., superoxide anion, hydrogen peroxide, hydroxyl free radical) can contribute to structural damage and the cultivation of infections (McGrath et al., 1999). In addition, ventilatory dysfunction as a result of structural lung damage, weaker respiratory muscles, dead space ventilation, and ventilatory reserve may contribute to the decline in exercise capacity in people with severe CF disease (Shei et al., 2019). To compensate for the increase in dead space and airway obstruction due to thick, tenacious, and sticky secretions seen in CF, an increase in minute ventilation during exercise often occurs (Cerny et al., 1982). Interestingly, a person that is free of lung disease rarely increases their peak minute ventilation above 65% of maximal voluntary ventilation. Although people with CF tend to increase their peak minute ventilation during exercise (Cerny et al., 1982), even those with

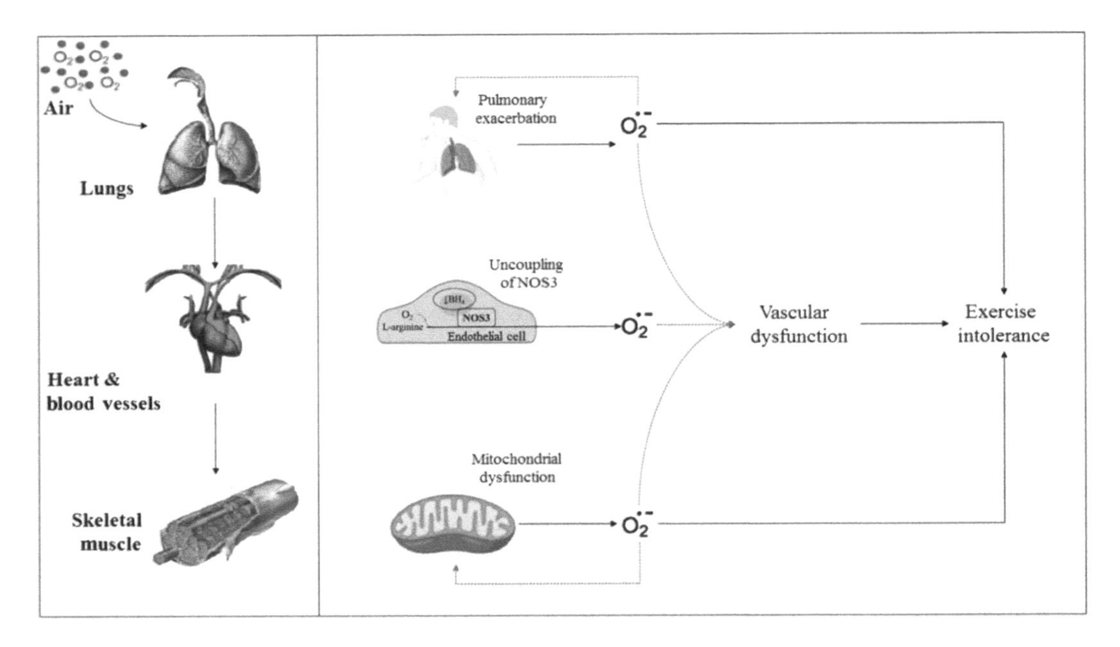

Figure 15.1 The four factors that can contribute to exercise intolerance and the primary sources of superoxide free radical production in CF that can contribute to exercise intolerance in CF.

severe CF lung disease are often able to maintain normal ventilation during maximal exercise (Cerny et al., 1982). Moreover, while the higher minute ventilation may help maintain adequate ventilation-perfusion (V/Q) and gas exchange during exercise, V/Q mismatching can still occur in people with severe lung disease (Godfrey and Mearns, 1971). Consequently, a mismatch in V/Q can result in oxyhemoglobin desaturation; however, oxyhemoglobin desaturation below 91% is rarely observed in people with an FEV1 ≥ 50% predicted (Nixon, 1996). The observations that CF patients with moderate-severe CF lung disease can maintain normal minute ventilation and proper oxyhemoglobin saturation during maximal exercise suggest that the lungs and pulmonary function may not be the limiting factor contributing to exercise intolerance in CF. Factors that are suspected to have significant effects on exercise tolerance in CF are discussed below.

Cardiovascular Function

Pulmonary vascular diseases have been identified in people with CF and often develop into frank pulmonary artery hypertension (PAH) over time. In fact, pulmonary artery enlargement, a consequence of PAH, occurs in approximately 50% of adults with moderate CF lung disease (Wells et al., 2016). Moreover, severe, chronic PAH can lead to Cor pulmonale, hypertrophy and dysfunction of the right ventricle, and a devastating cardiac consequence of CF (Moss, 1982). Indeed, PAH and changes in the physical structure of the heart can independently contribute to exercise intolerance in CF. Consequently, improvements in pulmonary vascular resistance using phosphodiesterase type 5 inhibition (PDE5) can not only improve PAH but also result in an increased exercise capacity (Montgomery et al., 2006). In addition to Cor pulmonale, left ventricular hypertrophy and systolic abnormalities in contractility identified by a decreased left ventricular strain pattern have been observed in CF (Pallin et al., 2018). Severe remodeling of the left ventricle can contribute to a reduction in stroke volume, which by itself can contribute to a lower exercise capacity (Pianosi and Pelech, 1996). These findings suggest that abnormal pulmonary vascular responses during exercise and cardiac remodeling can contribute to exercise intolerance in CF and reducing pulmonary vascular resistance can improve exercise intolerance in CF.

Vascular dysfunction is not limited to the pulmonary circulation, and people with CF exhibit systemic vascular endothelial dysfunction (Poore et al., 2012; Rodriguez-Miguelez et al., 2016). There is evidence of a reduced flow-mediated dilation (FMD), a reliable, noninvasive assessment of systemic conduit vascular endothelial function (Anderson et al., 1995) in young people with CF that have preserved spirometry function (Poore et al., 2012). Additionally, assessment of endothelial function was positively associated with pulmonary function and exercise parameters (peak ventilation, peak workload, and a modest relationship with $\dot{V}O_2$ peak) (Poore et al., 2012). Similarly, microvascular dysfunction is present in people with CF (Rodriguez-Miguelez et al., 2016). The microcirculation is critical for the exchange of O_2 to the exercising muscles and oxygen uptake kinetics are important to consider when investigating exercise intolerance. Two major hypotheses have been proposed regarding the rate-limiting control mechanisms of oxygen uptake kinetics during exercise: (1) the capacity of oxygen delivery to active muscle and (2) the ability of the oxygen to be utilized by the exercising muscle (Xu and Rhodes, 1999). Both $\dot{V}O_2$ response time and functional $\dot{V}O_2$ gain are measured during CPET and can be used to evaluate O_2 utilization by the muscle and O_2 delivery, respectively. People with CF experience a delayed $\dot{V}O_2$ response time at the onset of exercise and a lower functional $\dot{V}O_2$ gain (Fielding et al., 2015), indicating that there are impairments in both O_2 delivery (blood flow) and oxygen utilization at the skeletal muscle (uptake). In addition, impaired blood flow regulation during submaximal exercise has been observed in young people with CF and was accompanied by an increase in ROS production during exercise (Tucker et al., 2018). Specifically, alkoxyl free radical (α-phenyl-tert-butyl nitrone, PBN) was elevated in CF compared to healthy controls, and a negative relationship was observed between the change in retrograde velocity and oxidative stress (PBN) during exercise (Tucker et al., 2018). The increase in free radicals in people with CF is suggested to be due to both, the decrease in endogenous antioxidant buffering capacity and the excessive overload of ROS production during exercise.

Taken together, these studies support the notion that systemic conduit- and microvascular endothelial dysfunction are present in CF. Reductions in systemic vascular function can likely reduce O_2 delivery, and impairments in O_2 exchange can contribute to exercise intolerance. The negative association between changes in oxidative stress and brachial artery retrograde velocity during cycling suggests that exercise-induced ROS production may contribute to impaired blood regulation in CF during exercise. Treatments with pharmacological interventions to improve blood flow, such as sildenafil, have improved exercise capacity in CF (Rodriguez-Miguelez et al., 2019). These observations not only emphasize but also support the notion that vascular dysfunction and impaired oxygen uptake kinetics are large contributors to exercise intolerance in CF.

Although several mechanisms contribute to endothelial dysfunction, oxidative stress plays a critical role (Griendling and FitzGerald, 2003) in reducing endothelial-dependent vasodilation through uncoupling of nitric oxide synthase (NOS3) and the subsequent reduction in NO bioavailability (Roe and Ren, 2012). In fact, a reduction in tetrahydrobiopterin (BH4), a critical cofactor for endothelial NOS function (Vásquez-Vivar et al., 2001; Alp and Channon, 2004), can contribute to the uncoupling of endothelial NOS and production of superoxide rather than NO, thereby reducing NO bioavailability and increasing oxidative stress in the cell (Vásquez-Vivar et al., 2001). Accordingly, reducing systemic oxidative stress in CF may, in turn, increase NO bioavailability and improve endothelial function. Indeed, a significant improvement in endothelial function, evidenced by a 1.9% increase in FMD, was observed following antioxidant treatment (1,000 mg vitamin C, 600 IU vitamin E, and 600 mg α-lipoic acid); this was accompanied by a reduction in oxidative stress (Tucker et al., 2019). Ingestion of the antioxidant cocktail also reduced the circulating concentration of lycopene and lipid hydroperoxides, while preventing a reduction in α-tocopherol. In addition, treatment with oral BH4 enhances vascular endothelial function in CF (Jeong et al., 2019). The improvement in FMD was accompanied by a significant reduction in stimulated superoxide production in endothelial cells when cells were preincubated with patient plasma following treatment with BH4 compared to cells with pretreatment patient plasma. The reduction in stimulated superoxide in the endothelial cell was also accompanied by an increase in endothelial cell NO production.

Overall, the mounting evidence indicates that endothelial dysfunction is present in CF and can be attributed, at least in part, to elevated systemic oxidative stress. Reducing oxidative stress and improving the coupling of NOS3 can improve vascular endothelial function and blood flow regulation during exercise.

Skeletal Muscle Function

Skeletal muscle is a large metabolic tissue responsible for ATP metabolism which is a major source of energy production. Skeletal muscle mitochondria are a predominant generator of superoxide and hydrogen peroxide (Powers and Jackson, 2008; Dionisi et al., 1975). Elevated ROS disrupts the production and flow of the sarcoplasmic reticulum ATPase, which is essential in the production of ATP, and is often accompanied by mitochondria dysfunction in people with CF (Erickson et al., 2015; Shapiro, 1989; Liu et al., 2021). Elevated ROS damages cell DNA, protein, and lipids (Davies et al., 1982), thereby altering the structure and function of a cell. It is important to note, however, that a low concentration of ROS is required in skeletal muscle to regulate normal force production and cell signaling pathways (Powers and Jackson, 2008). Interestingly, cells from people with CF not only consume more oxygen at rest, but oxygen consumption is also greater during exercise (Shapiro, 1989) resulting in an increase in skeletal muscle ROS production. Nonetheless, the major consequences of elevated ROS in the skeletal muscle are increased muscle fatigue/weakness, decreased muscle force production, and impaired muscle strength, all of which are prevalent in people with CF (Hussey et al., 2002; Arikan et al., 2015; Elkin et al., 2000). There is a 50%–100% increase in ROS production during aerobic exercise due to the increase in O_2 consumption for energy production (Powers and Jackson, 2008), and the ability to maintain submaximal exercise is dependent on the oxidative capacity of the mitochondria (Willingham and McCully, 2017). Reducing ROS using antioxidants may improve skeletal muscle fatigue and muscle damage after submaximal muscle contraction (Jäger et al., 2019; Mason et al., 2020). Despite the evidence supporting a decline in skeletal muscle function in CF, the mechanisms have yet to be elucidated. Although the defect in skeletal muscle CFTR has been suggested to be an intrinsic factor (Divangahi et al., 2009), it is clear that systemic oxidative stress can play a role and may be exacerbated by accompanied muscle disuse, hypoxia, and corticosteroid use.

Although exercise increases ROS production, evidence suggests that regular exercise upregulates endogenous antioxidants (Vassalle et al., 2015). The muscle fiber contains both enzymatic and nonenzymatic antioxidants which are complex units that help regulate ROS (Powers and Jackson, 2008). Antioxidants such as superoxide dismutase, glutathione, uric acid, and bilirubin can be endogenously synthesized, while vitamin E, ascorbic acid, and carotenoids may be exogenously supplemented through dietary means (Powers and Jackson, 2008). Although supplementation with a high dose of Vitamin E (1000 IU/d) results in a modest change in exercise-induced oxidative stress (Sacheck et al., 2003), there is an urgent need to definitively ascertain if continued antioxidant supplementation is required or indeed detrimental in CF patients due to their elevated ROS status (Pingitore et al., 2015).

The findings discussed in this section provide evidence that skeletal muscle oxidative stress may contribute to exercise intolerance in CF. Mitochondria dysfunction results in a decline in energy production which results in reduced exercise capacity; skeletal muscle ROS is present and reducing ROS through antioxidant intervention may improve exercise tolerance in CF.

CONCLUSIONS

Oxidative stress is a key contributor to the pathogenesis of several diseases and is a common phenotype in people with CF. Although a moderate level of ROS is essential in keeping the optimal function and energy production in the cell, an excess accumulation can cause damage to the cells and result in the loss of pulmonary, cardiovascular, and skeletal muscle function. Exercise intolerance is also a distinguishing characteristic in CF and routine exercise testing has great clinical prognostic value. Although pulmonary dysfunction may not play a key role, the mechanisms that contribute to exercise intolerance in CF, to date, have yet to be fully elucidated. Indeed, there is evidence to support the negative role that oxidative stress has on cardiovascular and skeletal muscle function. Accordingly, elevated systemic oxidative stress may likely impact the function of many biological systems, contributing to exercise intolerance in people with CF.

REFERENCES

Alp, N. J. & Channon, K. M. 2004. Regulation of endothelial nitric oxide synthase by tetrahydrobiopterin in vascular disease. *Arterioscler Thromb Vasc Biol*, 24, 413–20.

Anderson, T. J., Uehata, A., Gerhard, M. D., Meredith, I. T., Knab, S., Delagrange, D., Lieberman, E. H., Ganz, P., Creager, M. A., Yeung, A. C., et al. 1995. Close relation of endothelial function in the human coronary and peripheral circulations. *J Am Coll Cardiol*, 26, 1235–41.

Arikan, H., Yatar, İ., Calik-Kutukcu, E., Aribas, Z., Saglam, M., Vardar-Yagli, N., Savci, S., Inal-Ince, D., Ozcelik, U. & Kiper, N. 2015. A comparison of respiratory and peripheral muscle strength, functional exercise capacity, activities of daily living and physical fitness in patients with cystic fibrosis and healthy subjects. *Res Dev Disabil*, 45–46, 147–56.

Centers for Disease Control and Prevention. 2020. Cystic Fibrosis [Online]. Available: https://www.cdc.gov/genomics/disease/cystic_fibrosis.htm [Accessed 12/29/2020 2020].

Cerny, F. J. 1989. Relative effects of bronchial drainage and exercise for in-hospital care of patients with cystic fibrosis. *Phys Ther*, 69, 633–9.

Cerny, F. J., Pullano, T. P. & Cropp, G. J. 1982. Cardiorespiratory adaptations to exercise in cystic fibrosis. *Am Rev Respir Dis*, 126, 217–20.

Childers, M., Eckel, G., Himmel, A. & Caldwell, J. 2007. A new model of cystic fibrosis pathology: Lack of transport of glutathione and its thiocyanate conjugates. *Med Hypotheses*, 68, 101–12.

Cystic Fibrosis News Today. Cystic Fibrosis Statistics [Online]. Available: https://cysticfibrosisnewstoday.com/cystic-fibrosis-statistics/ [Accessed 2/13/2021].

Davies, K. J., Quintanilha, A. T., Brooks, G. A. & Packer, L. 1982. Free radicals and tissue damage produced by exercise. *Biochem Biophys Res Commun*, 107, 1198–205.

Dionisi, O., Galeotti, T., Terranova, T. & Azzi, A. 1975. Superoxide radicals and hydrogen peroxide formation in mitochondria from normal and neoplastic tissues. *Biochim Biophys Acta (BBA) - Enzymol*, 403, 292–300.

Divangahi, M., Balghi, H., Danialou, G., Comtois, A. S., Demoule, A., Ernest, S., Haston, C., Robert, R., Hanrahan, J. W., Radzioch, D. & Petrof, B. J. 2009. Lack of Cftr in skeletal muscle predisposes to muscle wasting and diaphragm muscle pump failure in cystic fibrosis mice. *PloS Genet*, 5, e1000586.

Dwyer, T. J., Zainuldin, R., Daviskas, E., Bye, P. T. P. & Alison, J. A. 2017. Effects of treadmill exercise versus Flutter® on respiratory flow and sputum properties in adults with cystic fibrosis: A randomised, controlled, cross-over trial. *BMC Pulm Med*, 17, 14.

Elbasan, B., Tunali, N., Duzgun, I. & Ozcelik, U. 2012. Effects of chest physiotherapy and aerobic exercise training on physical fitness in young children with cystic fibrosis. *Ital J Pediatr*, 38, 2.

Elkin, S. L., Williams, L., Moore, M., Hodson, M. E. & Rutherford, O. M. 2000. Relationship of skeletal muscle mass, muscle strength and bone mineral density in adults with cystic fibrosis. *Clin Sci (Lond)*, 99, 309–14.

Erickson, M. L., Seigler, N., Mckie, K. T., Mccully, K. K. & Harris, R. A. 2015. Skeletal muscle oxidative capacity in patients with cystic fibrosis. *Exp Physiol*, 100, 545–52.

Fielding, J., Brantley, L., Seigler, N., Mckie, K. T., Davison, G. W. & Harris, R. A. 2015. Oxygen uptake kinetics and exercise capacity in children with cystic fibrosis. *Pediatr Pulmonol*, 50, 647–54.

Galli, F., Battistoni, A., Gambari, R., Pompella, A., Bragonzi, A., Pilolli, F., Iuliano, L., Piroddi, M., Dechecchi, M. C. & Cabrini, G. 2012. Oxidative stress and antioxidant therapy in cystic fibrosis. *Biochim Biophys Acta*, 1822, 690–713.

Godfrey, S. & Mearns, M. 1971. Pulmonary function and response to exercise in cystic fibrosis. *Arch Dis Childh*, 46, 144–51.

Griendling, K. K. & Fitzgerald, G. A. 2003. Oxidative stress and cardiovascular injury. *Circulation*, 108, 1912–16.

Hebestreit, H., Arets, H. G., Aurora, P., Boas, S., Cerny, F., Hulzebos, E. H., Karila, C., Lands, L. C., Lowman, J. D., Swisher, A., Urquhart, D. S. & European Cystic Fibrosis Exercise Working Group. 2015. Statement on exercise testing in cystic fibrosis. *Respiration*, 90, 332–51.

Hussey, J., Gormley, J., Leen, G. & Greally, P. 2002. Peripheral muscle strength in young males with cystic fibrosis. *J Cyst Fibros*, 1, 116–21.

Jäger, R., Purpura, M. & Kerksick, C. M. 2019. Eight weeks of a high dose of curcumin supplementation may attenuate performance decrements following muscle-damaging exercise. *Nutrients*, 11, 1692.

Jeong, J. H., Lee, N., Tucker, M. A., Rodriguez-Miguelez, P., Looney, J., Thomas, J., Derella, C. C., El-Marakby, A., Musall, J. B., Sullivan, J. C., Mckie, K. T., Forseen, C., Davison, G.

W. & Harris, R. A. 2019. Tetrahydrobiopterin improves endothelial function in patients with cystic fibrosis. *J Appl Physiol*, 126, 60–6.

Klijn, P. H., Oudshoorn, A., Van Der Ent, C. K., Van Der Net, J., Kimpen, J. L. & Helders, P. J. 2004. Effects of anaerobic training in children with cystic fibrosis: A randomized controlled study. *Chest*, 125, 1299–305.

Kriemler, S., Kieser, S., Junge, S., Ballmann, M., Hebestreit, A., Schindler, C., Stüssi, C. & Hebestreit, H. 2013. Effect of supervised training on FEV1 in cystic fibrosis: a randomised controlled trial. *J Cyst Fibros*, 12, 714–20.

Lands, L. C. & Hebestreit, H. 2015. Chapter 32-Exercise testing in CF, the what and how. In: Watson, R. R. (ed.) *Diet and Exercise in Cystic Fibrosis*. Boston, MA: Academic Press.

Liu, S., Yang, D., Yu, L., Aluo, Z., Zhang, Z., Qi, Y., Li, Y., Song, Z., Xu, G. & Zhou, L. 2021. Effects of lycopene on skeletal muscle-fiber type and high-fat diet-induced oxidative stress. *J Nutr Biochem*, 87, 108523.

Mason, S. A., Trewin, A. J., Parker, L. & Wadley, G. D. 2020. Antioxidant supplements and endurance exercise: Current evidence and mechanistic insights. *Redox Biol*, 35, 101471.

Mcgrath, L. T., Mallon, P., Dowey, L., Silke, B., Mcclean, E., Mcdonnell, M., Devine, A., Copeland, S. & Elborn, S. 1999. Oxidative stress during acute respiratory exacerbations in cystic fibrosis. *Thorax*, 54, 518–23.

Montgomery, G. S., Sagel, S. D., Taylor, A. L. & Abman, S. H. 2006. Effects of sildenafil on pulmonary hypertension and exercise tolerance in severe cystic fibrosis-related lung disease. *Pediatr Pulmonol*, 41, 383–5.

Moss, A. J. 1982. The cardiovascular system in cystic fibrosis. *Pediatrics*, 70, 728–41.

National Heart, Lung, and Blood Institute (NHLBI) 2018. Cystic Fibrosis [Online]. Available: https://www.nhlbi.nih.gov/health-topics/cystic-fibrosis [Accessed 12/29/2020 2020].

Nixon, P. A. 1996. Role of exercise in the evaluation and management of pulmonary disease in children and youth. *Med Sci Sports Exerc*, 28, 414–20.

Nixon, P. A., Orenstein, D. M., Kelsey, S. F. & Doershuk, C. F. 1992. The prognostic value of exercise testing in patients with cystic fibrosis. *N Engl J Med*, 327, 1785–1788.

Ntimbane, T., Comte, B., Mailhot, G., Berthiaume, Y., Poitout, V., Prentki, M., Rabasa-Lhoret, R. & Levy, E. 2009. Cystic fibrosis-related diabetes: From CFTR dysfunction to oxidative stress. *Clin Biochem Rev*, 30, 153–77.

Orenstein, D. M., Nixon, P. A., Ross, E. A. & Kaplan, R. M. 1989. The quality of well-being in cystic fibrosis. *Chest*, 95, 344–7.

Pallin, M., Keating, D., Kaye, D. M., Kotsimbos, T. & Wilson, J. W. 2018. Subclinical left ventricular dysfunction is influenced by genotype severity in patients with cystic fibrosis. *Clin Med Insights Circ Respir Pulm Med*, 12. doi: 10.1177/1179548418794154. eCollection 2018.

Petersson, J. & Glenny, R. W. 2014. Gas exchange and ventilation-perfusion relationships in the lung. *Eur Respir J*, 44, 1023–41.

Pianosi, P. & Pelech, A. 1996. Stroke volume during exercise in cystic fibrosis. *Am J Respir Crit Care Med*, 153, 1105–9.

Pingitore, A., Lima, G. P. P., Mastorci, F., Quinones, A., Iervasi, G. & Vassalle, C. 2015. Exercise and oxidative stress: Potential effects of antioxidant dietary strategies in sports. *Nutrition*, 31, 916–22.

Poore, S., Berry, B., Eidson, D., Mckie, K. & Harris, R. 2012. Evidence of vascular endothelial dysfunction in young patients with cystic fibrosis. *Chest*, 143, 939–45.

Powers, S. & Jackson, M. 2008. Exercise-induced oxidative stress: Cellular mechanisms and impact on muscle force production. *Physiol Rev*, 88, 1243–76.

Radlovic, N. 2012. Cystic fibrosis. *Srpski arhiv za celokupno lekarstvo*, 140, 244–9.

Rodriguez-Miguelez, P., Ishii, H., Seigler, N., Crandall, R., Thomas, J., Forseen, C., Mckie, K. T. & Harris, R. A. 2019. Sildenafil improves exercise capacity in patients with cystic fibrosis: A proof-of-concept clinical trial. *Ther Adv Chronic Dis*, 10. doi: 10.1177/2040622319887879. eCollection 2019.

Rodriguez-Miguelez, P., Thomas, J., Seigler, N., Crandall, R., Mckie, K. T., Forseen, C. & Harris, R. A. 2016. Evidence of microvascular dysfunction in patients with cystic fibrosis. *Am J Physiol Heart Circ Physiol*, 310, H1479–H1485.

Roe, N. D. & Ren, J. 2012. Nitric oxide synthase uncoupling: A therapeutic target in cardiovascular diseases. *Vascul Pharmacol*, 57, 168–72.

Rowbotham, N. J., Smith, S. J., Davies, G., Daniels, T., Elliott, Z. C., Gathercole, K., Rayner, O. C. & Smyth, A. R. 2020. Can exercise replace airway clearance techniques in cystic fibrosis? A survey of patients and healthcare professionals. *J Cyst Fibros*, 19, e19–e24.

Sacheck, J. M., Milbury, P. E., Cannon, J. G., Roubenoff, R. & Blumberg, J. B. 2003. Effect of vitamin E and eccentric exercise on selected biomarkers of oxidative stress in young and elderly men. *Free Radic Biol Med*, 34, 1575–88.

Schneiderman-Walker, J., Pollock, S. L., Corey, M., Wilkes, D. D., Canny, G. J., Pedder, L. & Reisman, J. J. 2000. A randomized controlled trial of a 3-year home exercise program in cystic fibrosis. *J Pediatr*, 136, 304–10.

Shapiro, B. L. 1989. Evidence for a mitochondrial lesion in cystic fibrosis. *Life Sci*, 44, 1327–34.

Shei, R.-J., Mackintosh, K. A., Peabody Lever, J. E., Mcnarry, M. A. & Krick, S. 2019. Exercise physiology across the lifespan in cystic fibrosis. *Front Physiol*, 10, 1382–1382.

Stallings, V. A., Stark, L. J., Robinson, K. A., Feranchak, A. P. & Quinton, H. 2008. Evidence-based practice recommendations for nutrition-related management of children and adults with cystic fibrosis and pancreatic insufficiency: Results of a systematic review. *J Am Diet Assoc*, 108, 832–9.

Tucker, M. A., Berry, B., Seigler, N., Davison, G. W., Quindry, J. C., Eidson, D., Mckie, K. T. & Harris, R. A. 2018. Blood flow regulation and oxidative stress during submaximal cycling exercise in patients with cystic fibrosis. *J Cyst Fibros*, 17, 256–263.

Tucker, M. A., Crandall, R., Seigler, N., Rodriguez-Miguelez, P., Mckie, K. T., Forseen, C., Thomas, J. & Harris, R. A. 2017. A single bout of maximal exercise improves lung function in patients with cystic fibrosis. *J Cyst Fibros*, 16, 752–758.

Tucker, M. A., Fox, B. M., Seigler, N., Rodriguez-Miguelez, P., Looney, J., Thomas, J., Mckie, K. T., Forseen, C., Davison, G. W. & Harris, R. A. 2019. Endothelial dysfunction in cystic fibrosis: Role of oxidative stress. *Oxid Med Cell Longev*, 2019, 1629638.

Urquhart, D. S. & Saynor, Z. L. 2018. Exercise testing in cystic fibrosis: Who and why? *Paediatr Respir Rev*, 27, 28–32.

van de Weert-Van Leeuwen, P. B., Hulzebos, H. J., Werkman, M. S., Michel, S., Vijftigschild, L. A. W., Van Meegen, M. A., Van Der Ent, C. K., Beekman, J. M. & Arets, H. G. M. 2014. Chronic inflammation and infection associate with a lower exercise training response in cystic fibrosis adolescents. *Respir Med*, 108, 445–452.

Vásquez-Vivar, J., Whitsett, J., Martásek, P., Hogg, N. & Kalyanaraman, B. 2001. Reaction of tetrahydrobiopterin with superoxide: EPR-kinetic analysis and characterization of the pteridine radical. *Free Radic Biol Med*, 31, 975–985.

Vassalle, C., Pingitore, A., De Giuseppe, R., Vigna, L. & Bamonti, F. 2015. Biomarkers Part II: Biomarkers to estimate bioefficacy of dietary/supplemental antioxidants in sport. In: Lamprecht, M. (ed.) *Antioxidants in Sport Nutrition*. Boca Raton, FL: CRC Press/Taylor & Francis© 2015 by Taylor & Francis Group, LLC.

Ward, N., Stiller, K. & Holland, A. E. 2019. Exercise is commonly used as a substitute for traditional airway clearance techniques by adults with cystic fibrosis in Australia: A survey. *J Physiother*, 65, 43–50.

Weibel, E. 2009. What makes a good lung? *Swiss Med Wkly*, 139(27–28), 375–86.

Wells, J. M., Farris, R. F., Gosdin, T. A., Dransfield, M. T., Wood, M. E., Bell, S. C. & Rowe, S. M. 2016. Pulmonary artery enlargement and cystic fibrosis pulmonary exacerbations: A cohort study. *Lancet Respir Med*, 4, 636–45.

Willingham, T. B. & Mccully, K. K. 2017. In vivo assessment of mitochondrial dysfunction in clinical populations using near-infrared spectroscopy. *Front Physiol*, 8, 689.

Wood, L. G., Fitzgerald, D. A., Lee, A. K. & Garg, M. L. 2003. Improved antioxidant and fatty acid status of patients with cystic fibrosis after antioxidant supplementation is linked to improved lung function. *Am J Clin Nutr*, 77, 150–9.

Xu, F. & Rhodes, E. C. 1999. Oxygen uptake kinetics during exercise. *Sports Med*, 27, 313–27.

Xue, R., Gu, H., Qiu, Y., Guo, Y., Korteweg, C., Huang, J. & Gu, J. 2016. Expression of cystic fibrosis transmembrane conductance regulator in ganglia of human gastrointestinal tract. *Sci Rep*, 6, 30926.

CHAPTER SIXTEEN

Ageing, Neurodegeneration and Alzheimer's Disease

THE UNDERLYING ROLE OF OXIDATIVE DISTRESS

Richard J. Elsworthy and Sarah Aldred

CONTENTS

Oxidative Distress in the Ageing Brain / 193
Oxidative Distress in Neurodegeneration: Insights from Alzheimer's Disease / 194
Evidence of Oxidative Distress in Alzheimer's Disease / 196
Energy Balance, Mitochondrial Dysfunction and Oxidative Distress in Alzheimer's Disease / 198
Therapeutic Intervention for Alzheimer's Disease from a Redox Biology Perspective / 199
Dietary Intervention for Alzheimer's Disease / 199
Physical Activity and Alzheimer's Disease / 201
Conclusion / 202
References / 202

OXIDATIVE DISTRESS IN THE AGEING BRAIN

Brain functionality declines with age and can largely be attributed to an accumulation of physiological changes that occur across the lifespan. These changes include genomic instability or DNA lesions (Barzilaiet al., 2017), stem cell depletion (Ermolaeva et al., 2018), dysfunctional cell-cell signalling and inflammation (Heneka et al., 2018), a loss of proteostasis (Hipp et al., 2019), compromised bioenergetics (Yin et al., 2016) and cellular senescence (Kritsilis et al., 2018). Dysregulation of such processes, which are critical for maintaining brain health, is connected by the phenomenon of redox change. Indeed, the accumulation of oxidised biomolecular products is considered a hallmark of the ageing brain. This accumulation of oxidative modification to DNA, proteins and lipids is termed oxidative distress a process which is capable of driving age-related disease (Luo et al., 2020).

Oxidative distress is defined as a disruption of redox homeostasis leading to biomolecular damage (Sies, 1985, 2015). This can arise due to an elevation in reactive oxygen species (ROS) generation or decline in repair systems which overwhelms antioxidant defence capacity (Beckman and Ames, 1998). The intersection of these three systems has been implicated in the ageing process and in a number of diseases such as Alzheimer's disease (AD), especially in cases where ageing is a major risk factor. The brain is a particularly susceptible organ to oxidative distress due to its high-energy demand and oxygen consumption, high content of poly-unsaturated fatty acids and a high level of iron which acts as a catalyst in oxidising reactions (Mecocci et al., 2018). Yet, ROS can act as 'pleiotropic signalling agents' (Sies and Jones, 2020) at physiological levels. This process, which is known as oxidative eustress, is essential for cell function

DOI: 10.1201/9781003051619-16

and survival. Such redox signalling is a property most strongly attributed hydrogen peroxide with a well-characterised mechanism of cysteine oxidation leading to allosteric protein conformational and ultimately functional change (Schieber and Chandel, 2014). The type of ROS generated and its local concentration are key factors in whether redox signalling or oxidative distress become apparent (Schieber and Chandel, 2014). The dynamic regulation of cell homeostasis via redox reactions is illustrated perfectly in the brain during maturation of neural precursor cells whereby ROS production is typically at elevated levels compared to mature neuronal cultures (Oswald et al., 2018). As signalling agents, ROS are critical for maintaining neuronal polarity, synaptic transmission and aiding cell proliferation (Oswald et al., 2018). In addition, cross-talk between neurons and astrocytes through distinct homeostatic mechanisms enables the maintenance of high functionality and redox balance (Baxter and Hardingham, 2016). For example, astrocytes can supply neurons with antioxidant protection through stimulation of antioxidant genes associated with the Nrf2 pathways, which is typically lower in neurons. Yet, if ROS generation goes unchecked, astrocytes can become reactive, further increasing oxidative distress, leading to neuroinflammation and neurodegeneration (Chen et al., 2020).

In the ageing brain, antioxidant and cellular waste removal systems become less effective (Daniele et al., 2018), and redox imbalance becomes apparent as a shift towards a pro-oxidative environment prevails. The oxidation of biomolecules can significantly impact cell survival as dysfunctioning biomolecules impact cellular processes essential for life such as metabolic function, perturbed signal transduction, or induce apoptosis and eventual cell death (Trachootham et al., 2008). In the brain, oxidative protein aggregation has been linked with neurodegeneration (Butterfield and Boyd-Kimball, 2018). Although brain ageing is an inevitable process, the rate of decline is variable. That is, some people who are matched for their chronological age do not share the same features associated with their biological or 'brain age' (Cole, 2020) (i.e., chronological and biological age are often asynchronous). This may, in part, explain why some individuals experience significant cognitive decline and neurodegeneration early as they age, whereas some people age with excellent cognitive health, free from apparent neurodegenerative disease.

OXIDATIVE DISTRESS IN NEURODEGENERATION: INSIGHTS FROM ALZHEIMER'S DISEASE

Old age is associated with an increased susceptibility of developing neurodegenerative diseases, the most prevalent of which is AD (Hou et al., 2019). The hallmark pathological features that typify the AD brain are the intracellular hyperphosphorylation of Tau protein 'tangles' and the extracellular accumulation insoluble Amyloid-β (Aβ) plaques (Alzheimer et al., 1995; Braak et al., 1998). This is in addition to a loss of brain volume and a reduction in synaptic function and formation, all of which lead to significant and progressive cognitive decline.

The identification of a small peptide named Aβ as the major constituent of the extracellular plaques was a driving factor in the development of the 'Amyloid-Cascade Hypothesis', which placed Aβ at the centre of AD research (Hardy and Higgins, 1992). This was supported by evidence of neighbouring neuritic and glial cytopathology in regions associated with memory and cognition near Aβ deposits (Selkoe and Hardy, 2016). Further, mutations within and close to the Aβ region in the *APP* gene are associated with aggressive autosomal dominant forms of familial AD (fAD) (Hunter and Brayne, 2018). The most common cause of fAD is missense mutations to the *PSEN1* and *PSEN2* genes, which are known to impact the catalytic subunit of the γ-secretase enzyme (Kelleher and Shen, 2017). γ-Secretase is responsible for a critical step in Aβ generation following beta amyloid cleaving enzyme-1 (BACE-1) cleavage of the amyloid-β precursor protein (AβPP). In many cases, *PSEN1/2* mutations increase Aβ production or increase the ratio of the longer Aβ1–42 to the more physiological Aβ1–40 peptide (Sun et al., 2017). Aβ1–42 can more rapidly aggregate at lipid membranes and form oligomeric Aβ species. These oligomers accumulate in endocytotic vesicles, and thus, an increase in the Aβ1–42:Aβ1–40 ratio has been linked to AD development (Zheng et al., 2017). Aβ1–42 has been shown to more easily propagate from smaller tetra- and hexameric oligomers to larger aggregates when compared to more resistant Aβ1–40 tetramers (Bernstein et al., 2009). The appearance of such 'intermediate' oligomeric forms of Aβ that are particularly toxic to neurons has provided a new area of investigation for the role of Aβ in the pathology of AD (Kim et al., 2003; Lesné et al., 2008). Further, Aβ oligomers have also been shown to diffuse from Aβ fibrillar plaques (Koffie et al., 2009) and exert cytotoxic effects which are associated with synapse loss and cognitive decline (Pickett et al., 2016). The processing of AβPP is not limited to BACE-1 cleavage and can be initiated by the enzyme 'a disintegrin and metalloproteinase-10' (ADAM10) (Kuhn et al., 2010) (Figure 16.1). ADAM10 cleavage of AβPP cuts through the Aβ-region and leads to shedding of soluble AβPPα. The sequential cleavage by γ-secretase releases a truncated p3 peptide, thus preventing the generation of Aβ (Kuhn et al., 2010). In AD, the expression and activity of ADAM10 are lower, which may increase the proteolytic processing of AβPP by BACE-1 and thus increase Aβ generation. Further, the soluble fragment AβPPα plays a role in neurogenesis, maintaining synaptic function and supporting the formation of neuronal networks (Kögel et al., 2012; Yuan et al., 2017), and is considered neuroprotective. Missense mutations to the *ADAM10* gene affecting the cysteine switch in the prodomain of the translated enzyme are also associated with an increased risk of AD (Kim et al., 2009). All of this evidence provides significant data for a role of Aβ in the progression of AD.

Although studying fAD has advanced our understanding of AD, these mutations only account for a small number of cases (estimated between 1% and 3%) and develop at a much more rapid and earlier age of onset. In contrast, the development of late-onset AD is associated with a multifactorial contribution of genetic risk and environmental influence (through epigenetics) that ultimately results in neurodegeneration. The involvement of Aβ as a single driving factor in late-onset AD has been questioned due to the discovery that a number of individuals can have elevated Aβ load in the brain but remain cognitively healthy (Haller et al., 2019; Mufson et al., 2016). There is also a lack of correlation between Aβ load and cognitive decline in individuals experiencing cognitive decline. Although likely to be a key component, it is highly unlikely that Aβ pathology alone is the cause of cognitive decline in AD.

Perhaps one of the key factors limiting our understanding of the mechanisms that cause AD is the inability to determine what the earliest pathological change is in the AD brain. One answer may be linked to oxidative distress (Nunomura et al., 2001).

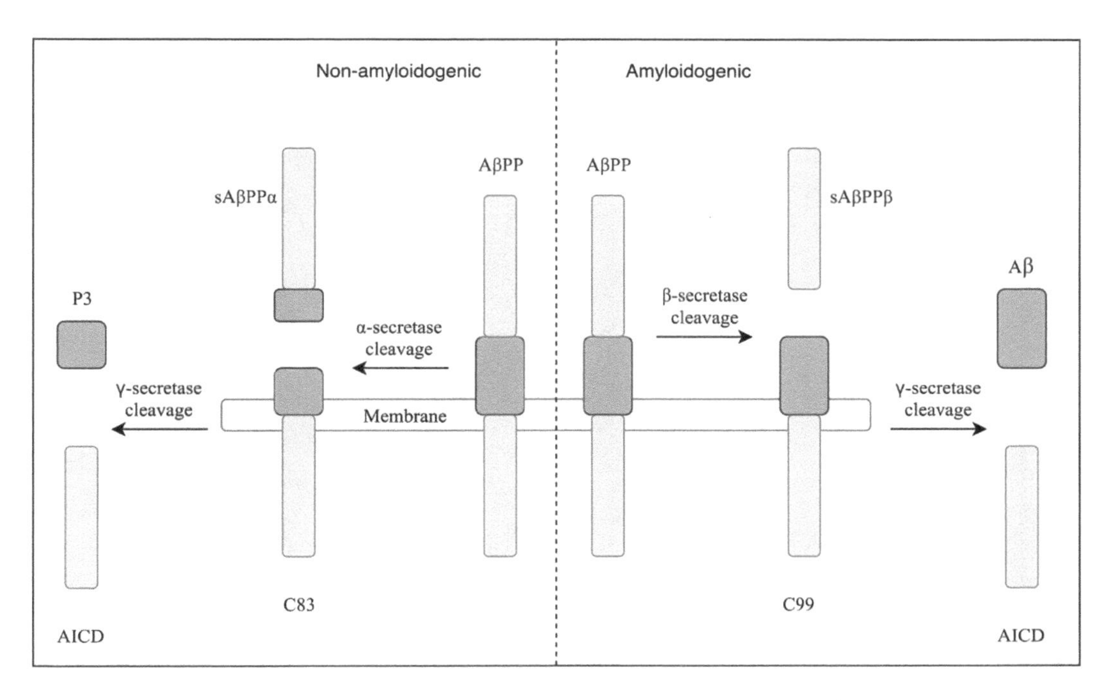

Figure 16.1 Diagrammatic representation of AβPP processing by secretase enzymes. Non-amyloidogenic processing is initiated by the α-secretase, ADAM10. ADAM10 cleavage of AβPP, followed by intramembrane γ-secretase cleavage, results in the release of neuroprotective sAβPPα preventing Aβ generation. Amyloidogenic AβPP processing is initiated by BACE-1, and subsequent γ-secretase cleavage, resulting in the generation of the Aβ peptide.

Evidence of Oxidative Distress in Alzheimer's Disease

As previously mentioned, the dynamic balance between ROS production and antioxidant defence is critical for physiological systems regulation and cell signalling (Sies, 1985, 2015; Sies and Jones, 2020). Therefore, any disturbance to redox homeostasis can have a significant impact on cellular functions and result in cell degeneration and death (Oswald et al., 2018). Such an imbalance in the AD brain is evidenced by increased lipid, protein and DNA (per)oxidation; a reduction in energy metabolism effected by oxidative modification to mitochondrial enzymes and increased transition metals capable of catalysing oxidation reactions (Butterfield and Halliwell, 2019). In fact, the incorporation of transition metals into Aβ fibril during assembly is a major source of ROS in AD (Cheignon et al., 2018). Further, the binding and insertion of Aβ oligomers can cause neuronal membrane integrity disruption and trigger lipid oxidising cascades (Figure 16.2). As such, with increasing Aβ load, the capacity for ROS production and subsequent oxidative distress is also increased (Cheignon et al., 2018; Markesbery,

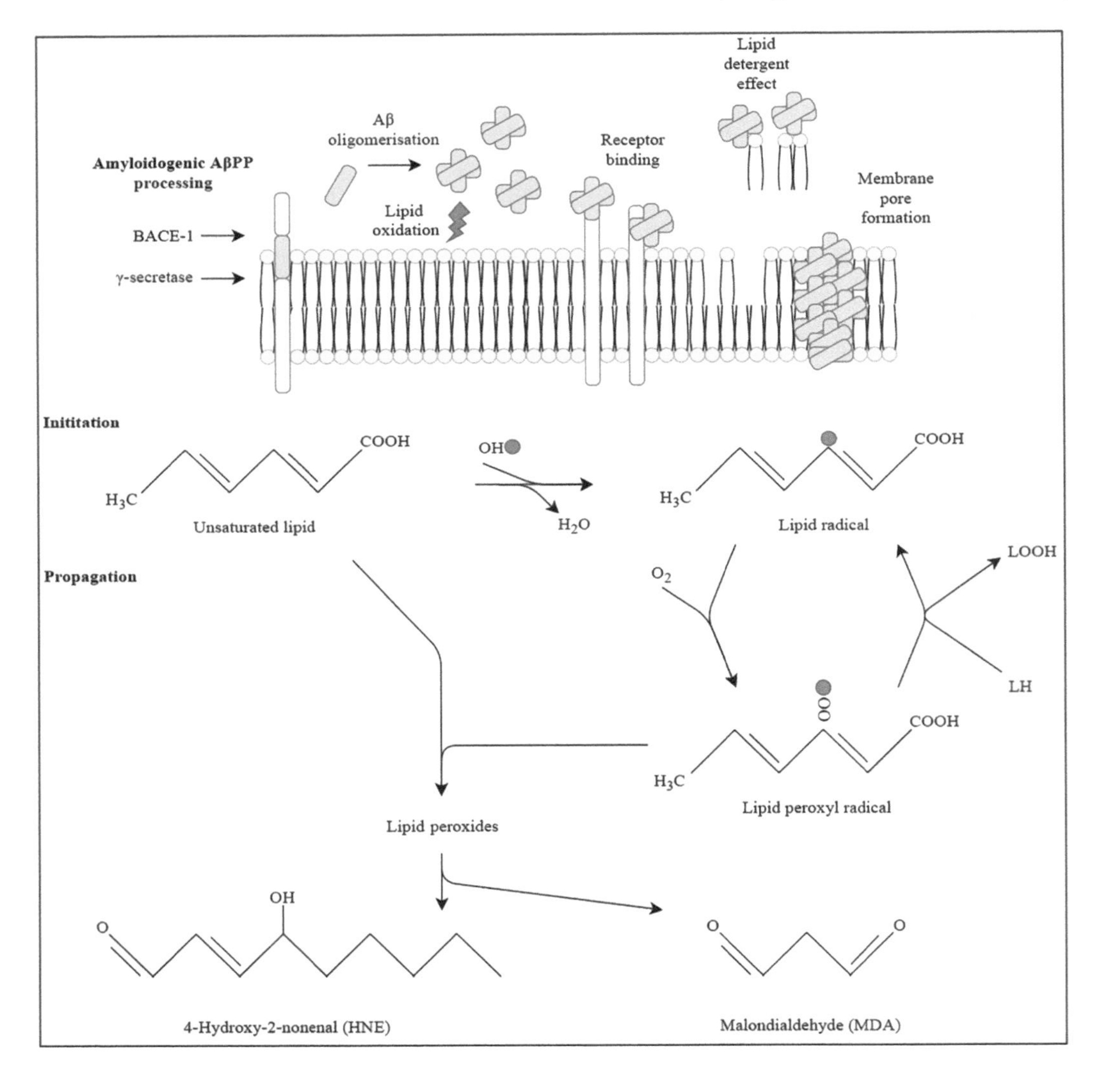

Figure 16.2 Aβ peptides generated following amyloidogenic AβPP processing can accumulate in lipid-rich regions or at membrane receptors forming oligomeric species. Aβ oligomers can disrupt the cell membrane, altering membrane architecture and permeability and facilitating the initiation of lipid peroxidation. Subsequent propagation steps can significantly increase lipid peroxidation products and trigger oxidation cascades leading to protein oxidation products.

1997). Each of these is individually able to create a pro-oxidant environment, but when they are combined, this may lead to a significant oxidative distress in AD pathogenesis. In fact, dysregulation of redox balance as either an initiating cause or crucial secondary event in the progression of neurodegeneration has been coined the 'Oxidative Stress Hypothesis' (Markesbery, 1997). This is supported by evidence of an elevation in 8-hydroxyguanosine and nitrotyrosine a marker of RNA and amino acid oxidation, respectively, in people with AD compared to cognitively healthy individuals. These markers of oxidation were elevated prior to the accumulation of Aβ and Tau, eventually decreasing as AD progresses. In addition, products of lipid peroxidation and nitrosylation have been shown to interfere with BACE-1 activity and subsequent Aβ generation (Arimon et al., 2015; Kwak et al., 2011; Nunomura et al., 2001).

Oxidation of proteins involved in Aβ clearance and degradation is evident in people with AD, suggesting oxidative distress may indeed be involved in early AD pathogenesis (Owen et al., 2010; Wang et al., 2003). However, the interaction between oxidative distress and Aβ accumulation is bidirectional in that oligomeric Aβ can trigger the propagation of oxidising cascades. This is perhaps most evident in lipid membranes. As previously mentioned, Aβ1–42 can readily form hydrophobic oligomers in lipid-rich membrane domains (Zheng et al., 2017) with the ability to insert and disrupt membrane structure. Aβ oligomers can then oxidise membrane lipids forming highly reactive electrophilic aldehydes, such as 4-hydroxynonenal (HNE), malondialdehyde and acrolein. HNE has been of particular interest, as elevated levels are associated with AD pathology (Benseny-Cases et al., 2014; Markesbery and Lovell, 1998; Scheff et al., 2016). Not only this, HNE-modified proteins are also elevated in AD (Butterfield et al., 2006; Perluigi et al., 2009). The addition of protein adducts such as HNE or a carbonyl group to membrane, cytosolic and mitochondrial proteins can subsequently result in protein dysfunction and may be critical in the progression of AD (Butterfield, 2020). Most notably, HNE bound to the low-density lipoprotein receptor-related protein 1 (LRP1) can significantly impair Aβ clearance. Consequently, the accumulation of Aβ in tandem with its oxidising potential may further oxidise LRP1, suggesting Aβ may impair its own clearance (Owen et al., 2010).

Lipid peroxidation has also been measured in AD by monitoring relatively stable products of fatty acid oxidation. The enzymatic and radical-mediated oxidation of arachidonic and docosahexaenoic acid produces prostaglandin-like F2-Isoprostanes and neuroprostanes respectively, which are commonly used for quantifying oxidative distress 'in vivo' (Yoshida et al., 2013). Although oxidative distress is widely regarded as a consistent feature of AD, there have been studies showing little to no change in total lipid and protein oxidation (Zabel et al., 2018). This apparent contradiction from research studies may be explained by differences in the time of sampling in relation to disease progression, or by the tissue or fluid compartment being analysed (Peña-Bautista et al., 2019). As previously mentioned, evidence of increased oxidative distress has been reported early in the pathogenies of AD but does not persist with disease progression (Nunomura et al., 2001). Therefore, rather than assess non-specific, global markers of oxidative distress, it is more informative to analyse oxidative modification to specific biomolecules at chosen stages of disease progression to better establish the role of oxidative distress in AD (Aldred et al., 2010; Sultana et al., 2010).

Cholesterol metabolism is widely considered to be a significant factor in AD pathology (Chang et al., 2017). This is perhaps best highlighted by the change in risk of developing AD in individuals with different *APOE* allelic variations. The *APOE* gene codes for the major cholesterol-carrying lipoprotein in the brain and is carried as a combination of three *APOE* alleles, ε2, ε3 and ε4 (Belloy et al., 2019). Carrying one or two copies of the *APOE* E4 allele significantly increases the risk of developing late-onset AD, when compared to the most common *APOE ε3* variant (Corder et al., 1993). *APOE* ε2 is associated with neuroprotection from late-onset AD (Corder et al., 1994). The presence of *APOE* ε4 is associated with Aβ oligomer production, neurofibrillary tangles, neuronal and synaptic degeneration, lipid bilayer disruption and oxidative stress (Butterfield and Mattson, 2020). Further, the loss of a direct HNE scavenging action in *APOE* ε4 compared to *APOE* ε3 and *APOE* ε2 has been proposed that could contribute to increase oxidative stress propagation (Butterfield and Mattson, 2020). Not only this, distinct lipid- and receptor-binding regions in *APOE* ε4 may increase neurotoxicity and facilitate mitochondrial dysfunction (Chang et al., 2005), processes which may lead to direct ROS generation (for 'mitochondrial dysfunction in AD' *see below*). In addition to *APOE*, there are a number of susceptibility genes involved in cholesterol pathways, such as *CLU, PICALM, BIN1* and *SORL1*, that increase the risk of developing AD (Dong et al., 2017). Not only this, but cholesterol and sphingolipid-rich lipid rafts on cellular membranes are key sites for Aβ generation and oligomerisation (Kim et al., 2006; Rushworth and Hooper, 2010). In contrast, depletion of cholesterol is associated with increased ADAM10 activity and non-amyloidogenic AβPP processing (Kojro et al., 2001; Matthews et al., 2017), and thus a decreased risk of AD. The brain is a cholesterol-rich organ that relies on endogenous synthesis, as free cholesterol is unable to cross the blood–brain barrier (Chang et al., 2017). The product of regulated enzymatic oxidation of cholesterol, termed oxysterol, is critical in the brain as it enables cholesterol flux across the blood–brain barrier helping to maintain cholesterol homeostasis. However, cholesterol is also susceptible to non-enzymatic oxidation, or autoxidation, which may reflect oxidative distress in the brain (Kulig et al., 2015). Both products of enzymatic and non-enzymatic oxysterol formation are elevated in AD and are related to disease progression (Testa et al., 2016). Further, elevation in the concentration of the products of cholesterol oxidation is considered a marker of reduced glucose metabolism, which is in itself a hallmark feature of the AD brain (Gamba et al., 2019). Mechanistically, adventitious autoxidation of membrane cholesterol may be a key step in the pathogenesis of AD and is mediated by its own effect on cell membrane integrity and impaired functionality of transmembrane proteins/receptors (Phan et al., 2013). For example, the inclusion of 7-ketocholesterol, an oxidation product of cholesterol, into cell membranes can prevent Aβ fibril formation. Instead, Aβ is maintained in an oligomeric state (Phan et al., 2018) which is capable of further oxidising membrane lipids. A positive feedback loop of oxidation events and Aβ oligomerisation can propagate, leading to significant and progressive neurodegeneration (Doig, 2018).

As the risk of developing AD progresses with advancing age, and ageing is associated with elevated oxidative distress, it is difficult to attribute to what extent oxidative stress leads to the initiation of the pathological process. Perhaps one explanation could be that propagating oxidative insults subject the brain to a new 'steady state' of redox balance, i.e., where higher baseline of ROS exists. Therefore, any further insult passes the 'antioxidant threshold' sooner and leads to neurodegeneration. A similar 'Two-Hit Hypothesis' has been proposed, whereby oxidative distress and/or abhorrent mitotic signalling can result in AD pathogenesis (Zhu et al., 2004). Whether oxidative distress plays a causal role in the onset of AD is unclear. However, it is clear that oxidative distress is highly coupled to brain energy metabolism, and both oxidative distress and perturbed brain energy metabolism are hallmarks of AD. It is therefore understandable that a significant amount of research has been directed towards identifying the pathways linking excessive ROS production and dysfunctional glucose metabolism in AD.

Energy Balance, Mitochondrial Dysfunction and Oxidative Distress in Alzheimer's Disease

Brain activity has particularly high energy demands and is heavily dependent on glucose metabolism. Efficient ATP production is critical for maintaining proper functioning (Butterfield and Halliwell, 2019). Over 75% of ATP produced in the brain is consumed by neurons (Hyder et al., 2013). A significant proportion of this is spent restoring neuronal membrane potentials, synthesising neurotransmitters, and in the axonal trafficking of biomolecules (Watts et al., 2018). The brain can also use alternative energy sources such as lactate or ketone bodies under certain conditions including, neurodevelopment, intense exercise and energy restriction whereby anaerobic systems are more heavily utilised (Mergenthaler et al., 2013). However, these alternate sources can only supplement the main metabolic process and cannot replace the need for glucose. In the AD brain, glucose metabolism is significantly compromised, and this is often detectable prior to the onset of clinically recognised symptoms (Croteau et al., 2018). Regions of hypometabolism in the brain are associated with a mixture of Aβ deposition and regions of atrophy (Grothe et al., 2016). There are a number of influencing factors thought to contribute to the loss of glucose oxidation in AD such as insulin resistance, proteostasis abnormalities, neuroinflammation, neuronal hyperactivity and mitochondrial dysfunction (for reviews see (Butterfield and Halliwell, 2019; Zilberter and Zilberter, 2017)).

Mitochondria are critical in the production of ATP as they are the site for oxidative phosphorylation. The oxidation of NADH and $FADH_2$, formed in glycolysis, fatty acid oxidation and the tricarboxylic acid cycle (TCA), is used to reduce ground state molecular dioxygen to water in the electron transport chain (ETC) which traverses the inner mitochondrial membrane (Zhao et al., 2019). During this process, protons are pumped into the intramembrane space to create a pH gradient and mitochondrial membrane potential (Belenguer et al., 2019), termed the proton-motive force (Mitchell, 1966). The entry of protons back into the matrix via the ATPase enzyme enables the phosphorylation of ADP to synthesis ATP (Belenguer et al., 2019). Mitochondrial oxidative phosphorylation accounts for a large portion of ATP synthesis in the brain, and therefore, a sufficient supply of metabolites is critical for effective cellular respiration. Although predominantly associated with their role in generating ATP, mitochondria are also involved in a number of other critical cellular processes, which include programmed cell death, calcium signalling, fatty acid oxidation and the innate immune response (Scott and Youle, 2010). Therefore, the development of mitochondrial dysfunction in an ageing brain would pose a significant challenge to cell function.

The mitochondrion is a significant source of ROS. Electron leakage from protein complexes I, II and III of the ETC facilitates the partial oxidation of oxygen, forming the superoxide radical ($O_2^{\cdot-}$). Under physiological conditions matrix, $O_2^{\cdot-}$ generation is comparatively low (steady state in the picomolar range) owing to the presence of the enzyme manganese superoxide dismutase. However, changes in the $NADH/NAD^+$ ratio and a high proton gradient, coupled with a lack of demand for ATP, can significantly increase $O_2^{\cdot-}$ and make the propagation of ROS cascades more likely (Cobley, 2020; Murphy, 2009). Superoxide dismutase is able to readily catalyse the conversion of $O_2^{\cdot-}$ to the non-radical, hydrogen peroxide (H_2O_2) which is able to diffuse across the mitochondrial membrane via aquaporins (Mondola et al., 2016). Under physiological conditions, H_2O_2 plays a critical role in redox signal transduction and thus contributes to oxidative eustress; however, the reduction of H_2O_2 and subsequent formation of the hydroxyl radical ($OH^{\bullet}$) – the most reactive radical species in the presence of iron and copper – are major steps in the propagation of oxidative damage to biomolecules (Gammella et al., 2016). These factors and thus the multifaceted role of mitochondria in maintaining cellular homeostasis have highlighted a likely role in neurodegeneration (Belenguer et al., 2019). Given the links between mitochondrial function, energy production and ROS generation, the importance of redox balance in the progression of AD is evident (Wang et al., 2014). The progressive decline in mitochondrial function in AD often results in defective ETC function (Santos et al., 2013). Global oxidative damage to mitochondrial DNA (mtDNA) has also been identified in AD, primarily by elevated 8-hydroxyadenine and 8-hydroxyguanine (Santos et al., 2013). As mtDNA is responsible for coding the subunits of the ETC and RNA that are essential for normal ETC function (Nissanka and Moraes, 2018), this can result in reduced ATP production and greater $O_2^{\cdot-}$ generation from the ETC, thus forming a cycle of increasing oxidation, known as the viscous cycle hypothesis. It should be noted, however, that not all mtDNA mutations increase ROS generation and, in fact, can prevent ROS formation (de Grey, 2005). For example, a mtDNA mutation effecting the synthesis of cytochrome b would limit complex III formation and subsequent superoxide generation (de Grey, 2005). Redox balance has also been implicated in the regulation of mitochondrial dynamics. Elevated ROS may interrupt normal fission and fusion events and reduce efficient mitophagy (Magi et al., 2016; Willems et al., 2015). Further, evidence of reduced mitochondrial recycling has been found in primary cells from people with AD (Martín-Maestro et al., 2017) leading to an accumulation of dysfunctional mitochondria. Overall, this can significantly impair sufficient ATP generation, increase ROS generation and subsequently decrease neuronal functioning. Therefore, focusing on methods of sustaining neuronal energy supply and balancing redox status are critical for preventing AD.

THERAPEUTIC INTERVENTION FOR ALZHEIMER'S DISEASE FROM A REDOX BIOLOGY PERSPECTIVE

Dietary Intervention for Alzheimer's Disease

There is a large body of research investigating the diet-disease relationship and how dietary interventions can reduce the prevalence of disease.

Excessive caloric intake, high (saturated) fat–high salt intake and consumption of micronutrient poor foods are some of the factors linked with an increased risk of metabolic and vascular diseases (Fernández-Sanz et al., 2019). Low nutritional status has also been associated with a person's risk of AD, with lower levels of nutrients associated with antioxidant properties in blood and brain of people with AD (de Wilde et al., 2017). Combined with the link of oxidative distress in the pathogenesis of AD, this has paved the way for a number of clinical trials aiming to restore redox balance through manipulating dietary intake (Persson et al., 2014; Verhaar et al., 2020). Further, the potential benefits of supplementing with vitamins and antioxidant compounds to delay the progression of AD has been of interest. However, the results of subsequent studies have been underwhelming and mixed.

Vitamin E refers to a group of compounds that are fat-soluble and known to possess antioxidant capacity. They are also regarded as important molecules with neuroprotective and anti-inflammatory biological properties. The solubility of vitamin E allows for an important role in the prevention of lipid membrane peroxidation (Brigelius-Flohé, 2009), a process which is pertinent in AD (Butterfield, 2020). Data from experimental models have shown that supplementation with vitamin E can improve cognitive function and reduce memory deficits, by modulating oxidative distress (Gugliandolo et al., 2017). Yet, these effects are not seen in larger clinical trials where little effect on cognitive function and memory has been reported (Brewer, 2010). In fact, at doses over 400IU/day, all-cause mortality may even increase (Miller et al., 2005). The inconsistency in effectiveness may be in part explained by the individual response to vitamin E supplementation, which has shown that people can be 'non-responders' (Lloret et al., 2009). In fact, dietary intake, age, gender, intestinal absorption, genetic polymorphisms and lifestyle factors may all contribute to an individual's response to vitamin E supplementation (Lloret et al., 2019). Questions over the appropriate dose to obtain an antioxidant response, and the timing of the intervention, may also explain mixed results. In any case, simply reducing oxidative distress by increasing antioxidant capacity in people with AD is highly unlikely to reverse neurodegeneration and may be better suited to target the early stages of the disease whereby preventing the initiation process of radical interaction is crucial. Although vitamin E has the potential to neutralise a number of ROS, this quenching reaction in turn produces the vitamin E radical which requires vitamin C to regenerate its antioxidant capability (Lloret et al., 2019). The vitamin E radical can also interact with lipid peroxides thus reaching the terminal oxidation reaction (Lloret et al., 2019).

To facilitate the removal of ROS, combinations of molecules with antioxidant properties have been tested in oral interventions. A combination of water-soluble vitamin C, the mineral selenium and vitamin E has been trialled as supplements. Whilst the results of these studies typically show a favourable action with regards to reducing the concentrations of oxidised biomolecules, they still fall short of improving clinical symptoms of AD (Arlt et al., 2012; Galasko et al., 2012; Kryscio et al., 2017). In fact, cognitive impairment has been seen to worsen with antioxidant treatment in some cases (Galasko et al., 2012). Perhaps aiming to quench ROS by globally increasing antioxidant capacity is not an effective strategy for treating AD. As previously mentioned, ROS have a pleiotropic role in the body, acting as cell signalling molecules critical to normal physiology (Sies and Jones, 2020). Further, it is critical that sufficient concentrations of antioxidants are achieved in areas that are highly susceptible to ROS damage. Therefore, using a more targeted approach to eliminating excessive ROS could hold more therapeutic benefits.

There a small number of such compounds being tested in ongoing clinical trials. One of which is the mitochondrial-targeted antioxidant, 'MitoQ'. Evidence of reduced oxidative stress, the prevention of synapse loss and neuropathy and preserved cognitive function has emerged from both 'in vitro' and 'in vivo' animal models using MitoQ (McManus et al., 2011, 2014; Young and Franklin, 2019). Supplementation with precursors to nicotinamide adenine dinucleotide (NAD) has also been the subject of supplementation studies in an attempt to restore the metabolic deficiencies associated with AD (Braidy et al., 2018; Xie et al., 2019). However, the usefulness of antioxidant supplementation alone is questionable. Whilst oxidative stress is undoubtably a prominent feature of AD, many antioxidant supplementation studies have been unsuccessful in trying to treat multifactorial or complex diseases due to problems with delivery of the antioxidant to the metabolic site where it is required. Instead, focusing on dietary interventions that can deliver the nutrient components to the desired site of action via normal metabolism, rebalance redox homeostasis via increased water and lipid-soluble antioxidants and target multiple aspects of disease pathology promises to be more beneficial.

Improving overall diet to include so-called 'superfoods' and towards a more nutrient-rich composition may have significant benefits for the ageing brain. Reduction of high saturated fat and sugar consumption and removal of excessive caloric surplus are known to reduce the risk of cerebrovascular and cardiometabolic dysfunction (Pan et al., 2018). Diets rich in antioxidants, monounsaturated fats such as omega 3, 6 and 9, high fibre and phytosterols are associated with improved cognition (Veurink et al., 2020). The Mediterranean diet matches this neuroprotective profile and has become of significant interest for reducing the risk of AD. The closer an individual's diet reflects that of the typical Mediterranean diet, the lower the risk of AD (Gardener et al., 2012). Further, the Mediterranean diet is associated with reduced markers of oxidative stress and inflammation highlighting an important role of restoring redox balance in dietary interventions (Dai et al., 2008).

Physical Activity and Alzheimer's Disease

Physical activity (PA) is widely regarded as an effective way to improve cerebrovascular and cardiometabolic health, maintain functional abilities and promote healthy ageing (Daskalopoulou et al., 2017). PA has also been investigated as a tool in the primary prevention of AD or as a disease slowing intervention (Morris et al., 2017). From this research, it has become apparent that increasing PA is more strongly associated with a reduced risk of other dementias such as vascular dementia, rather than AD by improving cerebral blood flow and metabolic risk factors (Trigiani and Hamel, 2017). This lack of evidence may stem from the difficulty in identifying the early phases or prodromal period of AD when PA interventions would be most effective. However, it is evident that promoting PA throughout the lifespan will convey some protective benefits for people with AD, whether this is by directly attenuating the core pathology or by reducing risk factors of AD (Hansson et al., 2019; Jia et al., 2019). Physical inactivity is linked with poor vascular function and an increase in the prevalence of risk factors associated with AD (Norton et al., 2014). Although the optimal 'dose' of PA required to elicit protective benefits for the brain is unknown, improved cognitive function is associated with increased PA in people with AD (Jia et al., 2019; Stephen et al., 2017). Yet, the effects of increased PA on the specific pathological features of AD are unclear. Differences in the Aβ and Tau profiles in people with AD are highly variable in the participant populations, and thus, it is difficult to assess the effectiveness of exercise in heterogenous conditions (Frederiksen et al., 2019; Liang et al., 2010). Further, the inconsistency in the efficacy of PA interventions in AD may be affected by the lack of a comparable approach in the intervention. Not only do the durations of PA interventions vary, but the mode, frequency and intensity of the activity also differ greatly. Many PA interventions may not be of sufficient duration to elicit the necessary responses for physiological benefits. This is evident when considering the almost paradoxical redox response to PA.

An acute bout of exercise is associated with an elevation in systemic ROS production and increased oxidative distress as a result of increased metabolic demand (Webb et al., 2017). Prolonged, high-intensity exercise produces a greater redox shift than that produced by lower intensity bouts (Powers et al., 2020). Thus, it seems as though avoiding exercise may be of benefit for some individuals where the bout of exercise would increase excess ROS generation and cause oxidative damage to biomolecules. However, the relatively low levels of ROS generated from exercise are unlikely to cause significant damage in the majority of individuals but instead would trigger beneficial redox-sensitive signalling responses (Webb et al., 2017). In fact, acute alterations in redox balance appear critical for optimal muscular contraction and signal for favourable exercise-induced adaptations (Powers et al., 2020). Although the major source of ROS generation from acute exercise is produced in distal muscle tissue, the increased cerebral blood flow and change in metabolic demand likely alter the redox state of the brain.

Older adults may have an impaired adaptive response to individual bouts of exercise compared to younger individuals (Nordin et al., 2014), perhaps due to an altered redox baseline, chronic exercise is advantageous across the lifespan (Done and Traustadóttir, 2016). In fact, repeated bouts of exercise cause an elevated expression of both oxidant buffering molecules and antioxidant enzyme activity (Powers et al., 2020). Furthermore, chronic exercise is associated with improved mitochondrial function, increased proteosome and protein degrading enzyme activity, reduced inflammatory signalling and neurotrophic factor release, such as BDNF, that may directly attenuate the pathological features of AD (Chen et al., 2016). Perhaps most crucially, these effects are seen in the brain, as well as peripheral tissues. By improving our knowledge of the redox-sensitive signalling mediated adaptive responses to specific exercise bouts, it will become possible to prescribe bespoke PA interventions that are more effective for people with AD. It is critical that gradually increasing overall PA provides a strong stimulus to induce adaptation yet avoids excessive tissue damage and muscle soreness that may manifest in a potentially vulnerable population. Thus, PA as an intervention can aid in the maintenance of oxidative eustress and the prevention of oxidative distress. Not only should exercise be recommended to improve the symptoms of AD, and reduce the prevalence of known risk factors, but also to enhance physical function, of which the latter can significantly and positively impact the quality of life.

CONCLUSION

The importance of investigating how redox balance is maintained (or disrupted) during ageing may prove pivotal in our ability to understand why the risk of neurodegeneration increases dramatically with age. Whether the generation of ROS, a reduction in repair mechanisms and antioxidant defence or a combination of each forms the causal perturbations underlying AD or contributes to the progression of the disease is not known. Yet, many of the hallmark features of the AD brain are linked by their ability to modulate redox balance, favouring oxidative distress. Attempts to reduce oxidative distress as a therapeutic intervention for AD have been unsuccessful; however, these are fundamentally flawed, often lacking target specificity and appropriate dosing. Instead, implementing lifestyle changes in mid-life, such as improved dietary intake and increased physical activity which have been shown to not only improve redox balance but modify a number of risk factors for AD, may be more beneficial. As with any treatment option, careful consideration is needed to optimise the beneficial physiological adaptations to dietary manipulation and exercise bouts to provide maximum protection against AD. One way that this may be achieved is by monitoring redox adaptation to interventions.

REFERENCES

Aldred, S., S. Bennett and P. Mecocci (2010). Increased low-density lipoprotein oxidation, but not total plasma protein oxidation, in Alzheimer's disease. *Clin Biochem* 43(3), 267–271.

Alzheimer, A., et al. (1995). An english translation of Alzheimer's 1907 paper, "Uber eine eigenartige Erkankung der Hirnrinde". *Clin Anat* 8(6), 429–431.

Arimon, M., et al. (2015). Oxidative stress and lipid peroxidation are upstream of amyloid pathology. *Neurobiol Dis* 84, 109–119.

Arlt, S., et al. (2012). Effect of one-year vitamin C- and E-supplementation on cerebrospinal fluid oxidation parameters and clinical course in Alzheimer's disease. *Neurochem Res* 37(12), 2706–2714.

Barzilai, A., B. Schumacher and Y. Shiloh (2017). Genome instability: Linking ageing and brain degeneration. *Mech Ageing Dev* 161(Pt A), 4–18.

Baxter, P. S. and G. E. Hardingham (2016). Adaptive regulation of the brain's antioxidant defences by neurons and astrocytes. *Free Radic Biol Med* 100, 147–152.

Beckman, K. B. and B. N. Ames (1998). The free radical theory of aging matures. *Physiol Rev* 78(2), 547–581.

Belenguer, P., et al. (2019). Mitochondria and the brain: Bioenergetics and beyond. *Neurotox Res* 36(2), 219–238.

Belloy, M. E., V. Napolioni and M. D. Greicius (2019). A quarter century of APOE and Alzheimer's disease: Progress to date and the path forward. *Neuron* 101(5), 820–838.

Benseny-Cases, N., et al. (2014). Microspectroscopy (μFTIR) reveals co-localization of lipid oxidation and amyloid plaques in human Alzheimer disease brains. *Anal Chem* 86(24), 12047–12054.

Bernstein, S. L., et al. (2009). Amyloid-β protein oligomerization and the importance of tetramers and dodecamers in the aetiology of Alzheimer's disease. *Nat Chem* 1(4), 326–331.

Braak, H., et al. (1998). Neuropathological hallmarks of Alzheimer's and Parkinson's diseases. *Prog Brain Res* 117, 267–285.

Braidy, N., R. Grant and P. S. Sachdev (2018). Nicotinamide adenine dinucleotide and its related precursors for the treatment of Alzheimer's disease. *Curr Opin Psychiatry* 31(2), 160–166.

Brewer, G. J. (2010). Why vitamin E therapy fails for treatment of Alzheimer's disease. *J Alzheimers Dis* 19(1), 27–30.

Brigelius-Flohé, R. (2009). Vitamin E: The shrew waiting to be tamed. *Free Radic Biol Med* 46(5), 543–554.

Butterfield, D. A. (2020). Brain lipid peroxidation and alzheimer disease: Synergy between the Butterfield and Mattson laboratories. *Ageing Res Rev* 64, 101049.

Butterfield, D. A. and D. Boyd-Kimball (2018). Redox proteomics and amyloid β-peptide: Insights into Alzheimer disease. *J Neurochem* 151, 459–487.

Butterfield, D. A. and B. Halliwell (2019). Oxidative stress, dysfunctional glucose metabolism and Alzheimer disease. *Nat Rev Neurosci* 20(3), 148–160.

Butterfield, D. A. and M. P. Mattson (2020). Apolipoprotein E and oxidative stress in brain with relevance to Alzheimer's disease. *Neurobiol Dis* 138, 104795.

Butterfield, D. A., et al. (2006). Elevated protein-bound levels of the lipid peroxidation product, 4-hydroxy-2-nonenal, in brain from persons with mild cognitive impairment. *Neurosci Lett* 397(3), 170–173.

Chang, S., et al. (2005). Lipid- and receptor-binding regions of apolipoprotein E4 fragments act in concert to cause mitochondrial dysfunction and neurotoxicity. *Proc Natl Acad Sci U S A* 102(51), 18694–18699.

Chang, T. Y., et al. (2017). Cellular cholesterol homeostasis and Alzheimer's disease. *J Lipid Res* 58(12), 2239–2254.

Cheignon, C., et al. (2018). Oxidative stress and the amyloid beta peptide in Alzheimer's disease. *Redox Biol* 14, 450–464.

Chen, W. W., X. Zhang and W. J. Huang (2016). Role of physical exercise in Alzheimer's disease. *Biomed Rep* 4(4), 403–407.

Chen, Y., et al. (2020). The role of astrocytes in oxidative stress of central nervous system: A mixed blessing. *Cell Prolif* 53(3), e12781.

Cobley, J. N. (2020). Mechanisms of mitochondrial ROS production in assisted reproduction: The known, the unknown, and the intriguing. *Antioxidants (Basel)* 9(10), 933.

Cole, J. H. (2020). Multimodality neuroimaging brain-age in UK biobank: Relationship to biomedical, lifestyle, and cognitive factors. *Neurobiol Aging* 92, 34–42.

Corder, E. H., et al. (1993). Gene dose of apolipoprotein E type 4 allele and the risk of Alzheimer's disease in late onset families. *Science* 261(5123), 921–923.

Corder, E. H., et al. (1994). Protective effect of apolipoprotein E type 2 allele for late onset Alzheimer disease. *Nat Genet* 7(2), 180–184.

Croteau, E., et al. (2018). A cross-sectional comparison of brain glucose and ketone metabolism in cognitively healthy older adults, mild cognitive impairment and early Alzheimer's disease. *Exp Gerontol* 107, 18–26.

Dai, J., et al. (2008). Association between adherence to the Mediterranean diet and oxidative stress. *Am J Clin Nutr* 88(5), 1364–1370.

Daniele, S., C. Giacomelli and C. Martini (2018). Brain ageing and neurodegenerative disease: The role of cellular waste management. *Biochem Pharmacol* 158, 207–216.

Daskalopoulou, C., et al. (2017). Physical activity and healthy ageing: A systematic review and meta-analysis of longitudinal cohort studies. *Ageing Res Rev* 38, 6–17.

de Grey, A. D. (2005). Reactive oxygen species production in the mitochondrial matrix: Implications for the mechanism of mitochondrial mutation accumulation. *Rejuvenation Res* 8(1), 13–17.

de Wilde, M. C., et al. (2017). Lower brain and blood nutrient status in Alzheimer's disease: Results from meta-analyses. *Alzheimers Dement (N Y)* 3(3), 416–431.

Doig, A. J. (2018). Positive feedback loops in Alzheimer's disease: The Alzheimer's feedback hypothesis. *J Alzheimers Dis* 66(1), 25–36.

Done, A. J. and T. Traustadóttir (2016). Aerobic exercise increases resistance to oxidative stress in sedentary older middle-aged adults. A pilot study. *Age (Dordr)* 38(5–6), 505–512.

Dong, H. K., et al. (2017). Integrated late onset Alzheimer's disease (LOAD) susceptibility genes: Cholesterol metabolism and trafficking perspectives. *Gene* 597, 10–16.

Ermolaeva, M., et al. (2018). Cellular and epigenetic drivers of stem cell ageing. *Nat Rev Mol Cell Biol* 19(9), 594–610.

Fernández-Sanz, P., D. Ruiz-Gabarre and V. García-Escudero (2019). Modulating effect of diet on Alzheimer's disease. *Diseases* 7(1), E12.

Frederiksen, K. S., et al. (2019). Moderate- to high-intensity exercise does not modify cortical β-amyloid in Alzheimer's disease. *Alzheimers Dement (N Y)* 5, 208–215.

Galasko, D. R., et al. (2012). Antioxidants for Alzheimer disease: A randomized clinical trial with cerebrospinal fluid biomarker measures. *Arch Neurol* 69(7), 836–841.

Gamba, P., et al. (2019). A crosstalk between brain cholesterol oxidation and glucose metabolism in Alzheimer's disease. *Front Neurosci* 13, 556.

Gammella, E., S. Recalcati and G. Cairo (2016). Dual role of ROS as signal and stress agents: Iron tips the balance in favor of toxic effects. *Oxid Med Cell Longev* 2016, 8629024.

Gardener, S., et al. (2012). Adherence to a mediterranean diet and Alzheimer's disease risk in an Australian population. *Transl Psychiatry* 2, e164.

Grothe, M. J., S. J. Teipel and A. S. D. N. Initiative (2016). Spatial patterns of atrophy, hypometabolism, and amyloid deposition in Alzheimer's disease correspond to dissociable functional brain networks. *Hum Brain Mapp* 37(1), 35–53.

Gugliandolo, A., P. Bramanti and E. Mazzon (2017). Role of vitamin E in the treatment of Alzheimer's disease: Evidence from animal models. *Int J Mol Sci* 18(12), 2504.

Haller, S., et al. (2019). Amyloid load, hippocampal volume loss, and diffusion tensor imaging changes in early phases of brain aging. *Front Neurosci* 13, 1228.

Hansson, O., et al. (2019). Midlife physical activity is associated with lower incidence of vascular dementia but not Alzheimer's disease. *Alzheimers Res Ther* 11(1), 87.

Hardy, J. A. and G. A. Higgins (1992). Alzheimer's disease: The amyloid cascade hypothesis. *Science* 256(5054), 184–185.

Heneka, M. T., R. M. McManus and E. Latz (2018). Inflammasome signalling in brain function and neurodegenerative disease. *Nat Rev Neurosci* 19(10), 610–621.

Hipp, M. S., P. Kasturi and F. U. Hartl (2019). The proteostasis network and its decline in ageing. *Nat Rev Mol Cell Biol* 20(7), 421–435.

Hou, Y., et al. (2019). Ageing as a risk factor for neurodegenerative disease. *Nat Rev Neurol* 15(10), 565–581.

Hunter, S. and C. Brayne (2018). Understanding the roles of mutations in the amyloid precursor protein in Alzheimer disease. *Mol Psychiatry* 23(1), 81–93.

Hyder, F., D. L. Rothman and M. R. Bennett (2013). Cortical energy demands of signaling and non-signaling components in brain are conserved across mammalian species and activity levels. *Proc Natl Acad Sci U S A* 110(9), 3549–3554.

Jia, R. X., et al. (2019). Effects of physical activity and exercise on the cognitive function of patients with Alzheimer disease: A meta-analysis. *BMC Geriatr* 19(1), 181.

Kelleher, R. J. and J. Shen (2017). Presenilin-1 mutations and Alzheimer's disease. *Proc Natl Acad Sci U S A* 114(4), 629–631.

Kim, H. J., et al. (2003). Selective neuronal degeneration induced by soluble oligomeric amyloid beta protein. *FASEB J* 17(1), 118–120.

Kim, M., et al. (2009). Potential late-onset Alzheimer's disease-associated mutations in the ADAM10 gene attenuate {alpha}-secretase activity. *Hum Mol Genet* 18(20), 3987–3996.

Kim, S. I., J. S. Yi and Y. G. Ko (2006). Amyloid beta oligomerization is induced by brain lipid rafts. *J Cell Biochem* 99(3), 878–889.

Koffie, R. M., et al. (2009). Oligomeric amyloid beta associates with postsynaptic densities and correlates with excitatory synapse loss near senile plaques. *Proc Natl Acad Sci U S A* 106(10), 4012–4017.

Kögel, D., T. Deller and C. Behl (2012). Roles of amyloid precursor protein family members in neuroprotection, stress signaling and aging. *Exp Brain Res* 217(3–4), 471–479.

Kojro, E., et al. (2001). Low cholesterol stimulates the nonamyloidogenic pathway by its effect on the alpha -secretase ADAM 10. *Proc Natl Acad Sci U S A* 98(10), 5815–5820.

Kritsilis, M., et al. (2018). Ageing, cellular senescence and neurodegenerative disease. *Int J Mol Sci* 19(10), 2937.

Kryscio, R. J., et al. (2017). Association of antioxidant supplement use and dementia in the prevention of Alzheimer's disease by vitamin E and selenium trial (PREADViSE). *JAMA Neurol* 74(5), 567–573.

Kuhn, P. H., et al. (2010). ADAM10 is the physiologically relevant, constitutive alpha-secretase of the amyloid precursor protein in primary neurons. *EMBO J* 29(17), 3020–3032.

Kulig, W., et al. (2015). Cholesterol under oxidative stress-How lipid membranes sense oxidation as cholesterol is being replaced by oxysterols. *Free Radic Biol Med* 84, 30–41.

Kwak, Y. D., et al. (2011). Differential regulation of BACE1 expression by oxidative and nitrosative signals. *Mol Neurodegener* 6, 17.

Lesné, S., L. Kotilinek and K. H. Ashe (2008). Plaque-bearing mice with reduced levels of oligomeric amyloid-beta assemblies have intact memory function. *Neuroscience* 151(3), 745–749.

Liang, K. Y., et al. (2010). Exercise and Alzheimer's disease biomarkers in cognitively normal older adults. *Ann Neurol* 68(3), 311–318.

Lloret, A., et al. (2009). Vitamin E paradox in Alzheimer's disease: It does not prevent loss of cognition and may even be detrimental. *J Alzheimers Dis* 17(1), 143–149.

Lloret, A., et al. (2019). The effectiveness of Vitamin E treatment in Alzheimer's disease. *Int J Mol Sci* 20(4), 879.

Luo, J., et al. (2020). Ageing, age-related diseases and oxidative stress: What to do next? *Ageing Res Rev* 57, 100982.

Magi, S., et al. (2016). Intracellular calcium dysregulation: Implications for Alzheimer's disease. *Biomed Res Int* 2016, 6701324.

Markesbery, W. R. (1997). Oxidative stress hypothesis in Alzheimer's disease. *Free Radic Biol Med* 23(1), 134–147.

Markesbery, W. R. and M. A. Lovell (1998). Four-hydroxynonenal, a product of lipid peroxidation, is increased in the brain in Alzheimer's disease. *Neurobiol Aging* 19(1), 33–36.

Martín-Maestro, P., et al. (2017). Slower dynamics and aged mitochondria in sporadic Alzheimer's disease. *Oxid Med Cell Longev* 2017, 9302761.

Matthews, A. L., et al. (2017). Scissor sisters: Regulation of ADAM10 by the TspanC8 tetraspanins. *Biochem Soc Trans* 45(3), 719–730.

McManus, M. J., M. P. Murphy and J. L. Franklin (2011). The mitochondria-targeted antioxidant MitoQ prevents loss of spatial memory retention and early neuropathology in a transgenic mouse model of Alzheimer's disease. *J Neurosci* 31(44), 15703–15715.

McManus, M. J., M. P. Murphy and J. L. Franklin (2014). Mitochondria-derived reactive oxygen species mediate caspase-dependent and -independent neuronal deaths. *Mol Cell Neurosci* 63, 13–23.

Mecocci, P., et al. (2018). A long journey into aging, brain aging, and Alzheimer's disease following the oxidative stress tracks. *J Alzheimers Dis* 62(3), 1319–1335.

Mergenthaler, P., et al. (2013). Sugar for the brain: The role of glucose in physiological and pathological brain function. *Trends Neurosci* 36(10), 587–597.

Miller, E. R., et al. (2005). Meta-analysis: High-dosage vitamin E supplementation may increase all-cause mortality. *Ann Intern Med* 142(1), 37–46.

Mitchell, P. (1966). Chemiosmotic coupling in oxidative and photosynthetic phosphorylation. *Biol Rev Camb Philos Soc* 41(3), 445–502.

Mondola, P., et al. (2016). The Cu, Zn superoxide dismutase: Not only a dismutase enzyme. *Front Physiol* 7, 594.

Morris, J. K., et al. (2017). Aerobic exercise for Alzheimer's disease: A randomized controlled pilot trial. *PLoS One* 12(2), e0170547.

Mufson, E. J., et al. (2016). Braak staging, plaque pathology, and APOE status in elderly persons without cognitive impairment. *Neurobiol Aging* 37, 147–153.

Murphy, M. P. (2009). How mitochondria produce reactive oxygen species. *Biochem J* 417(1), 1–13.

Nissanka, N. and C. T. Moraes (2018). Mitochondrial DNA damage and reactive oxygen species in neurodegenerative disease. *FEBS Lett* 592(5), 728–742.

Nordin, T. C., A. J. Done and T. Traustadóttir (2014). Acute exercise increases resistance to oxidative stress in young but not older adults. *Age (Dordr)* 36(6), 9727.

Norton, S., et al. (2014). Potential for primary prevention of Alzheimer's disease: An analysis of population-based data. *Lancet Neurol* 13(8), 788–794.

Nunomura, A., et al. (2001). Oxidative damage is the earliest event in Alzheimer disease. *J Neuropathol Exp Neurol* 60(8), 759–767.

Oswald, M. C. W., et al. (2018). Regulation of neuronal development and function by ROS. *FEBS Lett* 592(5), 679–691.

Owen, J. B., et al. (2010). Oxidative modification to LDL receptor-related protein 1 in hippocampus from subjects with Alzheimer disease: Implications for Aβ accumulation in AD brain. *Free Radic Biol Med* 49(11), 1798–1803.

Pan, A., et al. (2018). Diet and cardiovascular disease: Advances and challenges in population-based studies. *Cell Metab* 27(3), 489–496.

Peña-Bautista, C., et al. (2019). Free radicals in Alzheimer's disease: Lipid peroxidation biomarkers. *Clin Chim Acta* 491, 85–90.

Perluigi, M., et al. (2009). Redox proteomics identification of 4-hydroxynonenal-modified brain proteins in Alzheimer's disease: Role of lipid peroxidation in Alzheimer's disease pathogenesis. *Proteomics Clin Appl* 3(6), 682–693.

Persson, T., B. O. Popescu and A. Cedazo-Minguez (2014). Oxidative stress in Alzheimer's disease: Why did antioxidant therapy fail? *Oxid Med Cell Longev* 2014, 427318.

Phan, H. T., et al. (2013). The effect of oxysterols on the interaction of Alzheimer's amyloid beta with model membranes. *Biochim Biophys Acta* 1828(11), 2487–2495.

Phan, H. T. T., et al. (2018). Strikingly different effects of cholesterol and 7-ketocholesterol on lipid bilayer-mediated aggregation of amyloid beta (1–42). *Biochem Biophys Rep* 14, 98–103.

Pickett, E. K., et al. (2016). Non-fibrillar oligomeric amyloid-β within synapses. *J Alzheimers Dis* 53(3), 787–800.

Powers, S. K., et al. (2020). Exercise-induced oxidative stress: Friend or foe? *J Sport Health Sci* 9(5), 415–425.

Rushworth, J. V. and N. M. Hooper (2010). Lipid rafts: Linking Alzheimer's amyloid-β production, aggregation, and toxicity at neuronal membranes. *Int J Alzheimers Dis* 2011, 603052.

Samadi, M., et al. (2019). Dietary pattern in relation to the risk of Alzheimer's disease: A systematic review. *Neurol Sci* 40(10), 2031–2043.

Santos, R. X., et al. (2013). Mitochondrial DNA oxidative damage and repair in aging and Alzheimer's disease. *Antioxid Redox Signal* 18(18), 2444–2457.

Scheff, S. W., M. A. Ansari and E. J. Mufson (2016). Oxidative stress and hippocampal synaptic protein levels in elderly cognitively intact individuals with Alzheimer's disease pathology. *Neurobiol Aging* 42, 1–12.

Schieber, M. and N. S. Chandel (2014). ROS function in redox signaling and oxidative stress. *Curr Biol* 24(10), R453–462.

Scott, I. and R. J. Youle (2010). Mitochondrial fission and fusion. *Essays Biochem* 47, 85–98.

Selkoe, D. J. and J. Hardy (2016). The amyloid hypothesis of Alzheimer's disease at 25 years. *EMBO Mol Med* 8(6), 595–608.

Sies, H. (1985). Oxidative stress: Introductory remarks oxidative stress. *New York Acad J* 5, 1–8.

Sies, H. (2015). Oxidative stress: A concept in redox biology and medicine. *Redox Biol* 4, 180–183.

Sies, H. and D. P. Jones (2020). Reactive oxygen species (ROS) as pleiotropic physiological signalling agents. *Nat Rev Mol Cell Biol* 21(7), 363–383.

Stephen, R., et al. (2017). Physical activity and Alzheimer's disease: A systematic review. *J Gerontol A Biol Sci Med Sci* 72(6), 733–739.

Sultana, R., et al. (2010). Redox proteomic analysis of carbonylated brain proteins in mild cognitive impairment and early Alzheimer's disease. *Antioxid Redox Signal* 12(3), 327–336.

Sun, L., et al. (2017). Analysis of 138 pathogenic mutations in presenilin-1 on the in vitro production of Aβ42 and Aβ40 peptides by γ-secretase. *Proc Natl Acad Sci U S A* 114(4), E476–E485.

Testa, G., et al. (2016). Changes in brain oxysterols at different stages of Alzheimer's disease: Their involvement in neuroinflammation. *Redox Biol* 10, 24–33.

Trachootham, D., et al. (2008). Redox regulation of cell survival. *Antioxid Redox Signal* 10(8), 1343–1374.

Trigiani, L. J. and E. Hamel (2017). An endothelial link between the benefits of physical exercise in dementia. *J Cereb Blood Flow Metab* 37(8), 2649–2664.

Verhaar, B. J. H., et al. (2020). Nutritional status and structural brain changes in Alzheimer's disease: The NUDAD project. *Alzheimers Dement (Amst)* 12(1), e12063.

Veurink, G., G. Perry and S. K. Singh (2020). Role of antioxidants and a nutrient rich diet in Alzheimer's disease. *Open Biol* 10(6), 200084.

Wang, D. S., et al. (2003). Oxidized neprilysin in aging and Alzheimer's disease brains. *Biochem Biophys Res Commun* 310(1), 236–241.

Wang, X., et al. (2014). Oxidative stress and mitochondrial dysfunction in Alzheimer's disease. *Biochim Biophys Acta* 1842(8), 1240–1247.

Watts, M. E., R. Pocock and C. Claudianos (2018). Brain energy and oxygen metabolism: Emerging role in normal function and disease. *Front Mol Neurosci* 11, 216.

Webb, R., et al. (2017). The ability of exercise-associated oxidative stress to trigger redox-sensitive signalling responses. *Antioxidants (Basel)* 6(3), 63.

Willems, P. H., et al. (2015). Redox homeostasis and mitochondrial dynamics. *Cell Metab* 22(2), 207–218.

Xie, X., et al. (2019). Nicotinamide ribose ameliorates cognitive impairment of aged and Alzheimer's disease model mice. *Metab Brain Dis* 34(1), 353–366.

Yin, F., et al. (2016). Mitochondrial function in ageing: Coordination with signalling and transcriptional pathways. *J Physiol* 594(8), 2025–2042.

Yoshida, Y., A. Umeno and M. Shichiri (2013). Lipid peroxidation biomarkers for evaluating oxidative stress and assessing antioxidant capacity in vivo. *J Clin Biochem Nutr* 52(1), 9–16.

Young, M. L. and J. L. Franklin (2019). The mitochondria-targeted antioxidant MitoQ inhibits memory loss, neuropathology, and extends lifespan in aged 3xTg-AD mice. *Mol Cell Neurosci* 101, 103409.

Yuan, X. Z., et al. (2017). The role of ADAM10 in Alzheimer's disease. *J Alzheimers Dis* 58(2), 303–322.

Zabel, M., et al. (2018). Markers of oxidative damage to lipids, nucleic acids and proteins and antioxidant enzymes activities in Alzheimer's disease brain: A meta-analysis in human pathological specimens. *Free Radic Biol Med* 115, 351–360.

Zhao, R. Z., et al. (2019). Mitochondrial electron transport chain, ROS generation and uncoupling (review). *Int J Mol Med* 44(1), 3–15.

Zheng, W., M. Y. Tsai and P. G. Wolynes (2017). Comparing the aggregation free energy landscapes of amyloid beta(1–42) and amyloid beta(1–40). *J Am Chem Soc* 139(46), 16666–16676.

Zhu, X., et al. (2004). Alzheimer's disease: The two-hit hypothesis. *Lancet Neurol* 3(4), 219–226.

Zilberter, Y. and M. Zilberter (2017). The vicious circle of hypometabolism in neurodegenerative diseases: Ways and mechanisms of metabolic correction. *J Neurosci Res* 95(11), 2217–2235.

CHAPTER SEVENTEEN

Exercise, Metabolism and Oxidative Stress in the Epigenetic Landscape

Gareth W. Davison and Colum P. Walsh

CONTENTS

Introduction / 209
Metabolic Control of Epigenetic Mechanisms / 210
Metabolism / 210
One-Carbon Metabolism and DNA/Histone Methylation / 210
Histone Acetylation / 211
TCA Metabolite-Dependent Regulation of DNA Methylation and Histone Modification / 211
The Interplay between Exercise Stress and Epigenetics / 212
TCA Cycle Intermediates / 212
Lactate / 214
Oxidative Stress and ROS / 214
ROS and DNA Damage/Repair / 215
Modulation of Antioxidant Effectors / 217
Conclusion and Future Perspectives / 218
References / 219

INTRODUCTION

Epigenetic modification refers to heritable changes in gene regulation that occur without a change in the DNA nucleotide sequence (Bird, 2007) and primarily includes DNA cytosine methylation and histone post-translational modifications. While these heritable changes were once thought to only regulate tissue-specific transcription, X-chromosome inactivation and genomic imprinting, it is now known that environmental influences (i.e. exercise) can determine the degree of transient or persistent changes occurring as either altered DNA methylation or histone modifications (Davison et al., 2021). Together, these marks create 'the epigenetic landscape', allowing the genome to display unique properties and patterns within specific cell types (Guillaumet-Adkins et al., 2017).

Oxidative stress occurs as a consequence of disrupted redox signalling and ROS accumulation; this phenomenon increases with high-intensity exercise. ROS (e.g., $^{\bullet}OH$) leads to lipid, protein and DNA damage and can disrupt the epigenetic state of the cell. For example, ROS can influence the methylome through direct hydroxylation of 5-methylcytosine (5mC) (Lewandowska and Bartoszek, 2011), and via DNA oxidation and TET-mediated hydroxymethylation affecting DNA demethylation (Chia et al., 2011). ROS can also indirectly modulate epigenetics, since methylation and histone-modifying enzymes depend on essential metabolites, such as acetyl-CoA, α-ketoglutarate, Fe^{2+}, ascorbic acid and fumarate, indicating that epigenetic changes are tightly linked to global cellular metabolism (Guillaumet-Adkins et al., 2017; Davison et al., 2021), and any fluctuations

DOI: 10.1201/9781003051619-17

in these intermediates may lead a reprogramming of gene expression. Although metabolism and epigenetics trade many interacting factors, and it is well-known that high-intensity exercise increases ROS production and the number of by-products of macromolecular damage such as DNA oxidation, nevertheless the interplay between exercise, oxidative stress and epigenetics is incompletely understood. Herein, and primarily due to a lack of data, we take (in the main) a hypothetical approach to ascertain the complex interrelationship between exercise-induced metabolism/oxidative stress and epigenetics.

METABOLIC CONTROL OF EPIGENETIC MECHANISMS

Metabolism

Metabolism is defined as a set of chemical modifications within cells, catalysed by specific enzymatic reactions to maintain homeostasis in response to environmental conditions (Etchegaray and Mostoslavsky, 2016). Epigenetic modifications require intermediates of cell metabolism for enzymatic function; conversely, any change to cell metabolism can alter specific acetyltransferases and methyltransferases directly affecting epigenetic modification patterns (Keating and El-Osta, 2015). In addition to overall cell metabolic changes, the epigenome may also be modified by intracellular metabolite localisation; for example, the tricarboxylic acid cycle (TCA) cycle in mitochondria produces key regulatory molecules which are used as co-substrates for transcriptional and epigenetic processes (Shaughnessy et al., 2014; Keating and El-Osta, 2015). To the contrary, TCA enzymes may also block epigenetic processes. This interplay between metabolism and epigenetics has been termed 'metaboloepigenetics' (Donohoe and Bultman, 2012). The mechanics of metabolic intermediates controlling gene expression, however, is incompletely understood (Davison et al., 2021).

One-Carbon Metabolism and DNA/ Histone Methylation

One-carbon metabolism utilises nutrients for multiple biological functions (Mentch and Locasale, 2016). Two major components of one-carbon metabolism are the folate and methionine cycles: both act by transferring single carbon units to acceptor molecules. S-adenosylmethionine (SAM) provides the activated methyl (CH_3) donor group as the main substrate for methyltransferase reactions on DNA, as well as on arginine and lysine residues of histones, leading to DNA methylation and post-translational modification's (PTM) (Mentch and Locasale, 2016; Keating and El-Osta, 2015).

DNA methylation occurs in two contexts aided by three DNA methyltransferase (DNMT) enzymes. DNMT1 is a maintenance methyltransferase, which recognises hemimethylated DNA to form a symmetrically modified duplex during DNA replication, while DNMT3A and DNMT3B are primarily de novo enzymes that deposit CH_3 marks on previously unmodified cytosines (Okano et al., 1999; Grurnbaum et al., 1982). Methylation of DNA occurs with the addition of CH_3 at the C^5 position of the nucleoside cytosine, forming 5-methylcytosine (5mC) on CpG dinucleotides (80% of cytosine residues in CpG dinucleotides are methylated at position 5, Davison et al., 2021). The CpG dinucleotide is unique in that it only occurs at a low frequency, while simultaneously converging into CpG islands (CGI) extending for 300–3,000 base pairs, mainly in regulatory elements of genes such as enhancer and promoter regions (Hitchler and Domann, 2021). Located up- and downstream of CpG islands are CpG shores and CpG shelves which display greater tissue-specific methylation profiles (Seaborne and Sharples, 2020). Methylation is absent in CGI AT promoter regions with active transcription. However, when promoter-associated CGI are methylated, transcription is silenced (Bird, 2007). More recently, it has been shown by a number of genome-wide and functional studies that DNA methylation in the gene body also facilitates transcription (Wu et al., 2010; Neri et al., 2013; Irwin et al., 2014).

Methylation of histone H3 and H4 is via histone methyltransferase (HMT) enzymes that covalently add CH_3 from SAM onto the side-chain nitrogen atoms of mainly lysine and arginine residues (Vanzan et al., 2017; Wong et al., 2017). Lysine methyltransferases contain a conserved domain Su(var)3–9, Enhancer of zeste and Trithorax (SET) responsible for adding CH_3 specifically at H3K4, H3K9, H3K27 and H3K36 (Vanzan et al., 2017). The consequence of histone methylation is determined by the specific histone residue modified, the number of methyl groups

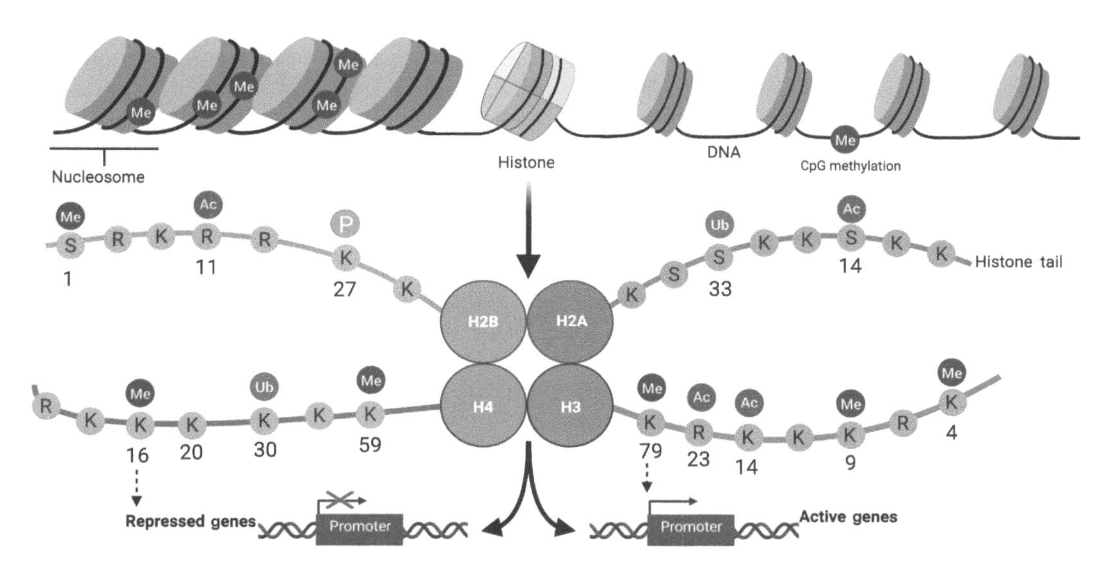

Figure 17.1 Chromatin and regulation of gene transcription. Gene transcription is regulated by chromatin accessibility. The nucleosome comprises 147 base pairs of tightly coiled DNA around each (two copies) of the four core histone proteins H2A, H2B, H3 and H4. R, Arginine; K, Lysine; S, Serine; Me, Methylation; Ac, Acetylation; Ub, Ubiquitination; P, Phosphorylation. H4K16me and H3K79me characterise a closed and open chromatin state, leading to repressed and active gene transcription respectively. (Taken from Davison et al. (2021).)

added (mono-, di- or tri-methylation), and the location within the N-terminal regions of either H3 or H4 (Kaelin and McKnight, 2013; Etchegaray and Mostoslavsky, 2016). These changes lead to a state of either euchromatin (lightly packed DNA promoting transcription) or heterochromatin (condensed DNA suppressing transcription; Figure 17.1) (Davison et al., 2021).

Histone Acetylation

Acetyl-CoA is involved in the transfer of an acetyl group to lysine amino acids on the N-terminal tails of histone to yield acetylation (Martinez-Reynes and Chandel, 2020; Vanzan et al., 2017). Acetylation occurs on each of the four histones via HAT enzymes, and HATs can be sensitive to fluxes in intracellular acetyl-CoA concentration (Kaelin and McKnight, 2013; Wong et al., 2017). Histone acetylation regulates gene transcription. For example, acetylation (i.e. H3K9ac, H3K27ac) at specific gene loci is implicated in transcriptional activation by opening chromatin structure, through a combination of (1) histone acetylation neutralising their positive charge which may weaken the interaction of the nucleosome with the DNA, leading to the opening of chromatin and active transcription and (2) histone acetylation acting as a docking station for the recruitment of transcription regulators (Vanzan et al., 2017; Wang et al., 2018; Davison et al., 2021).

TCA Metabolite-Dependent Regulation of DNA Methylation and Histone Modification

TCA cycle metabolites can control DNA and histone demethylation. Cytosine residues are demethylated by the oxidation of 5-methylcytosine (5mC) to 5-hydroxymethylcytosine (5hmC), catalysed by the ten-eleven translocation (TET 1–3) dioxygenase proteins, followed by decarboxylation to 5-formylcytosine (5fC) and 5-carboxylcytosine (5caC). 5caC is subsequently excised by thymine DNA glycosylase (TDG's) to restore unmethylated cytosine (He et al., 2011; Rodriguez et al., 2017). TET proteins control these sequential oxidation steps, and in doing so use α-Ketoglutarate (α-KG), oxygen and Fe^{2+} as co-factors (Figure 17.2). α-KG is a by-product of oxidative phosphorylation in the TCA cycle and is produced directly from isocitrate via isocitrate dehydrogenase 2 (IDH2) and 3 (IDH3). Its involvement in TET activity is dependent on the intracellular ratio of α-KG to succinate, fumarate or 2-hydroxyglutarate (2-HG) (Martinez-Reynes and Chandel, 2020). Changes in these ratios in the mitochondria can therefore block TET activity, potentially leading to hypermethylation of DNA (Davison et al., 2021). α-KG-related hydroxylation reactions further depend on the regeneration of Fe^{2+} and involve L-ascorbic acid (vitamin C) by inducing the reduction of oxidised Fe^{3+} to Fe^{2+} to restore TET activity (Martinez-Reynes and Chandel, 2020). Due to mutations in the L-gulonolactone oxidase (*Gulo*) gene in the primate lineage rendering it into a non-functional pseudogene (*GULOP*), humans have the inability to synthesize vitamin C in vivo and are therefore required to supplement through dietary sources (Etchegaray and Mostoslavsky, 2016). The role of ascorbate as a key co-factor for TET dioxygenases to sustain and complete the hydroxylation of 5mC to 5hmC is supported by Blaschke et al. (2013), who showed rapid and global DNA demethylation in mouse embryonic stem cells in response to the addition of vitamin C to the cell media. As such, TET proteins can respond under physiological changes (e.g. hypoxia) to agonists such as α-KG, Fe^{2+}, ascorbic acid or antagonist molecules such as fumarate and succinate and are likely to also be responsive to an altered and dysfunctional metabolism (Kaelin and McKnight, 2013; Davison et al., 2021).

Histone demethylation can have a repressive or activating role, depending on context, and occurs through lysine-specific demethylases (LSD) and the JumonjiC (JmjC) domain proteins (Lu and Thompson, 2012). LSD1 incorporates the co-factor flavin adenine dinucleotide (FAD^+), which is reduced to $FADH_2$ (Miranda-Goncalves et al., 2018) following demethylation of H3K4me1, H3K4me2, H3K9me1 and H3K9me2 (Etchegaray and Mostoslavsky, 2016). The paralog LSD2 also utilises FAD^+ to demethylate H3K4me1 and H3K4me2 (Karytinos et al., 2009). Similar to the workings of the TET isoenzymes, the *circa* 30 JmjC domain-contain histone demethylases rely on Fe^{2+} and α-KG co-factors to aid the removal of methyl groups from arginine and trimethylated lysine (H3K4, H3K9, H3K27, H3K36 and H4K20) in an oxidative-style reaction producing hydroxymethyl lysine. JmjC is also known to be inhibited by the TCA intermediates fumarate and succinate (Wang et al., 2018; Rodriguez et al., 2017; Etchegaray and Mostoslavsky, 2016; Davison et al., 2021).

Histone deacetylation, on the other hand, leads to a closed chromatin configuration and gene silencing. Deacetylation reactions are metabolically responsive (Wang et al., 2018; Miranda-Goncalves et al., 2018), where the glycolytic substrate nicotinamide adenine dinucleotide (NAD) acts as a redox cofactor in the deacetylation activity of sirtuins (HDAC enzymes). There are seven sirtuins in mammalian cells, with SIRT1, 2, 6 and 7 localised to the nucleus (Miranda-Goncalves et al., 2018; Imai et al., 2000). As NAD is subject to oxidation in normal cellular metabolism, it yields NAD^+ and NADH, and any adjustment in the NAD^+/NADH ratio can subsequently change sirtuin activity (Figure 17.2). For instance, when cells have net positive charged (e.g. due to an increased glucose flux), the NAD^+/NADH ratio drops, and this metabolic sensor inhibits sirtuin activity to regulate gene expression (Wong et al., 2017). In contrast, when cell NAD^+ concentration is elevated (increased NAD^+/NADH ratio), sirtuin activation occurs. So, when cells are deprived of ATP and other metabolic substrates, NAD^+ levels become elevated and SIRT1, in particular, is upregulated leading to histone (H3K9ac and H3K14ac) deacetylation and a potential up-regulation of metabolic enzymes (Canto et al., 2009; Davison et al., 2021; Etchegaray and Mostoslavsky, 2016).

THE INTERPLAY BETWEEN EXERCISE STRESS AND EPIGENETICS

TCA Cycle Intermediates

Metabolism provides energy in the form of adenosine triphosphate (ATP) aligned to glycolysis and activation of the TCA cycle to meet the demands of exercising skeletal muscle. Exercise bioenergetics may have profound effects on the epigenome by influencing the availability of metabolites such as α-KG, and the relative flux of metabolites through the TCA cycle as indicated in the preceding sections. The primary evidence linking exercise-induced TCA cycle intermediates and epigenetic modifications involve the key mitochondrial metabolite citrate. The nuclear form of acetyl-CoA is a central generator of

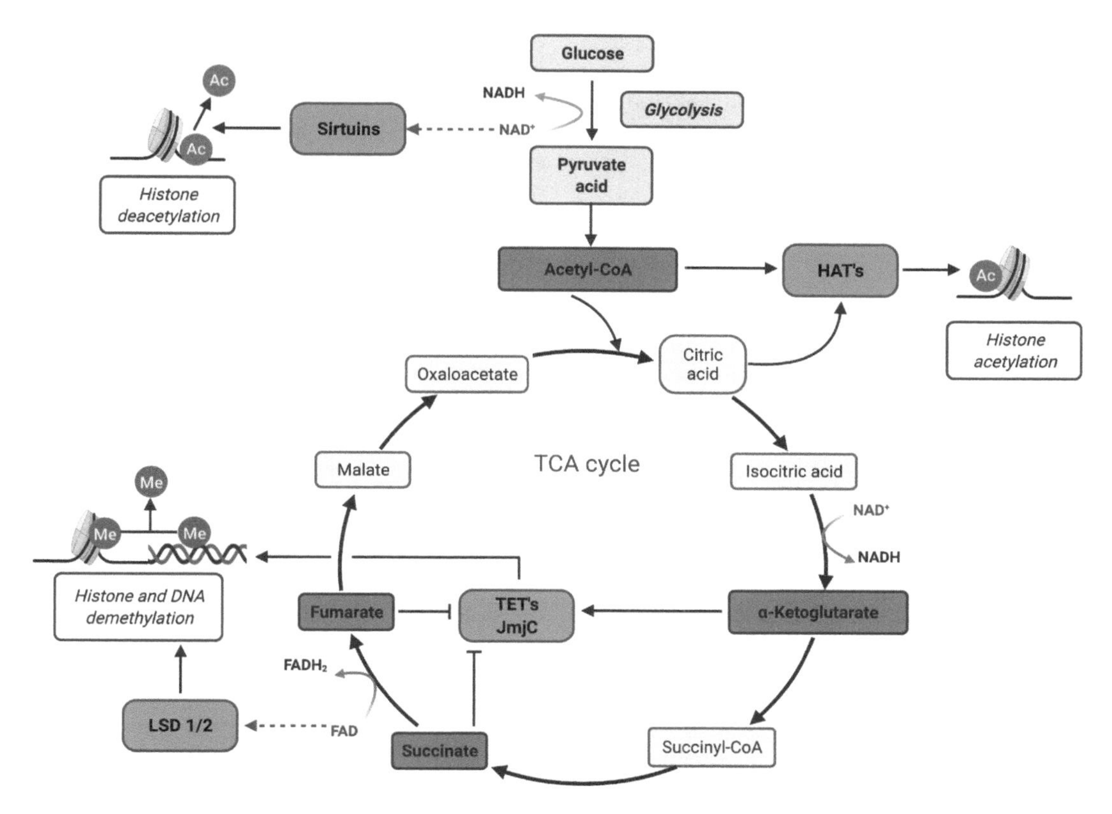

Figure 17.2 TCA cycle and glycolytic regulation of chromatin and DNA (De) methylation. Products of glycolysis (cytosol) and the tricarboxylic acid cycle (TCA; mitochondria) are required for chromatin and DNA modifications. Acetyl-CoA synthesised from pyruvate acid and fatty acid oxidation provides substrate for histone acetyltransferases (HATs). Histone deacetylation requires sirtuin molecules via the co-factor nicotinamide adenine dinucleotide (NAD). Lysine-specific demethylases (LSD) utilise flavin adenine nucleotide (FAD), while α-ketoglutarate controls the ten-eleven translocation (TET) enzymes and Jumonji C domain-containing (JmjC) demethylases. Fumarate and succinate block TET and JmjC molecules. Key intermediate metabolites involved in the control of histone and DNA modification are shown in red. Molecules affected by key metabolites are shown in orange. (Modified from Davison et al. (2021).)

acetyl molecules in histone acetylation induction. Mitochondrially derived citrate supplies nuclear acetyl-CoA through ATP citrate lyase (ACLY), drawing citrate from the cytoplasm into the nucleus to yield acetyl-CoA and oxaloacetate (Figure 17.3). ACYL is important in the nexus between the metabolic cell state and epigenetic modifications at the level of the histone and may become a rate-limiting step in the transcriptional competency of the cell (Seaborne and Sharples, 2020).

In a skeletal muscle context, ACLY silencing impairs myoblast and satellite cell differentiation, accompanied by a decrease in the gene expression of fast myosin heavy chain isoforms, and the myogenic regulatory factor MyoD. Conversely, overexpression of ACLY enhances MyoD expression via hyperacetylation of H3K9, 14 and 27 (Das et al., 2017). In this regard, it is theoretically plausible that exercise stress may stimulate the pathway involving ACYL induction and the conversion of citrate to acetyl-CoA. This metabolic flux may lead to hyperacetylation of key skeletal muscle remodelling transcripts (Seaborne and Sharples, 2020). Using a whole-genome sequencing approach, Sailani et al. (2019) determined that lifelong exercise can positively modify the epigenome to protect skeletal muscle cells against a reduced metabolic capacity and oxidative stress. DNA methylation was differentially lower (in exercise vs sedentary) in 714 promoters for genes encoding critical enzymes relating to glycogen metabolism, glycolysis, and the TCA cycle. In general, promoter regions of genes involved in the oxidative stress response were hypomethylated in the exercise group. Collectively, these hypomethylation profiles suggest that lifelong exercise can manipulate the capacity for glycolysis and TCA cycle flux and protect cells from age-related damage and deterioration brought on by oxidative stress.

Lactate

Lactate is a metabolite produced during glycolysis, where glucose is converted to two pyruvate molecules; these can either be fluxed into lactate or transported into mitochondria forming acetyl-CoA (Izzo and Wellen, 2019). Strenuous exercise activates the cell glycolytic pathway to elevate lactate, where an increase in exercise intensity leads to a greater increase in lactate production. Lactate has a distinct and direct role in cell epigenetics, where it acts as an endogenous HDAC inhibitor, leading to histone hyperacetylation with corresponding dysregulation of gene expression. In this sense, lactate is an important transcriptional regulator, linking the metabolic state of the cell to gene transcription (Latham et al., 2012). More recently, lactate has been implicated as a precursor substrate to an epigenetic modification termed lactylation. Zhang et al. (2019) identified 28 lactylation sites on core histone lysine residues that directly stimulated gene transcription from chromatin. Through a combination of functional experiments and using isotopically labelled glucose ($^{13}C_6$-glucose), it was postulated that lysine lactylation depends solely on glucose flow through glycolysis – i.e. as glycolysis increases, intracellular lactate rises, paralleling an increase in histone lactylation. Functionally, lysine lactylation occurred mainly in promoter regions of protein-coding genes and aligned to corresponding gene expression of these transcripts, suggesting a protranscriptional characteristic of the modification (Liberti and Locasale, 2020). Whilst this novel data is intriguing, caveats remain: (1) the data warrants independent validation; (2) whilst the authors demonstrate that lactyl-CoA is a lactyl-group donor for lysine lactylation, the enzymes that produce lactyl-CoA from lactate and the cell concentration of lactyl-CoA remain unknown (Izzo and Wellen, 2019; Davison et al., 2021). Undoubtedly, more work is required to ascertain the definitive link between intracellular lactate production and its ability to modify the epigenome. However, it does raise the intriguing possibility that exercise-induced lactate production can subsequently regulate the epigenetic landscape of the genome and the transcriptional capacity of coding genes. Moreover, lactate molecules can shuttle between cells and tissues, as such lactate production as a function of exercise may facilitate crosstalk between cells and tissue contributing to a broader adaptive responsiveness within the biological milieu (Seaborne and Sharples, 2020).

Oxidative Stress and ROS

Oxidative stress (defined by the modification of DNA, RNA, lipids and proteins) influences chromatin structure, DNA methylation, enzymatic and non-enzymatic post-translational modifications of histones and DNA-binding proteins. The effect of oxidative stress on aforementioned chromatin alterations mediates several cellular modifications, including modulation of gene expression, cell death, cell survival and mutagenesis; the latter of which are disease-driving mechanisms in human pathologies (Kreuz and Fischle, 2016). The complex interlink between oxidative stress and epigenetics is summarised in Table 17.1.

High-intensity strenuous exercise leads to a state of oxidative stress exacerbated by the production of reactive oxygen species (ROS; typical examples include $O_2^{\bullet-}$, H_2O_2 and $^{\bullet}OH$). ROS are generated during exercise and in exercise recovery as by-products of systemic and cellular metabolism, and while the mechanisms of ROS production are comprehensively outlined elsewhere, in brief, prototypic NADPH-oxidases (NOX) are a primary source of superoxide anions ($O_2^{\bullet-}$). In particular, NOX2 with its access to the cytosolic domain and the localisation of NOX4 to mitochondria are often cited as potential suspects in exercise-induced production of $O_2^{\bullet-}$ and H_2O_2 (NOX4 only) (Williamson and Davison, 2020). Mitochondrial generation of $O_2^{\bullet-}$ and H_2O_2 involves electron leakage from donor redox centres of the electron transport chain (ETC) yielding a state of uni- or bivalent reduction of oxygen. There are at least 11 known mitochondrial sites of $O_2^{\bullet-}$ and H_2O_2 production encompassing NADH/NAD$^+$ and ubiquinol/ubiquinone (UQH2/UQ); each with different production capabilities (Williamson and Davison, 2020). Oxidative phosphorylation in mitochondria may only be a significant source of ROS production following exercise, as $O_2^{\bullet-}$ seems to be greater in State 4 respiration (i.e. rest) compared with State 3 (i.e. exercise) (Henriquez-Olguin

TABLE 17.1

Overview of oxidative stress and epigenetic nexus.

Oxidative stress and chromatin structure

- Aligned to a particular cell type, oxidative stress causes heterochromatin loss and cell death or protective stabilisation of heterochromatin.
- Direct oxidative modifications of histones alter chromatin structure and induce changes in transcription and DNA replication.

Oxidative stress and DNA methylation

- Oxidative stress-induced inhibition of DNA methyltransferases and relocalisation of DNA methyltransferases to CpG islands induce global DNA hypomethylation and local promoter hypomethylation.
- Inhibition of Fe(II)- and α-KG-dependent TET DNA demethylases by oxidation, fumarate or succinate leads to altered DNA methylation.

Oxidative stress and histone modification

- Oxidative stress inhibits Fe(II)- and α-KG-dependent histone demethylases of the JmjC family and potentially activates histone hypermethylation.
- Oxidative stress directly inhibits histone deacetylases and can also stimulate their gene expression alongside locally increasing NAD^+ to activate SIRT deacetylases.

Oxidative stress regulates DNA-associated proteins

- Oxidative stress sequesters the corepressor TRIM28 from chromatin by binding to tyrosyl-t-RNA synthetase or oxidised HP1γ leading to activation of gene expression.
- Oxidised HMGB1 loses its DNA-bending activity and its ability to displace histone H1, possibility inhibiting gene expression and stabilising chromatin. Aligned to this, it gains a DNA end-joining function contributing to DNA repair following oxidative stress.
- The second messenger PI(5)P is upregulated by oxidative stress and modulates the functions of the corepressor ING2, the general transcription factor TAF3 and the chromatin-modifying protein UHRF1 – controlling transcription at multiple levels.

SOURCE: Modified from Kreuz and Fischle (2016).

et al., 2020; Goncalves et al., 2015), and it is possible that the net production of $O_2^{\cdot-}$ and H_2O_2 after exercise causes persistent damage over time. Mitochondria can influence epigenetic signalling through H_2O_2 production with the premise that any epigenetic alterations in response to ROS may in turn lead to altered expression of genes that regulate mitochondrial metabolism (Shaughnessy et al., 2014). Moreover, the lysine-specific histone Demethylases 1 and 2 (LSD1, LSD2) are known to demethylate the mono- and di-methylated Lys4 and Lys9 of histone H3 via an FAD-dependent oxidative reaction involving H_2O_2 (Forneris et al., 2008). Intriguingly, the LSD1-associated production of H_2O_2 promotes the production of 8-oxodG, suggesting that although roaming free iron is lowered by the pool of nuclear ferritin, H_2O_2 may still lead to $^{\bullet}OH$ production which probably involves iron-containing complexes associated with the DNA structure (Gorini et al., 2021). The work of Barres et al. (2012), to our knowledge, is the only experiment to quantify epigenetics and ROS concurrently following exercise up to 80% VO_{2max} Exercise remodelled human skeletal muscle promoter methylation by causing a state of global hypomethylation with a dose-dependent mRNA expression in a select number of genes: PGC-1a, PDK4 and PPAR-δ. However, promoter methylation in differentiated L6 rat myotubes was unaltered by 1 mM H_2O_2 exposure (Note: this represents an abnormal physiological dose), even though ROS increased mRNA expression, suggesting in this instance, that DNA methylation does not exclusively control exercise-induced gene expression.

Furthermore, ROS production can oxidise signalling proteins and transcription factors, and also important components of the epigenetic machinery, including DNA and RNA. Various modifications (e.g. to thiol groups) cause conformational changes that can affect enzymatic activity, protein-protein interaction and protein sub-cellular localisation (Dimauro et al., 2020).

ROS and DNA Damage/Repair

Our laboratory has consistently demonstrated that high-intensity exercise leads to guanine nucleotide oxidation (Fogarty et al., 2013) and DNA single- and double-strand breaks; the latter quantified using a marker of histone phosphorylation, γ-H2AX combined with 53BP1 (Williamson and Davison, 2020). Other work by Williamson et al. (2020) shows that high-intensity (95% heart rate max) exercise increases nuclear (single-cell gel electrophoresis comet assay) and mitochondrial DNA damage (Long Amplicon-Quantitative Polymerase Chain Reaction, LA-qPCR assay) across lymphocytes and muscle respectively. DNA bases are directly modified by ROS, where the hydroxyl free radical ($^{\bullet}OH$) can play an integral role through its production by Fenton reactions that involve the reduction of H_2O_2 by either ferrous ($k \sim 76\ M^{-1}s^{-1}$) or copper ions ($k \sim 4.7 \times 10^3\ M^{-1}s^{-1}$) (Davison et al., 2016, 2021). $^{\bullet-}OH$-mediated DNA changes are initiated by hydrogen abstraction or by $^{\bullet}OH$ interfering with a DNA base (Davison, 2016); an example being the oxidation of 5mC to 5hmC leading to demethylation at CpG sites (Madugundu et al., 2014; Dimauro et al., 2020; Figure 17.3). While not shown in an exercise context per se, ROS-induced DNA damage may interfere with epigenetic processes. $^{\bullet}OH$ production can directly modify DNA methylation through oxidation of guanosine to 8-oxo-2′-deoxyguanosine (8-oxodG). 8-OxodG in normal cells represents ~1 in 10^6 guanines, but this can increase by an order of magnitude under conditions of oxidative stress (Fleming and Burrows, 2020a), with an increased steady-state level of 8-oxodG lesions leading to stalled replication forks and C:G to A:T transverse mutations (van Loon et al., 2010). Usually, 8-oxodG is recognised and removed from an OG:C base pair in duplex DNA by 8-oxoguanine DNA glycosylase (OGG1) through cleavage of the glycosidic bond between the ribose and the base, releasing 8-oxodG and generating an apurinic (AP) site in DNA. This short-patch base excision repair (BER) mechanism further involves AP site cleavage by AP-endonuclease-1 (APE1) to re-synthesise undamaged DNA using the cytosine base opposite as a template for insertion of an undamaged G nucleotide (Fleming and Burrows, 2020a). However, on 8-oxodG accumulation, methylation of adjacent cytosines is attenuated, leading to a state of hypomethylation and transcriptional activation (Le and Fujimori, 2012). Zhou et al. (2016) postulate that OGG1 promotes DNA demethylation by recruitment of TET1 to the oxidised lesion, thus suggesting a model in which oxidative stress recruits OGG1/TET1 complex proteins to 8-oxodG, facilitating the conversion of 5mC through to 5caC close to sites of ROS-induced damage. However, apart from thee reports there is currently not a great deal of additional evidence of APE1 or OGG1 being linked to DNA demethylation in a wider context.

Further interplay between DNA repair, oxidatively generated 8-oxodG and transcriptional activity has been documented in mammalian genes, with Fleming et al. (2017a) demonstrating that DNA damage in G-quadruplex-forming sequences (PQS) of a promoter can lead to a 300% increase in gene expression. This suggests that 8-oxodG represents an epigenetic-like modification, while G-quadruplex-forming sequences serve as sensors of oxidative stress. G-quadruplexes are G-rich sequences that can fold in DNA structures and comprise four or more continuous runs of oligo-G (specific sites that are optimally reactive toward one-electron oxidation), interspersed with one or more loops between the G run nucleotides. The GGG runs in one strand of DNA duplex can fold, forming layers in G-tetrads, and are exposed sites for G oxidation by a one-electron mechanism. The carbonate radical anion ($CO_3^{\bullet-}$) is a potent one-electron oxidant formed from carbon dioxide and peroxynitrite, and also directly from bicarbonate; muscle bicarbonate concentration can rise multiple fold during high-intensity exercise. While $CO_3^{\bullet-}$ ($pKa < 0$) is more selective and less oxidising than $^{\bullet}OH$ ($E^0 = 1.78\ V$ vs. $2.3\ V$ at pH 7.0), it can initiate many damaging reactions in a biological system (Medinas et al., 2007; Di Meo and Venditti, 2020).

A G-quadruplex-forming sequence in a promoter can serve as a redox sensor by changing from standard DNA duplex to a G-4 fold following just one G oxidation (Fleming and Burrows, 2020b). Although the majority of human gene promoters are G rich, with *circa* 40% containing potential G-4 sequences (Huppert and Balasubramanian, 2007), the presence of 8-oxodG alone may not trigger a shift from duplex to quadruplex. However, when OGG-1 responds to an increased generation of 8-oxodG, the AP site destabilises the duplex structure, leading to refolding to the quadruplex if the AP is in a loop while retaining a stable G4. If 8-oxodG is present following G-4 folding, OGG-1 may bind for signalling purposes, and alongside the binding of a redox effector factor molecule (e.g. *APE1*), recruitment of transcription factors may occur (Fleming and Burrows, 2020a). Interestingly, removal of 8-oxodG from a DNA coding (sense) strand by OGG-1-mediated BER upregulated transcription, while for 8-oxodG residing in the template (antisense) strand of *VEGF*, transcription was downregulated by a BER-independent mechanism (Fleming et al., 2017b). In the context of cellular adaptation to oxidative stress generated by high-intensity exercise, this data is salient. 8-OxodG in a promoter region impacts the transcriptional pre-initiation complex that can lead to the coupling of DNA repair and initiation of transcription, while 8-oxodG in a coding region may interfere with progression of the transcriptional elongation complex leading to stalling or truncated transcription (Fleming et al., 2017b).

In a similar line of investigation, Sulkowski et al. (2020) ascertained a connection between 2-HG, succinate, fumarate and DNA repair. The TCA cycle metabolites inhibit the lysine demethylase KDM4B, resulting in aberrant hypermethylation of H3K9 at loci in close proximity to DNA double-strand breaks. This conceals a local H3K9 trimethylation signal that is essential for the proper functioning of the DNA homology-dependent repair (HDR) pathway. Consequently, the recruitment of salient HDR transcriptional regulators – Tip60 and ATM – is impaired at DNA break sites leading to reduced end resection. Of note is the increased TCA cycle activity of succinate and fumarate commonly observed in exercise (Brugnara et al., 2012), and it is thus plausible that a disruption in chromatin signalling and DNA repair, as outlined, may be exacerbated in certain forms of exercise that activate the oxidative phosphorylation pathway.

Whilst facets of the aforementioned are hypothetical mainly from an exercise perspective, its postulates should be subject to scientific enquiry through rigorous experimental testing in an exercise setting.

Modulation of Antioxidant Effectors

In mammalian cell types, the production of ROS is regulated by enzymatic and non-enzymatic antioxidants. ROS, particularly H_2O_2, are important signalling molecules acting to induce transcription factor activation (e.g. Nrf2) and subsequent gene expression of a host of intracellular antioxidant enzymes (Figure 17.3). Gene expression of antioxidant enzymes can also be directly modulated by DNA methylation and post-translational histone modifications. For instance, the epigenetic regulation (methylation and histone changes) of superoxide dismutase 2 (*SOD2*) can affect promoter and enhancer sites, alongside other upstream regulatory elements (Cyr et al., 2013). Although the inter-relationship between exercise and epigenetic regulation of antioxidant enzymes is still in its infancy (Dimauro et al., 2020), Nguyen et al. (2019) found that glutathione peroxidase 1 (*Gpx1*) was hypomethylated following 3 months of exercise. This alteration was observed in the second exon of the *Gpx1* gene and occurred alongside an increase in mRNA expression. In a more recent study, Sailani et al. (2019) demonstrated that protein levels of key antioxidant enzymes catalase (*CAT*) and superoxide dismutase 2 (*SOD2*) were enhanced with exercise, corresponding to a hypomethylation of *SOD2* and *CAT* promoters (20% and 18%).

Strenuous exercise reduces systemic and intracellular non-enzymatic antioxidants, including ascorbic acid. Indeed, our laboratory has shown that oral ascorbic acid supplementation (1,000 mg) provides an effective prophylaxis against exercise-induced oxidative stress (Davison et al., 2008). From an epigenetics perspective, ascorbic acid is a known co-factor regulating TET isoenzymes, and in vivo data demonstrate that ascorbic acid enhances TET-mediated 5mC oxidation in mice (Yin et al., 2013; Blaschke et al., 2013). It is thus conceivable, where high-intensity exercise compromises ascorbic acid concentration, that a transient inactivation of TET may ensue, leading to a more hypermethylated state. Equally so, cells with normal ascorbic acid concentration may also respond to high-dose ascorbic acid supplementation leading to a change in methylation via the overactivation of intracellular TET enzyme activity. In this context, we recently demonstrated in a small-scale randomized controlled trial ($n = 24$) that high-intensity exercise (1 hour cycle ergometer at 70% VO_{2max}) did indeed lead to gains in DNA methylation genome-wide, as assessed using Illumina EPIC arrays (unpublished). While 10 g daily ascorbic acid supplementation for 4 weeks on its own did not appear to lead to any significant demethylation, however, when combined with the exercise we observed a significant loss of

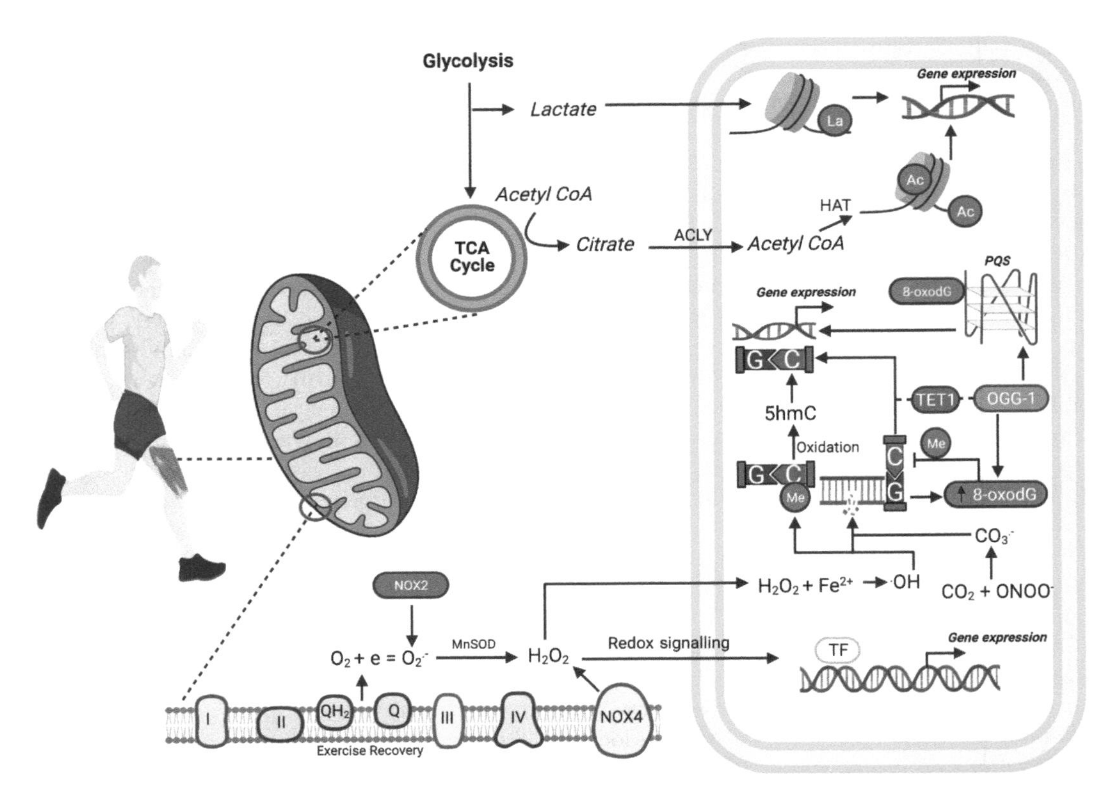

Figure 17.3 Interplay between exercise, epigenetics and gene transcription. Hypothetical schematic of exercise and mitochondria-nucleus crosstalk. Exercise leads to ROS production. $^{\bullet}OH$ generation increases 5mC oxidation/5hmC. Hydroxyl ($^{\bullet}OH$) and carbonate ($CO_3^{\bullet-}$) radicals attack G (guanine nucleotides) leading to 8-oxodG accumulation (also in DNA quadruplexes, PQS), demethylated C (cytosine) and OGG-1 activation causing transcriptional changes. OGG-1 recruits TET1 leading to hypomethylation and binds to PQS complex causing gene expression. Metabolic substrates lactate (La) and citrate control histone lactylation and acetylation respectively leading to gene expression. H_2O_2 acts as a redox signalling molecule leading to gene expression of antioxidant enzymes. TCA, tricarboxylic acid cycle; Fe^{2+}, iron; Me, methyl group; Ac, acetyl group; HAT, histone acetyltransferase; ACLY, ATP citrate lyase; NOX, NADPH oxidase.

DNA methylation genome-wide, consistent with the above hypothesis.

CONCLUSION AND FUTURE PERSPECTIVES

Unambiguous crosstalk exists between metabolism, oxidative stress and epigenetics; however, the extent to which exercise *per se* can influence and interfere with this crosstalk is unclear. High-intensity exercise enhances the production of metabolic intermediates such as lactate, fumarate, succinate and α-KG, while ROS (hydroxyl radicals mainly) can damage DNA guanine nucleotides. The link between exercise, DNA damage and epigenetics, in particular, deserves a comprehensive enquiry. Moreover, ROS have the potential to play a significant role in shaping the epigenetic landscape, not only through their adverse (oxidation) control of DNA methylation and PTM's, but also as modulators of epigenetic mechanisms through redox signalling, with essential implications for localised transcriptional regulation. Finally, there is a need to ascertain the role of exercise-induced ROS and their potential to oxidise the several thiol-based transcription factors aligned to epigenetic regulation. To this end, this narrative should be viewed as a starting point to explore the multidirectional connections between exercise, metabolism and oxidative stress in the epigenetic landscape.

REFERENCES

Barres, R., Yan, Y., Egan, B. et al. (2012) Acute exercise remodels promoter methylation in human skeletal muscle. *Cell Metabolism* 15, 405–411.

Bird, A. (2007) Perceptions of epigenetics. *Nature* 447, 396–398.

Blaschke, K., Ebata, K.T., Karimi, M.M. et al. (2013) Vitamin C induces TET-dependent DNA demethylation and a blastocyste-like state in ES cells. *Nature* 500, 222–226.

Brugnara, L., Vinaixa, M., Murillo, S., Sarmino, S., Rodriguez, M.A., Beltran, A., Davison, G.W., Correig, X., Novials, A. (2012) Metabolomics approach for analyzing the effects of exercise in subjects with type 1 diabetes mellitus. *PLoS One* 7(7), e40600.

Canto, C., Gerhart-Hines, Z., Feige, J.N. et al. (2009) AMPK regulates energy expenditure by modulating NAD^+ metabolism and SIRT1 activity. *Nature* 458, 1056–1060.

Chia, N., Wang, L., Lu, X. et al. (2011) Hypothesis: Environmental regulation of 5-hydroxymethylcytosine by oxidative stress. *Epigenetics* 6, 853–856.

Cyr, A.R., Hitchler, M.J., Domann, F.E. (2013) Regulation of SOD2 in cancer by histone modifications and CpG methylation: Closing the loop between redox biology and epigenetics. *Antioxidants & Redox Signaling* 18, 1946–1955.

Das, S., Morvan, F., Morozzi, G. et al. (2017) ATP citrate lyase regulates myofiber differentiation and increases regeneration by altering histone acetylation. *Cell Reports* 21, 11.

Davison, G.W. (2016) Exercise and oxidative damage in nucleoid DNA quantified using single cell gel electrophoresis: Present and future application. *Frontiers Physiology* 7, 249. doi:10.3389/fphys.2016.00249.

Davison, G.W., Ashton, T., George, L., Young, I.S., McEneny, J., Davies, B., Jackson, S.K., Peters, J., Bailey, D.M. (2008) Molecular detection of exercise-induced free radicals following ascorbate prophylaxis in type 1 diabetes mellitus: A randomised controlled trial. *Diabetologia* 51, 2049–2059.

Davison, G.W., Irwin, R.E., Walsh, C.P. (2021) The metabolic-epigenetic nexus in type 2 diabetes mellitus. *Free Radical Biology and Medicine* 170, 194–206.

Di Meo, S., Venditti, P. (2020) Evolution of the knowledge of free radicals and other oxidants. *Oxidative Medicine and Cellular Longevity* 2020, 9829176.

Dimauro, I., Paronetto, M.P., Caporossi, D. (2020) Exercise, redox homeostasis and the epigenetic landscape. *Redox Biology* 35, 101477.

Donohoe, D.R., Bultman, S.J. (2012) Metaboloepigenetics: Interrelationships between energy metabolism and epigenetics control of gene expression. *Journal of Cellular Physiology* 227, 3169–3177.

Etchegaray, J.-P., Mostoslavsky, R. (2016) Interplay between metabolism and epigenetics: A nuclear adaptation to environmental changes. *Molecular Cell* 62, 695–711.

Fleming, A.M. Burrows, C.J. (2020b) Interplay of guanine oxidation and G-quadruplex folding in gene promoters. *Journal of the American Chemical Society* 142, 1115–1136.

Fleming, A.M., Burrows, C.J. (2020a) On the irrelevancy of hydroxyl radical to DNA damage from oxidative stress and implications for epigenetics. *Chemical Society Reviews* 49, 6524–6528.

Fleming, A.M., Ding, Y., Burrows, C.J. (2017a) Oxidative damage is epigenetic by regulating gene transcription via base excision repair. *PNAS* 114, 2604–2609.

Fleming, A.M., Zhu, J., Ding, Y., Burrows, C.J. (2017b) 8-Oxo-7,8-dihydroguanine in the context of a gene promoter G-quadruplex is an on–off switch for transcription. *ACS Chemical Biology* 12, 2417–2426.

Fogarty, M., De Vito, G., Hughes, C.M., Burke, G., Brown, J.C., McEneny, J., Brown, D., McClean, C., and Davison, G.W. (2013) Effects of α-lipoic acid on mtDNA damage following isolated muscle contractions. *Medicine and Science in Sport and Exercise* 45, 1469–1477.

Forneris, F., Binda, C., Battaglioli, E., Mattevi, A. (2008) LSD1: Oxidative chemistry for multifaceted functions in chromatin regulation. *Trends in Biochemical Sciences* 33, 181–189.

Goncalves, R.L.S., Quinlan, C.L., Perevoshchikova, I.V. et al. (2015) Sites of superoxide and hydrogen peroxide production by muscle mitochondria assessed *ex Vivo* under conditions mimicking rest and exercise. *The Journal of Biological Chemistry* 290, 209–227.

Gorini, F., Scala, G., Cooke, M.S. et al. (2021) Towards a comprehensive view of 8-oxo-7,8-dihydro-2'-deoxyguanosine: Highlighting the intertwined roles of DNA damage and epigenetics in genomic instability. *DNA Repair* 97, 103027.

Grurnbaum, Y., Cedar, H., Razin, A. (1982) Substrate and sequence specificity of a eukaryotic DNA methylase. *Nature* 295, 620–622.

Guillaumet-Adkins, M., Yañez, Y., Peris-Diaz, M.D. et al. (2017) Epigenetics and oxidative stress in aging. *Oxidative Medicine and Cellular Aging*, 2017. doi: 10.1155/2017/9175806.

He, Y.-F., Li, B.-Z., Li, Z. et al. (2011) Tet-mediated formation of 5-carboxylcytosine and its excision by TDG in Mammalian DNA. *Science* 333, 1303–1307.

Henriquez-Olguin, C., Meneses-Valdes, R., Jensen, TE. (2020) Compartmentalized muscle redox signals controlling exercise metabolism: Current state, future challenges. *Redox Biology* 35, 101473.

Hitchler, M.J., Domann, F.E. (2021) The epigenetic and morphogenetic effects of molecular oxygen and its derived reactive species in development. *Free Radical Biology and Medicine*. doi: 10.1016/j.freeradbiomed.2021.01.008.

Huppert, J.L., Balasubramanian, S. (2007) G-quadruplexes in promoters throughout the human genome. *Nucleic Acids Research* 25, 406–413.

Imai, S., Armstrong, C.M., Kaeberlein, M. et al. (2000) Transcriptional silencing and longevity Sir2 is and NAD- dependent histone deacetylase. *Nature* 403, 795–800.

Irwin, R.E., Thakur, A., O'Neill, K.M., Walsh, C.P. (2014) 5-hydroxymethylation marks a class of neuronal gene regulated by intragenic methylcytosine levels. *Genomics* 104, 383–292.

Izzo, L.T., Wellen, K.E. (2019) Histone lactylation links metabolism and gene regulation. *Nature* 574, 492–493.

Kaelin, W.G., McKnight, S.L. (2013) Influence of metabolism on epigenetics and disease. *Cell* 153, 56–69.

Karytinos, A., Forneris, F., Profumo, A. et al. (2009) A novel mammalian flavin-dependent histone demethylase. *Journal of Biological Chemistry* 284, 17775–17782.

Keating, S.T., El-Osta, A. (2015) Epigenetics and metabolism. *Circulation Research* 116, 715–736.

Kreuz, S., Fischle, W. (2016) Oxidative stress signalling to chromatin in health and disease. *Epigenomics* 8, 843–862.

Latham, T., Mackay, L., Sproul, D. et al. (2012) Lactate, a product of glycolytic metabolism, inhibits histone deacetylase activity and promotes changes in gene expression. *Nucleic Acids Research* 40, 4794–4803.

Le, D.D., Fujimori, D.G. (2012) Protein and nucleic acid methylating enzymes: Mechanisms and regulation. *Current Opinion in Chemical Biology* 16, 507–515.

Lewandowska, J., Bartoszek, A. (2011) DNA methylation in cancer development, diagnosis and therapy: Multiple opportunities for genotoxic agents to act as methylome disruptors or remediators. *Mutagenesis* 26, 475–487.

Liberti, M.V., Locasale, J.W. (2020) Histone lactylation: A new role for glucose metabolism. *Trends in Biochemical Sciences* 45, 179–82.

Lu, C., Thompson, C.B. (2012) Metabolic regulation of epigenetics. *Cell Metabolism* 16, 9–17.

Madugundu, G.S., Cadet, J., Wagner, J.R. (2014) Hydroxyl-radical-induced oxidation of 5-methylcytosine in isolated and cellular DNA. *Nucleic Acids Research* 42, 7450–7460.

Martinez-Reynes, I., Chandel, N.S. (2020) Mitochondrial TCA cycle metabolites control physiology and disease. *Nature Communications* 11, 102.

Medinas, D.B., Cerchiaro, G., Trindade, D.F., Augusto, O. (2007) The carbonate radical and related oxidants derived from bicarbonate buffer. *IUBMB Life* 59, 255–262.

Mentch, S.J., Locasale, J.W. (2016) One carbon metabolism and epigenetics: Understanding the specificity. *Annals of the New York Academy of Sciences* 1363, 91–98.

Miranda-Goncalves, V., Lamerinhas, A., Henrique, R., et al. (2018) Metabolism and epigenetic interplay in cancer: Regulation and putative therapeutic targets. *Frontiers in Genetics* 9, 429. doi: 10:.3389/fgene.2018.00427.

Neri, F., Krepelova, A., Incarnato, D., Maldotti, M., Parlato, F., Galvagni, F., Matarese, H. (2013) DNMT3L antagonizes DNA methylation at bivalent promoters and favours DNA methylation at gene bodies in ESCs. *Cell* 155, 121–134.

Nguyen, A., Duquette, N., Mamarbachi, M., Thorin, E. (2019). Epigenetic regulatory effect of exercise on glutathione peroxidase 1 expression in the skeletal muscle of severely dyslipidemic mice. *PLoS One* 11, e0151526.

Okano, M., Bell, D.W., Haber, D.A. et al. (1999) DNA methyltransferases Dnmt3a and Dnmt3b are essential for de novo methylation and mammalian development. *Cell* 99, 247–257.

Rodriguez, H., Rafehi, H., Bhave, M. et al. (2017) Metabolism and chromatin dynamics in health and disease. *Molecular Aspects of Medicine* 54, 1–15.

Sailani, M.R., Halling, J.F., Møller, H.D., Lee, H., Plomgaard, P., Pilegaard, H., Snyder, M.P., Regenberg, B. (2019) Lifelong physical activity is associated with promoter hypomethylation of genes involved in metabolism, myogenesis, contractile properties and oxidative stress resistance in aged human skeletal muscle. *Scientific Reports* 9, 3272.

Seaborne, R.A., Sharples, A.P. (2020) The interplay between exercise metabolism, epigenetics, and skeletal muscle remodeling. *Exercise and Sport Sciences Reviews* 48, 188–200.

Shaughnessy, D.T., McAllister, K., Worth, L. et al. (2014) Mitochondria, energetics, epigenetics, and cellular responses to stress. *Environmental Health Perspectives* 122, 1271–1278.

Sulkowski, P.L., Oeck, S., Dow, J. et al. (2020) Oncometabolities suppress DNA repair by distrupting local chromatin signalling. *Nature* 582, 586–591.

van Loon, B., Markkanen, E., Hubscher, U. (2010) Oxygen as a friend and enemy: How to combat the mutational potential of 8-oxo-guanine. *DNA Repair* 9, 604–616.

Vanzan, L., Sklias, A., Herceg, Z., Murr, R. (2017) Mechanisms of histone modifications. In: Tollefsbol, T.O. (Ed.) *Handbook of Epigenetics*, 2nd Edn. Academic Press: New York, 25–39.

Wang, Z. Long, H. Chang, C. et al. (2018) Crosstalk between metabolism and epigenetic modifications in autoimmune diseases: A comprehensive overview. *Cellular and Molecular Life Sciences* 75, 3353–3369.

Williamson, J., Davison, G.W. (2020) Targeted antioxidants in exercise-induced mitochondrial oxidative stress: Emphasis on DNA damage. *Antioxidants* 9, 1142.

Williamson, J., Hughes, C.M., Cobley, J.N., Davison, G.W. (2020) The mitochondrial-targeted antioxidant MitoQ, attenuates exercise-induced mitochondrial DNA damage. *Redox Biology* 36, 101673.

Wong, C.C., Qian, Y., Yu, J. (2017) Interplay between epigenetics and metabolism in oncogenesis: Mechanisms and therapeutic approaches. *Oncogene* 36, 3359–3374.

Wu, H., Coskun, V., Tao, J., Xie, W., Ge, W., Yoshikawa, K., Li, E., Zhang, Y., Sun, Y.E. (2010) DNMT3a-dependent nonpromoter DNA methylation facilitates transcription of nurosgenic genes. *Science* 329, 444–448.

Yin, R., Mao, S.Q., Zhao, B. et al. (2103) Ascorbic acid enhances Tet-mediated 5-methylcytosine oxidation and promotes DNA demethylation in mammals. *Journal of the American Chemical Society* 135, 10396–10403.

Zhang, D., Tang, Z., Huang, H. et al. (2019) Metabolic regulation of gene expression by histone lactylation. *Nature* 574, 575–583.

Zhou, X., Zhuang, Z., Wang, W. et al. (2016) OGG1 is essential in oxidative stress induced DNA demethylation. *Cellular Signalling* 28, 1163–1171.

INDEX

Note: **Bold** page numbers refer to tables and *italic* page numbers refer to figures.

acetyl-CoA 211, *213*, 213–214
acetyl-CoA through ATP citrate lyase (ACLY) 213, *218*
activated methyl (CH_3) donor group 210
acute oxidative eustress 75
adaptive redox response 76
    antioxidant enzymes 76–79
    basal redox status role 75–76
    chronic oxidative distress 72–73
    Nrf2 response to exercise training 74
    ROS generation sites 72–73
adenosine triphosphate (ATP) 199, 212
    mitochondrial respiration 51
    muscle contraction 24
    skeletal muscle function 188
    TCA cycle intermediates 212
AD pathology 197
ageing brain 193–194, 201
air-blood barrier 44, 44
airway clearance technique (ACT) 184–185
alveolar-capillary region 43, 44
Alzheimer's disease
    dietary intervention for 199–201
    energy balance 198–199
    mitochondrial dysfunction 198–199
    oxidative distress 194–198
    physical activity 201–202
    therapeutic intervention 199
American College of Sports Medicine (ACSM) 168
amyloid-cascade hypothesis 194
amyloid cleaving enzyme-1 (BACE-1) 194
amyloid-β (Aβ) 194–198, 201
amyloid-β precursor protein (AβPP) 194, *195*, *196*, 198
anaerobic capacity 185
antioxidant response elements (ARE) 73, 125, 129
antioxidants 2, 117
    capacity 25–26
    defence systems 1, 29–30, 117, 124–125
    effectors modulation 217–218
    enzyme activity 17, 19, 76–79
    supplementation 29
antioxidant supplements
    antioxidant defenses 124–125
    context 118–120
    heterogeneity of 117
    methodologies used 117–118
    mitochondrial biogenesis 125–126
    muscle hypertrophy/strength 126–127
    oxidative stress 128–130
    performance
        endurance performance 130
        substrate metabolism 127–128
    vascular function 128–130
APOE allelic variations 197
arachidonic acid
    non-enzymatic peroxidation 4
ARE *see* antioxidant response element (ARE)
arterial oxygen saturation (SaO_2) 44
ataxia telangiectasia mutated (ATM) 64
athletic performance 115, 117, 119
ATP synthase 28, 59, 199; *see also* adenosine triphosphate (ATP)

basal redox status role 75–76
base excision repair (BER) mechanism 216
binary logic gate 13
bioavailability 46, 123, 126, 131, 154, 158
biogenesis pathway 62, 126
biological systems 2, 5, 13, 188, 216
biomarkers 4–5, 16, 104, 153

calmodulin-dependent protein kinase (CAMK) 125
CaMKII activity 28
cancer
    oxidative stress 168–174
    physical activity 167–168
capillary density 49
carbon dioxide (CO_2) 43, 44, 51, 185, 216
5-carboxylcytosine (5caC) 211
cardiopulmonary exercise testing (CPET) 184
cardiovascular disease 103
cardiovascular function 186–188
carotenoid astaxanthin 127

catalase (CAT) 124
catechol-O-methyltransferase (COMT) 156
C–C Chemokine Receptor-7 (CCR7) 88
cellular metabolism 24, 79, 90, 125, 209, 212, 214
cellular signalling pathways 23, 48
cholesterol metabolism 197
chronic erythrocyte oxidative stress 43
chronic inflammatory disease 85
chronic low-grade oxidative stress 72
chronic oxidative distress 72–73
chronic supplementation 109, 119, 128
chronic training stimulus 13
classical cytosolic intracellular 62
click-PEG workflows 17–18, 18
continuous endurance exercise performance 141–142
Copper 4
CpG islands (CGI) 210
c-reactive protein 159
cysteine modifications 7
cysteine oxidation 88, 194
2-cysteine peroxiredoxin (PRDX) isoforms 17
cysteine thiols role 88
cystic fibrosis (CF)
  cardiopulmonary exercise testing 184
  cardiovascular function 186–188
  exercise, benefits of 184–185
  inflammation stress 183–184
  lung disease 186
  mechanistic insight 185
  overview of 183
  oxidative stress 183–184
  pulmonary function 185–186
  skeletal muscle function 188
cystic fibrosis transmembrane conductance regulator (CFTR) gene 183
cytochrome C oxidase subunit IV (COXIV) 126
CyTOF™ mass cytometry 90
cytosolic 2-Cys PRDX isoform activity 19
cytosolic glutathione (GSH) 13

D-amino-acid oxidase 3
delayed onset muscle soreness (DOMS) 123
deoxyribonucleic acid (DNA) 103
  histone methylation 210–211
  methylation 210, 211–212, 217
  methyltransferase 210
  nucleotide sequence 209
  oxidation 98
2-deoxyribose moiety 104
D-glyceraldehyde-3-phosphate 18
DHE *see* dihydroethidium (DHE)
2,6-diamino-5-formamido4-hydroxypyrimidine (FAPy-G) 106
2′,7′-dichlorofluorescin (DCF) **12**
dietary nitrate supplementation
  on continuous endurance exercise performance 141–142
  ergogenic effect, mechanisms for 143–145
  on highintensity exercise performance 142–143
diffusion-controlled rate 3
dihydroethidium (DHE) 11
dinitrosyliron complexes (DNICs) 138
DNA damage 71
  contributing variables 109
  exercise-induced RONS sources 106–107
  Hydroxyl attack 106
  mtDNA damage 108–109
  nDNA damage 107–108
  pathological signals 104
DNA methyltransferase (DNMT) 210
DNMT *see* DNA methyltransferase (DNMT)
DOMS *see* delayed onset muscle soreness (DOMS)
double-strand breaks (DSB) 108
DSB *see* double-strand breaks (DSB)
dual-specificity phosphatases (DUSPs) 30
DUOX isoforms 31
DUSPs *see* dual-specificity phosphatases (DUSPs)

electron paramagnetic resonance (EPR) spectroscopy 1, 5, 12
electron transferring flavoprotein dehydrogenase (ETFDH) 59
electron transport chain (ETC) 59, 199, 214
electrophilic Michael acceptors 154
ELISA methods 16, 97, 98
elite cyclists 119
endothelial-dependent vasodilation 187
endothelial dysfunction 49, 128, 186, 187
endothelial function 46, 49, 129, 155, 187
endothelial NO synthase (eNOS) 128, 137
endothelium-derived vasodilator 137
endurance performance 130–131, 142, 158
energy balance 198–199
energy metabolism 47, 196
epigenetic landscape 209
  antioxidant effectors modulation 217–218
  DNA/histone methylation 210–211
  DNA methylation 211–212
  histone acetylation 211
  lactate 214
  metabolism 210
  one-carbon metabolism 210–211
  oxidative stress and ROS 214–215
  ROS and DNA damage/repair 215–217
  TCA cycle intermediates 212–214
  TCA cycle metabolites 211–212
EPR spectroscopy *see* electron paramagnetic resonance (EPR) spectroscopy
EPR spin trapping 13
ergogenic effect, mechanisms for 143–145
erythrocyte glutathione 48, 48
erythrocyte reduced glutathione (GSH) 88
erythrocytes
  computational model 48
  energetics 46–48
  oxygen 45
  redox network 45–46
ETC *see* electron transport chain (ETC)
exercise
  benefits of 184–185
  oxidative stress 5
  performance 156–158
  recovery 158–160
  redox biology 48
  RONS damages 110, *110*
  training 62–63
exercise immunology
  couple redox biology 85–91
  global oxidation 86–87
  and redox biology 86

redox reactions 86
single-cell analysis 87
single cell approaches 89
'thiol' group 88
extracellular vesicles (EVs) 90, 170, *171*

family of transcription factors (FoxO) 7
fast troponin I isoform (TnI(f)) 28
fast-twitch glycolytic fibres 25
fatty acid translocase (FAT/CD36) 128
fatty acid utilization 29
Fenton reactions 4, 106, 216
Fick equation 42
F_2-isoprostane 4, 75
flow-mediated dilation (FMD) 186
fluorescein-5 maleimide (F5M) 88
FMD *see* flow-mediated dilation (FMD)
'food first' philosophy 120
footprint oxidative damage 12
forced expiratory volume in 1 second (FEV1) 185
forced vital capacity (FVC) 185
5-formylcytosine (5fC) 211
free radical-mediated protein damage 5
free radical production 1, 5, 13, 61, 115, 118, 160, 183

gastrointestinal (GI) tract 154
gene regulation 209
gene transcription 7, 62, 76, 77, *77*, 211, 214
glial cytopathology 194
global hypomethylation 215
global tyrosine nitration 16
glucose homeostasis 28, 62
glucose metabolism 46, 198
glutathione peroxidase 1 (Gpx1) 217
glutathione peroxidase (GP_X) 124
glutathione redox state **12**
glutathione synthetase transferases (GSTs) 77
glycolysis-associated enzymopathies 46
G-quadruplex-forming sequences (PQS) 216
GSH *see* cytosolic glutathione (GSH)
γ-H2AX detection 108

healthy redox homeostasis 75
hemoglobin (cHb) 42, 45
hemoglobin autooxidation 46
highintensity exercise performance 142–143
high-performance liquid chromatography mass spectrometry (HPLC-MS-MS) 14
histone acetylation 211
histone demethylation 212
histone methyltransferase (HMT) 210
HMT *see* histone methyltransferase (HMT)
homologydependent repair (HDR) pathway 217
HPLC-MS-MS *see* high-performance liquid chromatography mass spectrometry (HPLC-MS-MS)
humoral immunity 86
hydrazine biotin 14
hydrogen bond network 15
hydrogen peroxide (H_2O_2) 3, 11, 42, 52, 60, 76, 127
hydroperoxyl radical (HO_2) 2, 3
8-hydroxy-2′-deoxyguanosine 5, 107
2-hydroxyglutarate (2-HG) 212
hydroxyl-derived reactions 106
hydroxyl free radical (OH) 4, 104
5-hydroxymethylcytosine (5hmC) 211
4-hydroxynonenal (HNE) 32, 197
13-hydroxyoctadecadienoic acid (HODE) 32
HyPer family 63
hypertrophy 30–31
hypochlorous acid 86
hypometabolism 198
hypoxia modulation 172

immune cell mobilisation 88, 89
immune natural killer (NK) cells 172
immune system 85, 86, 89, 90
immunoblotting 15
immunological assays 16–18, *17*, 118
immunometabolism 86, 90
immuno-spin trapping 13
inducible NO synthase (iNOS) 137
inflammation stress 183–184
intermyofibrillar mitochondria produces 51
intracellular cytokine analysis 89
intracellular location 6
intracellular muscle adaptation 6
intracellular muscle biology 2
intramuscular adaptation 24
intramuscular free radical 6
inverse electron demand Diels–Alder (IEDDA) 17
Ironman triathlons 108
isocitrate dehydrogenase 2 (IDH2) 212

JumonjiC (JmjC) 212

Keap1 oxidation 73
α-Ketoglutarate (α-KG) 211
knowledge gap 30

lactate 90, 119, 170, 214
lactormone 32
L-arginine supplementation 138
L-cysteine 141
L-gulono-gamma-lactone oxidase 125
lipid hydroperoxides 48
lipid peroxidation 4, 16, 86
lipid-soluble antioxidants 201
lipoprotein receptor-related protein 1 (LRP1) 197
lipoxygenases (LOX) 72
local tissue oxygenation 49
lung clearance index (LCI) 185
lungs 43–45
lysine amino acids 211
lysine-specific demethylases (LSD) 212

macromolecular damage 5, 210
mammalian target of rapamycin (mTOR) 126
manganese superoxide dismutase (MnSOD) 15
mechanistic insight 185
mechano-growth factor (MGF) 30
messenger RNA (mRNA) 96
metabolic equivalent of task (MET) 168
metabolic phenotypes 89
metal ions sequestration 4
5-methylcytosine (5mC) 209, 211
MGF *see* mechano-growth factor (MGF)
microcirculation
regulation 49, 50
structure 48–49
micro-RNA (miRNA) 96

microvascular dysfunction 187
mitochondria 5
biogenesis 29, 125–126
cellular (redox) processes 51–53
dissolved oxygen 51
DNA damage **105**
dysfunction 198–199
genome 65, 103
nucleoids 64
oxygen flow and consumption 51
respiratory rate 95
ROS production 61
via myoglobin-mediated delivery 51
mitochondrial-derived peptide (MDP) 65, 110
mitochondrialderived ROS act 63
mitochondrial derived peptides 64
mtDNA damage 63–64
peroxiredoxins 64
mitochondrial glycerophosphate dehydrogenase (mGPDH) 59
mitochondrial open-reading frame of the 12S rRNA-c (MOTS-c) 65
mitochondrial oxidative phosphorylation (OXPHOS) 59
mitochondrial redox regulation
and energy metabolism 59
reactive oxygen species 59–60
skeletal muscle mitochondrial ROS production 60–62
mitochondrial-targeted superoxide detection dyes (MitoSOX) 63
mitochondria-targeted quinone (MitoQ) 13, 73, 200
MitoQ *see* mitochondria-targeted quinone (MitoQ)
MnSOD *see* manganese superoxide dismutase (MnSOD)
MnSOD nitration analysis 16
5′ monophosphate-activated protein kinase (AMPK) 124
mtDNA damage 63, 108–109
muscle 49–51
"communicates" 24
contraction 24–25
function 123
hypertrophy/strength 126–127
recovery 130
muscle adaptations 28–29
antioxidant defence 29–30
hypertrophy 30–31
mitochondrial biogenesis 29
muscle-damaging protocol 88
muscle-function modification 24
myeloid-derived suppressor cells (MDSCs) 172
myosin composition 25

N-acetyl cysteine (NAC) 74, 88, 118, 125, 160
NAD *see* nicotinamide adenine dinucleotide (NAD)
NADPH *see* nicotinamide adenine dinucleotide phosphate (NADPH)
NADPH oxidase enzymes (NOXs) 24
NADPH oxidase (NOX) isoforms 62, 72
nDNA damage 107–108
quantification of 104
near-infrared spectroscopy (NIRS) 158
negative feedback mechanism 76
neuronal NO synthase (nNOS) 137
NF-κB activation 30
nicotinamide adenine dinucleotide (NAD) 200, 212
nicotinamide adenine dinucleotide phosphate (NADPH) 24, 90, 156
nitric oxide (NO) 4, 42, 86, 117, 137
nitric oxide biochemistry 137
cost-effective approach 138
dietary nitrate supplementation 141–142
nitratenitrite-nitric oxide pathway 138–141
nitric oxide synthase (NOS3) 187, 188
nitrogen molecule 4
nitrosonium ion 138
nitrosothiols 49
3-nitrotyrosine (3-NT) 15
NLR family pyrin domain containing 3 (NLPR3) 155
NO *see* nitric oxide (NO)
non-enzymatic antioxidants 117
non-omic approaches 15
NO_3-rich saliva 138
NO synthase (NOS) enzymes 137
NO• synthases (eNOS) 49
NOX2-derived ROS 27
NOXs *see* NADPH oxidase enzymes (NOXs)
n-6 poly-unsaturated fatty acids 4
Nrf2 activation 73, 74
Nrf2 response to exercise training 74
N-terminal kinase (JNK) 125
nuclear erythroid 2-related factor (Nrf-2) 7
nuclear factor-erythroid-2-related factor 2 (Nrf2) 154
nuclear factor κB (NF-κB) 124
nuclear genome 104
nucleotide metabolism 4

omics workflows, oxidative damage 13, 15
onset of muscle soreness (DOMS) 119
open metabolic system 2
oral microflora catalyse 138
organic solvent 5
oxidation of low-density lipoprotein (ox-LDL) 128
oxidative attack 5
oxidative damage 5, 19
biological meaning 16
measurement approaches 13
measuring reactive species 12–13
novel approaches 13–16
oxidative eustress 11–12
oxidative distress 60
in ageing brain 193–194
concept of 2
in neurodegeneration 194–195
oxidative eustress 11–12, 86, 193
appreciation for 89
oxidative eustress–muscle adaptation relationship 6
oxidative macromolecule damage 13
oxidative phosphorylation (OXPHOS) complexes 59, 64–65, 211
oxidative RNA modifications 96
oxidative stress (OS) 4, 48, 60, 86, 115, 118, 128–130, 167, 183–184
definition of 2
description of 2
overview of **215**
Oxidative Stress Hypothesis 197
Oximouse database 17
8-oxo-2′-deoxyguanosine (8-oxodG) 216
8-oxo-7,8-dihydro-2′-deoxyguanosine (8-oxodG) 96–98
oxygen 2
in biological systems 2, **2**
consumption rate 95
rate-limiting control mechanisms 187

oxygen-centred free radical 2
oxygen delivery (DO_2)
 definition of 42
oxygen transport
 erythrocytes
  computational model 48
  energetics 46–48
  oxygen 45
  redox network 45–46
  Fick equation 42
 lungs 43–45
 microcirculation
  regulation 49, 50
  structure 48–49
 mitochondria
  cellular (redox) processes 51–53
  oxygen flow and consumption 51
 muscle 49–51
 quantitative snapshot 42–43
oxyhemoglobin desaturation 186

PAH *see* pulmonary artery hypertension (PAH)
Paper-to-Podium (P-2-P) framework 116, *116*, 117
PBMCs *see* peripheral blood mononuclear cells (PBMCs)
pentane gas 4
pentose phosphate pathway (PPP) 17, 46
peripheral blood mononuclear cells (PBMCs) 86
peroxiredoxins (Prxs) 76–77
peroxiredoxin/thioredoxin (PRDX/TRDX) 124
peroxomonocarbonate ($HOOCO_2$) 18
peroxynitrite (ONOO–) 127
peroxynitrite anion (ONOO–) 4
PGC-1α activity 29
(poly)phenols
 action, mechanisms of 153–156
 and exercise performance 156–158
 exercise recovery 158
 and exercise recovery 158–160
 "periodised" approach 158
 practical application 160
α-phenyl-*tert*-butylnitrone (PBN) 5
phosphatase and tensin homolog (PTEN) 30
phosphate creatine system 25
phosphodiesterase type 5 inhibition (PDE5) 186
physical activity (PA) 201–202
 in cancer survival 167–168
physical exercise session 25
phytochemical-rich fruit 157
Piper Fatigue Inventory Scale 170
pleiotropic signaling molecules 42
pleiotropic signalling agents 193
polyunsaturated fatty acids 46
post-translational modification (PTM) 210
PPP *see* pentose phosphate pathway (PPP)
prdx isoform dimers 19
pro-inflammatory cytokines 159
proliferator-activated receptor-γ coactivator 1-α
  (PGC1α) 156
"proof-of-concept" 48
prooxidant–antioxidant balance 2
pro-oxidant environment 31, 197
prooxidative muscular environment 23
protein 5
 carbonylation 86
 carbonyls 5
 cysteine residues 26–27
 glutathionylation 77
 oxidation 5
 synthesis 30, 96
protein phosphatase 2A (PP2A) 30
protein-protein interaction 215
proton-motive force 199
Prx enzymes inhibition 76–77
PTEN *see* phosphatase and tensin homolog (PTEN)
PTM *see* post-translational modification (PTM)
pulmonary artery hypertension (PAH) 186
pulmonary function 185–186
putative mechanisms 140

quantitative snapshot 42–43

radiation therapy 172
radiotherapy response 172
rapid glucose transport (GLUT1) 90
reactive nitrogen intermediates (RNIs) 138
reactive nitrogen species (RNS) 4, 31
reactive oxygen and nitrogen species (RONS) 30, 42, 86,
  103, 153
reactive oxygen species (ROS) 1, 2, 23, 60, 71, 95, 123,
  167, 193
 antioxidant capacity 25–26
 and contractile function 27–28
 glucose homeostasis 28
 muscle adaptations 28–29
 muscle contraction 24–25
 ROS waves concept 32
 skeletal muscle 6, 25–26
 tissue specific adaptation 6
reactive species 12–13
redox biomarker 13
redox controls systems 7
"Redox Exerkine Signaling" dilemma 74
redox homeostasis 193
redox proteomic approaches 17
redox reactions 86
redox-sensitive protein 28
redox-sensitive signalling pathways 29
redox-sensitive signalling proteins 64
redox signalling
 antioxidant enzyme activity 19
 novel immunological assays 16–18
 protein cysteine residues 26–27
"reduced" monomeric band 17
reduction of lipid hydroperoxides (R-OOH) 48
reductive stress 33, 86
rehabilitation 167
respiratory system 44
"reverse electron transfer" (RET) 60
RNA oxidation
 acute exercise 98
 biological consequences **96**
 biological functions 95
 chemical modifications 96
 epitranscriptomic changes 95–97
 non-coding types 95
 pathophysiological roles 96
 in vivo setting 97
RNS *see* reactive nitrogen species (RNS)
RONS *see* reactive oxygen nitrogen species (RONS)
ROS *see* reactive oxygen species (ROS)

ROS and DNA damage/repair 215–217
ROS generation sites 72–73
ROS-rich environment 23–24
    local and systemic ROS waves 31–33
ryanodine receptor-Ca^{2+} release channel 1 (RyR1) 27

S-adenosylmethionine (SAM) 210
sarcoplasmic reticulum (SR) 27
sarcoplasmic reticulum Ca^{2+} ATPases (SERCAs) 25, 27
Schiff base formation 16
second-pass metabolism 138
γ-secretase enzyme 194
S-glutathionylation 27, 28
signal transduction pathways 24
single-cell analysis 87
single cell approaches 89
single-cell gel electrophoresis assay 4
single unpaired electron 4
skeletal muscle function 188
skeletal muscle mitochondria 61
skeletal muscle mitochondrial ROS production
    during exercise 61–62
    at rest 60–61
"slow-twitch oxidative" fibres 25
small humanin-like-peptide 6 (SHLP6) 65
S-nitrosation 11
S-nitrosocysteine (CysNO) 141
S-nitrosoglutathione (GSNO) 141
SOD *see* superoxide dismutase (SOD)
sports performance 115
spurious carboxylesterase 13
SR *see* sarcoplasmic reticulum (SR)
steady-state redox balance 2
stem cell depletion 193
S-transnitrosylation 138
substrate metabolism 127–128
Su(var)3–9, Enhancer of zeste and Trithorax (SET) 210
sulphenic acid formation 11
sulphhydryl groups 26–27
superfoods 201
superoxide anion 2–3
superoxide dismutase 2 (SOD2) 217
superoxide dismutase (SOD) 1, 3, 28, 30, 124, 154
superoxide production 60
"switch" approach 17
"switch on/off" mechanism 27
systemic inflammation 183
systemic oxidative stress 61
systemic redox assays 20

TAC assay *see* total antioxidant capacity (TAC) assay
Tau protein 194
TBARS assay *see* thiobarbituric acid-reactive substance
        (TBARS) assay
TCA *see* tricarboxylic acid cycle (TCA)
T cell function 89
T cell receptor (TCR) 90
ten-eleven translocation (TET 1–3) 211
tetrahydrobiopterin (BH4) 187
thiobarbituric acid-reactive substance (TBARS) assay 4, 12,
        12, 118
thiol oxidation 18, 88
thiol protein 18
thymine DNA glycosylase (TDG) 211
time trial performance test 142
total antioxidant capacity (TAC) assay **12,** 19, 118, 170
transcription factor A mitochondrial (Tfam) 126
transient receptor potential (TRP) 27
tricarboxylic acid cycle (TCA) 199, 210
    chromatin, glycolytic regulation 213
    DNA (De) methylation 213
    intermediates 212–214
    metabolites 211–212
    oxidative phosphorylation 212
Two-Hit Hypothesis 198
type 2 diabetes (T2D) 129
tyrosine 34 (Y34) nitration 15

ubiquinol/ubiquinone (UQH2/UQ) 59, 61, 214
ubiquitin proteasome activity 17
unpaired electron process 4

vascular dysfunction 186
vascular function 128–130
ventilation-perfusion (V/Q) 186
Vienna Active Aging Study 98
vitamin C supplementation 6, 75, 130
vitamin E supplementation 6, 119
VO_2 max 130

Western blot/ELISA 16
Western blotting 17
whole-body oxygen consumption (VO_2) 42
World Anti-Doping Agency (WADA) 118

xanthine dehydrogenase (XDH) 25
xanthine oxidase activity 6, 62, 72
xanthine oxidoreductase (XOR) 25, 138
XDH *see* xanthine dehydrogenase (XDH)
XOR *see* xanthine oxidoreductase (XOR)

Yo-Yo Intermittent Recovery-1 test 142–143